D0386340

ANALYTICAL GEOCHEMISTRY

SERIES

Methods in Geochemistry and Geophysics

1. A.S. RITCHIE
CHROMATOGRAPHY IN GEOLOGY

2. R. BOWEN
PALEOTEMPERATURE ANALYSIS

3. D.S. PARASNIS
MINING GEOPHYSICS

4. I. ADLER
X-RAY EMISSION SPECTROGRAPHY IN GEOLOGY

6. A.J. EASTON
SILICATE ANALYSIS BY WET CHEMICAL METHODS

7. E.E. ANGINO AND G.K. BILLINGS
ATOMIC ABSORPTION SPECTROMETRY IN GEOLOGY

8. A. VOLBORTH
ELEMENTAL ANALYSIS IN GEOCHEMISTRY A. MAJOR ELEMENTS

9. P.K. BHATTACHARYA AND H.P. PATRA
DIRECT CURRENT GEOELECTRIC SOUNDING

10. J.A.S. ADAMS AND P. GASPARINI
GAMMA-RAY SPECTROMETRY OF ROCKS

11. W. ERNST
GEOCHEMICAL FACIES ANALYSIS

Methods in Geochemistry and Geophysics

5

ANALYTICAL GEOCHEMISTRY

BY

THE LORD ENERGLYN,
M. Sc., Ph.D., M.I.Min.E., C.Eng., D.Sc.,
M.I.M.M., F.L.S., F.R.G.S., F.G.S.

Head of Department of Geology,
The University of Nottingham,
University Park, Nottingham,
Great Britain

AND

L. BREALEY,
B.Sc., Ph.D., F.R.I.C., M.Inst.F.

Department of Geology,
The University of Nottingham,
University Park, Nottingham,
Great Britain

ELSEVIER PUBLISHING COMPANY
AMSTERDAM / LONDON / NEW YORK
1971

ELSEVIER PUBLISHING COMPANY
335 JAN VAN GALENSTRAAT
P.O. BOX 211, AMSTERDAM, THE NETHERLANDS

ELSEVIER PUBLISHING COMPANY LTD.
BARKING, ESSEX, ENGLAND

AMERICAN ELSEVIER PUBLISHING COMPANY, INC.
52 VANDERBILT AVENUE
NEW YORK, NEW YORK 10017

ISBN 0-444-40826-6
LIBRARY OF CONGRESS CARD NUMBER 76-103358

PRINTED IN THE NETHERLANDS

V

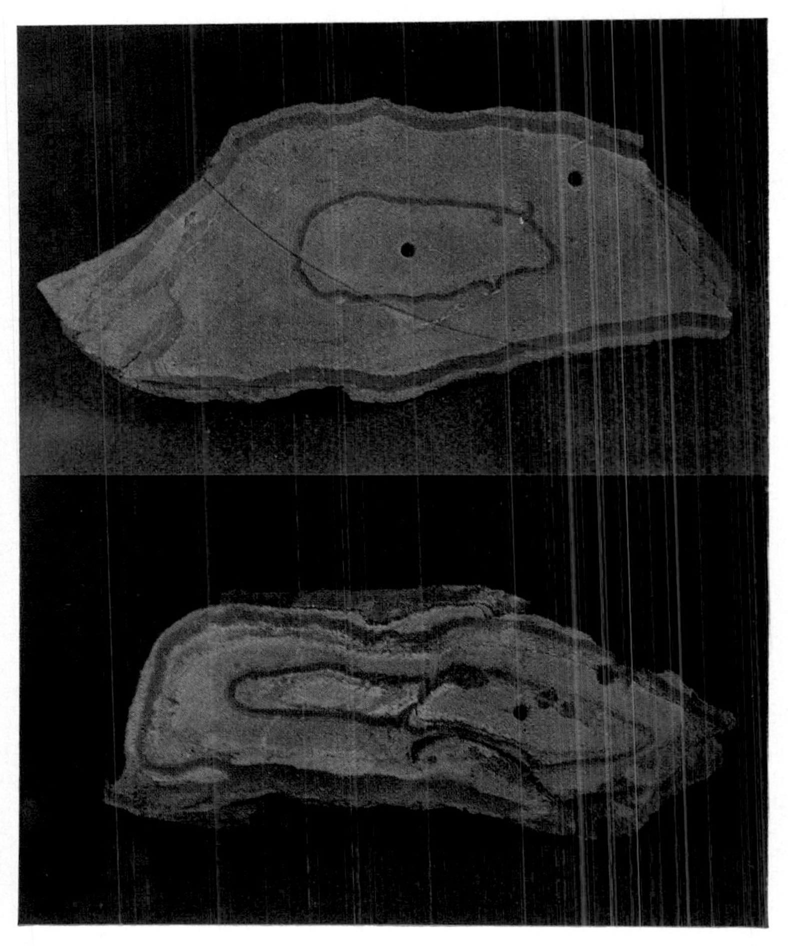

Simple and multiple diffusion structures, such as these samples of boxstone from the Northampton Ironstone, serve to illustrate the interaction of geochemical processes which can now be resolved by modern methods of analysis.

Preface

This is an attempt by a geologist and a chemist to produce a form of introductory manual to techniques which are commonly required for the analysis of rocks and minerals. We are aware of the limitations of such a book, but our intention is to introduce some analytical techniques to those geologists who did not study chemistry as undergraduates. Its objective is to guide the geologist rather than to attempt to instruct him in chemistry.

The drafting of the diagrams and tables was carried out by Mrs. Denise Stones to whom we are also indebted for checking the manuscript and the proofs. The typing was carried out by Miss I. R. Bruce. To our colleagues Dr. Dennis Field, Dr. R. B. Elliott and Dr. P. K. Harvey, we are most grateful for analytical data and advice in the compilation of the book. On the other hand, we do not commit them to geochemical observations and deductions for which we are entirely responsible. We would also like to acknowledge the debt we owe to the large number of analysts whose work has been quoted in this text. In particular we should like to thank Professor David Williams and the Institution of Mining and Metallurgy, London, for their permission to include in this book a large part of the paper on "chromographic analysis" under the title of "membrane colorimetry".

Contents

CHAPTER 1

Introduction to Geochemistry

INTRODUCTION

From the stage when our planet developed a solid envelope—the so-called primordial crust—the progress of mineral development was probably attained by an ordered arrangement of elements. Thereafter the pattern of geochemical events is in part now readable but elsewhere the design has yet to be unfolded. This is the quest of the geochemist. His task is rendered difficult by the impossibility of testing earth bound reactions on a laboratory scale—reactions which have been engendered by temperatures and pressures which have changed in intensity from time to time. On the other hand the challenge imposed by these parameters excites the imagination and extends the ingenuity of the geologists prepared to use chemical and physical techniques to extend our knowledge of the petrogenesis of the crust of the earth.

In the initial stages of crustal formation it seems reasonable to assume that elements assembled themselves into a metallic phase and then developed through sulphides into silicates. In other words a process analogous to that which takes place in those metallurgical melts in which oxygen is in excess of sulphur. This broad concept is supported by our existing knowledge of the differentiation processes which have led to the consolidation of magmas. There does appear to be reasonable evidence to suggest that the same controlling factors used in synthesising pure crystalline compounds in the laboratory must have exercised their influence over the formation of impure minerals. Consequently, it is reasonable to assume that the structure of minerals is largely determined by the same electro-chemical and geometric factors which have been evaluated for synthetic compounds. Therefore, it can be assumed that since oxygen is the most important ion in crustal rocks, its geometric envelopment of cations dominates the crystalline structure of a large number of rock-forming minerals. The number of oxygen ions which can surround an element to form a cation determines the co-ordination number of the cation. This may be as low as 3 or as high as 12. Obviously the smaller the cation the lower will be its co-ordination number. Consequently, the study of ionic size by means of X-ray diffraction has proved invaluable in the understanding of crystal systems. In developing geometric dispositions there is a tendency for the cations and anions to assume a configuration possessing a minimum potential energy consistent with the neutralisation of ionic charges. Consequently, crystal structures must satisfy geometrical, electrical and potential energy conditions. Such threefold control naturally eliminates the possibility of certain compounds occurring as natural minerals. This is well illustrated by felspars—the alumino-silicates of potassium, sodium and calcium. On valency values alone one might expect the other alkalis to be capable of replacement by lithium, but owing to the wide differences of atomic radii involved, this is not possible. On the other hand, rubidium and barium can replace potassium whilst magnesium and ferrous iron are incapable of replacing calcium. In other words, the formation of crystalline minerals is dominated by the size of the ions whilst the valency or combining power of the ions is of lesser importance. This is particularly true for the silicates. Moreover, since geometric factors predominate to such a great extent in mineralization, the principle of isomorphism becomes a matter of much greater significance to the geochemist than to the chemist concerned with the preparation of pure chemical compounds in the laboratory. This accounts for the fact that compounds of analogous composition are not always isomorphous.

Difference in ionic size may control the establishment of certain crystal forms. For example, carbonate minerals are either trigonal or orthorhombic. The lower trigonal symmetry is achieved only by those cations whose radius is greater than 1 Å. Consequently, the carbonates of strontium, lead, silver and barium form trigonal crystals. Conversely the carbonates of copper, nickel, manganese, magnesium and ferrous iron achieve the higher symmetry of orthorhombic crystals as their atomic radii are less than 1Å. Calcium carbonate is an interesting borderline case, as the ionic radius of calcium is 0.99 Å. Under differing conditions it is capable of crystallizing as trigonal crystals of calcite or as orthorhombic aragonite (EVANS, 1964).

CHEMICAL CLASSIFICATION OF IGNEOUS AND METAMORPHIC ROCKS

A considerable number of methods has been postulated for the petrographic and petrological interpretation of chemical analyses. Many of them have proved to be too complicated for the purposes of depicting changes which have taken place during the crystallization of magmas and the metamorphism of all types of igneous and sedimentary rocks. JOHANNSEN (1931) has summarized and commented on a number of these methods and has shown that for most practical purposes only two of these methods have proved to be of value to most petrologists and geochemists. One is known as the "C.I.P.W." or "American" classification and the other is generally referred to as the "Niggli Equivalent Norm".

Four American petrographers, C. Whitman Cross, Joseph P. Iddings, Louis V. Pirsson and Henry A. Washington, produced a classification of analytical data which has become known by their surname initials as the C.I.P.W. classification (CROSS et al., 1902). Their classification was based on calculations which were very similar to those devised three years earlier by OSANN (1899), but the classification differed in the use of "normative" minerals instead of the molecules represented by the analyses. This was an advance which resulted in a classification of igneous rocks in terms of "standard" or "normative" mineral molecules. These were pure molecules representing the principal chemical characteristics of the actual rock-forming minerals—the so-called modal minerals. Among these modal minerals it was recognised that certain oxides/elements are mainly located by molecules which are incapable of change regardless of the chemical composition of rock. Thus P_2O_5 is embodied in apatite, and ZrO_2 appears only in zircon. On the other hand, some oxides/elements may enter different molecules according to the amounts of other associated constituents. For example, soda may occur in albite or nepheline whilst potash may be located in orthoclase, leucite, or kaliophilite, according to the amount of silica which was potentially available during the process of crystallization of the magma. Consequently, a modal or numerical analysis of the rock in thin section determines to a large extent the mechanics of assembling the normative values from the list of oxides obtained by chemical analysis. A list of the normative mineral molecules used in the C.I.P.W. classification is shown in Table I, using the following molecular weights for the important oxides: SiO_2 (60.06), Al_2O_3 (101.94), Fe_2O_3 (159.68), FeO (71.84), MgO (40.32), CaO (56.07), Na_2O (62.0), K_2O (94.2), TiO_2 (79.9), MnO (70.93), P_2O_5 (142.06). The classification itself is shown in Table II.

The C.I.P.W. method of calculating normative minerals from chemical analyses is summarized as follows:

(*1*) The percentage weight of the oxides is divided by their molecular weights in order to obtain a so-called "molecular number".

(*2*) The amounts of MnO and NiO are combined with that of FeO and so eliminated from the picture.

(*3*) The amounts of BaO and SrO are likewise added to CaO for eliminative purposes.

TABLE I

NORMATIVE MINERAL MOLECULES FOR C.I.P.W. CLASSIFICATION

Mineral	Group		Molecule
Salic group			
Quartz (Q)			SiO_2
Corundum (C)			Al_2O_3
Zircon (Z)			$ZrO_2.SiO_2$
Orthoclase (or)	F		$K_2O.Al_2O_3.6SiO_2$
Albite (ab)	F		$Na_2O.Al_2O_3.6SiO_2$
Anorthite (an)	F		$CaO.Al_2O_3.2SiO_2$
Leucite (lc)	L		$K_2O.Al_2O_3.4SiO_2$
Nephelite (ne)	L		$Na_2O.Al_2O_3.2SiO_2$
Kaliophilite (kp)	L		$K_2O.Al_2O_3.2SiO_2$
Halite (hl)			NaCl
Thenardite (th)			$Na_2O.SO_3$
Sodium carbonate (nc)			$Na_2O.CO_2$
Femic group			
Acmite (ac)	P		$Na_2O.Fe_2O_3.4SiO_2$
Sodium metasilicate (ns)	P		$Na_2O.SiO_2$
Potassium metasilicate (ks)	P		$K_2O.SiO_2$
Diopside (di)	P		$CaO(Mg, Fe)O.2SiO_2$
Wollastonite (wo)	P		$CaO.SiO_2$
Hypersthene (hy)	P		$(Mg, Fe)O.SiO_2$
Olivine (ol)	O		$2(Mg,Fe)O.SiO_2$
Calcium orthosilicate (cs)	O		$2CaO.SiO_2$
Magnetite (mt)	H	M	$FeO.Fe_2O_3$
Chromite (cm)	H	M	$FeO.Cr_2O_3$
Hematite (hm)	H	M	Fe_2O_3
Ilmenite (il)	T	M	$FeO.TiO_2$
Titanite (tn)	T	M	$CaO.TiO_2.SiO_2$
Perofskite (pf)	T		$CaO.TiO_2$
Rutile (ru)	T		TiO_2
Apatite (ap)	A		$3(3CaO.P_2O_5).CaF_2$
Fluorite (fr)	A		CaF_2
Pyrite (pr)	A		FeS_2
Calcite (cc)	A		$CaO.CO_2$

TABLE II

THE C.I.P.W. CLASSIFICATION

Class:		Class 1. Persalic–Ratio of sal: fem greater than 7.0 Class 2. Dosalic–Ratio of sal: fem between 7.0 and 1.667 Class 3. Sulfemic–Ratio of sal: fem between 0.60 and 0.143 Class 5. Perfemic–Ratio of sal: fem less than 0.143
Order:	$\frac{\text{Quartz(Q) or Lenads (L)}}{\text{Felspars (F)}}$	Order I. $Q \gg F$ Order 2. $Q > F$ Order 3. $Q = F$ Order 4. $F > Q$ Order 5. $F > Q$ or L Order 6. $F > L$ Order 7. $F = L$ Order 8. $L > F$ Order 9. $L \gg F$
Rang:	$\frac{\text{mol. } Na_2O + \text{mol. } K_2O}{\text{mol. } CaO}$	Rang 1. $(Na_2O + K_2O) \gg CaO$ Rang 2. $(Na_2O + K_2O) > CaO$ Rang 3. $(Na_2O + K_2O) = CaO$ Rang 4. $CaO > (Na_2O + K_2O)$ Rang 5. $CaO \gg (Na_2O + K_2O)$
Subrang:	$\frac{\text{mol. } K_2O}{\text{mol. } Na_2O}$	Subrang 1. $K_2O \gg Na_2O$ Subrang 2. $K_2O > Na_2O$ Subrang 3. $K_2O = Na_2O$ Subrang 4. $K_2O < Na_2O$ Subrang 5. $K_2O \ll Na_2O$

(*4*) Oxides which form specific minerals identified as modal minerals in their section are allocated as follows:

Apatite: CaO equal to 3.33 that of P_2O_5 (or 3.00 P_2O_5 and 0.33 F, if the latter is present).

Halite: Na_2O equal to that of Cl_2.

Thernardite: Na_2O equal to that of SO_3 when the rock contains minerals of the Hauynite group.

Pyrite: FeO equal to 0.5 S.

Chromite: FeO equal to Cr_2O_3.

Ilmenite: FeO equal to TiO_2.

Titanite: CaO equal to TiO_2 plus an equal amount of

SiO_2, if lime remains after providing for anorthite.

Rutile: excess TiO_2.

Fluorite: CaO equal to F.

Calcite: CaO equal to CO_2.

Cancrinite: Na_2O equal to CO_2.

Zircon: total of ZrO_2 plus an equal amount of SiO_2.

(5) The distribution of alumina is controlled as follows:

(*a*) Where alumina is in excess, collectively or separately, of potash, soda and lime:

Orthoclase: Al_2O_3 equal to K_2O plus $6SiO_2$.

Albite: Al_2O_3 equal to Na_2O plus $6SiO_2$.

Anorthite: Al_2O_3 equal to CaO plus $2SiO_2$.

Corundum: if Al_2O_3 is still in excess.

(*b*) Where alumina is exceeded collectively or individually by the amounts of potash, soda and lime, the excess alkalis are modulated as follows: K_2O is calculated as potassium metasilicate ($K_2O.SiO_2$); Na_2O is calculated as acmite by the addition of an equal amount of Fe_2O_3 plus 4 : 1 SiO_2. If there is still an excess of Na_2O over Fe_2O_3 it is calculated as sodium metasilicate ($Na_2O.SiO_2$). If there is an excess of Fe_2O_3 over the remaining Na_2O this is assigned to magnetite with the addition of an equal amount of FeO derived from the excess remaining after the formation of pyrite, chromite and ilmenite. Any remaining Fe_2O_3 is calculated as haematite, whilst CaO is calculated as diopside by adding an equal amount of MgO + FeO plus an equal amount of SiO_2; or it is reserved provisionally for wollastonite.

(*6*) Any excess of MgO + FeO, after the formation of diopside, is allocated to hypersthene.

So far, it has been assumed that the calculations are dealing with analyses of rocks possessing adequate amounts of SiO_2. In those analyses possessing insufficient quantities of silica to complete the foregoing normative calculations a kind of reverse process of subtraction and reassembly of oxides must take place in order to derive a stoichiometric picture from the molecular numbers. The number of changes could be various and even pointless unless modal control is exercised over the choice of normative minerals. In general terms the following rules tend to guide the analyst in dealing with low values of silica:

(1) If there is not enough silica to equal half the amount of MgO + FeO to form hypersthene then these oxides are assembled as a molecule of olivine.

(2) Likewise the CaO and TiO_2 is calculated as perovskite and not as titanite.

(*3*) The sum of SiO_2 needed to form the metasilicates, zircon, acmite, diopside, wollastonite, orthoclase and albite is deducted from the total silica.

(*4*) If there is an excess of silica more than twice, but less than six times that of Na_2O to form provisional albite, then this is distributed between albite and nepheline as follows:

$$\text{albite molecules} = \frac{SiO_2 + 2\,Na_2O}{4}$$

$$\text{nepheline molecules} = Na_2O - \text{albite molecules}$$

(5) If the silica is less than twice the amount of available Na_2O the soda is alloted to nepheline and the K_2O is distributed between orthoclase and leucite as an outcome of the following changes: olivine is substituted for hypersthene, perovskite is substituted for titanite, nepheline is substituted for albite, orthoclase molecules = $^1/_2$ (SiO − $4K_2O$), and leucite molecules = K_2O − orthoclase molecules.

(*6*) If the SiO_2 is less than four times the K_2O, then the CaO, MgO and FeO must be redistributed. The most common case is where there is no wollastonite or not enough to satisfy the deficiency in SiO_2, after allocating all the K_2O to form nepheline, anorthite, acmite and olivine and possibly to form zircon and sodium metasilicate. If in rare cases there is insufficient SiO_2 to form leucite the presence of the kaliophilite molecule must be assumed to be present in the norm.

To illustrate the application of these rules to the evaluation of analyses in terms of the C.I.P.W. classification the following example has been chosen to illustrate the way in which the C.I.P.W. normative mineral molecules can be used to enhance petrological research. In the application of these rules the choice of values for the molecular weights of the oxide is important. On balance it now seems advisable to base these on the atomic weights of the most stable isotopes of the elements. The weight of molecules derived from the division of the percentage values of the oxides by these molecular weights will be found in practice to yield a more reliable set of values than the employment of approximate values. On the other hand it will be found convenient to use conventional approximations of the molecular weights for the normative minerals in arriving at the percentage values of the salic and femic groups which determine the position of the analysis in the C.I.P.W. classification (Table II). The example illustrated by Table III serves to illustrate the application of the foregoing rules to the evaluation of a chemical analysis in terms of the C.I.P.W. classification. The chemical analysis is of a quartz–andesine–biotite gneiss (Table III).

The "Niggli Equivalent Norm" is now widely used as a means of relating rocks to their petrogenetic setting.

NIGGLI (1920, 1936, 1938, 1950) was clear about the objectives which should be achieved from the production of a chemical analysis of a rock. It should not be limited to the classification of that rock—the principal objective of the C.I.P.W. classification. Rock analyses should be used to interpret the geochemical processes by which various assemblages of minerals can form under thermodynamic control from similar groups of oxides. Consequently, Niggli expanded the list of so-called normative minerals to include those minerals of complex structures which were omitted from the C.I.P.W. classification. By including the sesquioxide-bearing pyroxenes, the hornblendes, the micas and the accessory minerals he was able to rationalize the analytical picture

TABLE III

CHEMICAL ANALYSIS OF A QUARTZ–ANDESINE–BIOTITE GNEISS

Oxide	*Wt.%*	*Mol. wt.*	*No. mol.*	*ap*	*il*	*or*	*ab*	*an*	*C*	*mt*	*en*	*fs*	*Q*
SiO_2	76.41	60.09	1.2715			00.1740	00.3042	0.0714			0.0042	0.0012	00.7150
Al_2O_3	12.67	101.96	0.1290			00.0290	00.0507	0.0357	0.0086				
Fe_2O_3	0.83	189.70	0.0052							0.0056			
FeO	0.82	71.85	0.0114		0.0026					0.0056			
MgO	0.18	40.32	0.0045								0.0030	0.0015	
CaO	2.04	56.08	0.0363	0.0006				0.0357					
Na_2O	3.14	61.98	0.0507				00.0507						
K_2O	2.74	94.20	0.0290			00.0290							
TiO_2	0.21	79.90	0.0026		0.0026								
MnO	0.03	70.94	0.0004										
P_2O_5	0.03	141.06	0.0002	0.0002									
Approx. mol. weights				336	152	556	524	278	102	232	100	140	60
% Norm. Minerals				0.0670	0.3950	16.1240	26.5700	9.920	0.8800	1.4100	1.3300	0.4040	42.9000

Rock type: granite gneiss. *Locality*: the Island in Callaherick Lough, Co. Galway, Eire. (Unpublished analysis by P. K. Harvey, University of Nottingham.)

Classification:

Quartz = 42.90	Q		*Class*: $\frac{\text{sal}}{\text{fem}} = \frac{96.394}{2.08} = 46.3$ (Class 1)
Orthoclase = 16.124			
Albite = 26.57	52.614 F	sal. 96.394	*Order*: $\frac{\text{Quartz}}{\text{F}} = \frac{42.90}{52.614} = 0.8$ (Order 4)
Anorthite = 9.92			
Corundum = 0.88	C		*Rang*: $\frac{Na_2O + K_2O}{Ca} = \frac{0.0507 + 0.0290}{0.0363} = 2.1$ (Rang 2)
Enstatite = 1.33			
Forsterite = 0.404	1.734 P		*Subrang*: $\frac{K_2O}{Na_2O} = \frac{0.0290}{0.0507} = 0.57$ (Subrang 4)
Magnetite = 1.41			
Ilmenite = 0.395	1.805 M	fem. 2.08	
Apatite = 0.67	A		

more clearly in terms of the normative components of a rock. By copying the practice of conventional chemists he attached symbols to his normative components in much the same way as was portrayed in the C.I.P.W. classification. By their use he attempted to satisfy the following conditions.

(*1*) To obtain an evaluation of an analysis which would be directly comparable with the observed or modal compositions of the rock.

(*2*) To simplify the calculation of the oxides into normative molecules and thereby enable one not only to classify a rock analysis but also to place it in its position in a petrogenetic reaction series.

(*3*) To introduce analyses of actual minerals into the calculations instead of a limited number of idealized formulae, such as those used in the C.I.P.W. classification.

The so-called "Niggli values" are obtained by the choice of the correct units—the equivalent weights of the minerals involved. The size of these equivalent weight units is to some extent arbitrary, since the molecular concept of classical chemistry is not entirely valid for crystalline minerals. The only factor which is decisive in the choice of the equivalent weight is the ratio of the elements involved in the empirical formulae of the minerals concerned. For example, in the chemical composition of orthoclase the ratio of Si:K:Al:O is 3:1:1:8. This ratio could be used to postulate orthoclase (usually abbreviated to 1 Or) as either $K.AlO_2.(SiO_2)_3$, $K_2O.Al_2O_3.6SiO_2$ or $^1/_3$ (Si_3O_8AlK). These possibilities were conventionalized by Niggli by assembling the normative molecule so that it contained equivalent numbers of atoms of the elements Si, Al, Fe, Mn, Mg, Ca, Na, K, Ti, P, Zr, S, etc. These he referred to as "conformable formula units" and for practical purposes the sum of these elements was calculated to unity. This conclusion was based on the geochemical fact that all electropositive elements (cations) contained in rock-forming minerals have relatively low atomic numbers and that their values are closely grouped to one another. Moreover, the light elements like hydrogen and the heavy elements play an insignificant part in the petrogenesis of the lithosphere and can thus be excluded from the computation of chemical analyses. Into this, since the majority of the rock-forming minerals are silicates, the combination of elements with silicon and oxygen lessens the effect which differences in their atomic weights and atomic numbers have upon the molecular or equivalent weights of the mineralogical molecule as a whole. Thus, Niggli was able to show that it was possible to designate numerical indices to minerals by means of the summation of their electropositive elements. For example if the formula for orthoclase is taken as $K_2O.Al_2O_3.6SiO_2$ then the sum of the electropositive elements is 10. If this is reduced to unity, then the symbolic use of Or is employed (in the same sense as the "or" of the C.I.P.W. classification) to represent orthoclase in a reaction series then Or $= \frac{1}{10}$ ($K_2O.Al_2O_3$. $6SiO_2$) and would represent an equivalent weight of 55.65. Additional examples are as follows:

Albite (1 Ab) = $^1/_{10}$ ($Na_2O.Al_2O_3.6SiO_2$). Equiv. wt. = 52.43.

Anorthite (1 An) = $^1/_5$ ($CaO.Al_2O_3.2SiO_2$). Equiv. wt. = 55.63.

Nepheline (1 Ne) = $^1/_6$ ($Na_2O.Al_2O_3.2SiO_2$). Equiv. wt. = 47.34.

Forsterite (1 Fo) = $^1/_3$ ($2MgO.SiO_2$). Equiv. wt. = 46.90.

Enstatite (1 En) = $^1/_2$ ($MgO.SiO_2$). Equiv. wt. = 50.19.

Quartz (1 Qu) = 1 (SiO_2). Equiv. wt. = 60.06.

Using this notation petrogenic equations assume a very simple form, as is indicated by the following example:

$$Na_2O.Al_2O_3.\ 2SiO_2 + 4SiO_2 = Na_2O.Al_2O_3.6SiO_2$$
$$6\ Ne + 4\ Q = 10\ Ab$$

In reactions involving H_2O and CO_2 these should be

TABLE IV

NIGGLI VALUES

Computable equivalent weights

SiO_2,60.1; 1/2 Al_2O_3,51; 1/2 Fe_2O_3,79.8; FeO, 71.8: MgO, 40.3; CaO, 56.1; 1/2Na_2O, 31; 1/2 K_2O, 47.1; TiO_2,79.9; 1/2 P_2O_5,71

	Normative components	*Equiv. wt.*
Quartz	$1 Q = 1SiO_2$	60.1
Felspars		
Orthoclase, potassium felspar	$1 Or = 1/10 (6SiO_2.Al_2O_3.K_2O)$	55.6
Albite, sodium felspar	$1 Ab = 1/10 (6SiO_2.Al_2O_3.Na_2O)$	52.4
Anorthite, calcium felspar	$1 An = 1/5 (2SiO_2.Al_2O_3.CaO)$	55.6
Felspathoids		
Kaliophylite, potassium-nepheline	$1 KP = 1/6 (2SiO_2.Al_2O_3.K_2O)$	52.7
Nepheline	$1 Ne = 1/6 (2SiO_2.Al_2O_3.Na_2O)$	47.3
Leucite	$1 Lc = 1/6 (3SiO_2.Al_2O_3.K_2O)$	54.5
Analcite	$1 Anc = 1/8 (4SiO_2.Al_2O_3.Na_2O.\ 2H_2O)$	55.0
Gehlenite	$1 Ge = 1/5 (SiO_2.Al_2O_3.2CaO)$	54.8
Na-Gehlenite	$1 Na\text{-}Ge = 1/5 (2SiO_2.\ 1/2\ Al_2O_3.CaO.\ 1/2\ Na_2O)$	51.6
Fe-Gehlenite	$1 Fe\text{-}Ge = 1/5 (SiO_2.AlFeO_3.\ 2CaO)$	60.6
Akermanite	$1 Ak = 1/5 (2SiO_2.MgO.\ 2\ CaO)$	54.5
Fe-Akermanite	$1 Fe\text{-}Ak = 1/5 (2SiO_2.FeO.2\ CaO)$	60.8
Sodalite	$1 Sod = 1/20 (6SiO_2.3\ Al_2O_3.\ 3Na_2O.2\ NaCl)$	48.4
Noselite	$1 Nos = 1/21 (6SiO_2.\ 3Al_2O_3.\ 3Na_2O.Na_2O.SO_3)$	47.3
Hauynite	$1 Hau = 1/22 (6SiO_2.3Al_2O_3.3Na_2O.2CaO.\ 2SO_3)$	51.2
Cancrinite	$1 Canc = 1/20 (6SiO_2.3Al_2O_3.3Na_2O.2CaO.2CO_2)$	52.6
Olivine group		
Forsterite	$1 Fo = 1/3 (SiO_2.2MgO)$	46.9
Fayalite	$1 Fa = 1/3 (SiO_2.2FeO)$	67.9
Monticellite	$1 Mont = 1/3 (SiO_2.MgO.CaO)$	52.1
Tephroite	$1 Tephr = 1/3 (SiO_2.2MnO)$	67.3
Augites and hornblendes		
Enstatite	$1 En = 1/2 (SiO_2.MgO)$	50.2
Hypersthene	$1 Hy = 1/2 (SiO_2.FeO)$	65.9
Wollastonite	$1 Wo = 1/2 (SiO_2.CaO)$	58.1
Diopside	$1 Di = 1/4 (2SiO_2.MgO.CaO)$	54.1

TABLE IV (continued)

	Normative components	Equiv. wt
Hedenbergite	1 Hed = 1/4 ($2SiO_2.FeO.CaO$)	62.0
Tschermaks compound	1 Ts = 1/4 ($SiO_2.Al_2O_3.CaO$)	54.5
Aegirine (Acmite)	1 Ac = 1/8 ($4SiO_2.Fe_2O_3.Na_2O$)	57.7
K-Aegirine (K-Acmite)	1 K-Ac = 1/8 ($4SiO_2.Fe_2O_3.K_2O$)	61.8
Jadeite	1 Jd = 1/8 ($4SiO_2.Al_2O_3.Na_2O$)	50.5
Spodumene	1 Spod = 1/8 ($4SiO_2.Al_2O_3.LiO_2$)	46.5
Common hornblende	1 Ho = 1/15 ($7SiO_2.Al_2O_3.(4Mg,Fe)O.2CaO.H_2O$)	54.3–62.7
Riebeckite$_1$	1 Rb_1 = 1/15 ($8SiO_2.Fe_2O_3.3FeO.Na_2O.H_2O$)	62.3
Riebeckite$_2$	1 Rb_2 = 1/32 ($16SiO_2.Fe_2O_3.8FeO.3Na_2O.2H_2O$)	59.9
Riebeckite$_3$	1 Rb_3 = 1/32 ($16SiO_2.2Fe_2O_3.6FeO.3Na_2O.H_2O$)	60.4
Micas		
Mg-Biotite	1 Mg-Bi = 1/16 ($6SiO_2.Al_2O_3.6MgO.K_2O.2H_2O$)	52.1
Fe-Biotite	1 Fe-Bi = 1/16 ($6SiO_2.Al_2O_3.6FeO.K_2O.2H_2O$)	64.0
Muscovite$_1$	1 Ms_1 = 1/14 ($6SiO_2.3Al_2O_3.K_2O.2H_2O$)	56.9
Muscovite$_2$ (phengitic)	1 Ms_2 = 1/14 ($7SiO_2.2Al_2O_3.(Mg,Fe)O.K_2O.H_2O$)	56.8–59.0
Lepidolite	1 Lep = 1/32 ($12SiO_2.5Al_2O_3.3Li_2O.4KF.2H_2O$)	49.6
Zinnwaldite	1 Zwd = 1/16 ($6SiO_2.2Al_2O_3.2FeO.$ 2 $LiF.2KF$)	54.7
Scapolite group		
Chloride-marialite	1 Ma = 1/32 ($18SiO_2.3Al_2O_3.3Na_2O.$ 2NaCl)	52.8
Carbonate-marialite	1 Ma_2 = 1/32 ($18SiO_2.3Al_2O_3.4Na_2O.CO_2$)	52.4
Sulphate-marialite	1 Ma_3 = 1/33 ($18SiO_2.3Al_2O_3.4Na_2O.SO_3$)	52.0
Chloride-meionite	1 Me_1 = 1/16 ($6SiO_2.2Al_2O_3.3CaO.CaCl_2$)	59.1
Carbonate-meionite	1 Me_2 = 1/16 ($6SiO_2.3Al_2O_3.4CaO.CO_2$)	58.4
Sulphate-meionite	1 Me_3 = 1/17 ($6SiO_2.3Al_2O_3.4CaO.SO_3$)	57.1
Accesories, oxides, etc.		
Ca-phosphate	1 Cp = 1/5 ($P_2O_5.3$ CaO)	62.0
Apatite	1 Ap = 1/16 ($3P_2O_5.9CaO.CaF_2$)	63.0
Titanite	1 Tit = 1/3 ($SiO_2.TiO_2.CaO$)	65.3
Rutile	1 Ru = 1 TiO_2	79.9
Perovskite	1 Pf = 1/2 ($TiO_2.CaO$)	68.0
Magnetite	1 Mt = 1/3 ($Fe_2O_3.FeO$)	77.1
Ilmenite	1 Ilm = 1/2 ($TiO_2.FeO$)	75.9
Hematite	1 Hm = 1/2 Fe_2O_3	79.8
Chromite	1 Cm = 1/3 ($Cr_2O_3.FeO$)	74.6
Fluorite	1 Fr = 1 CaF_2	78.1
Pyrite	1 Pr = 1/3 FeS_2	40.0
Corundum	1 C = 1/2 Al_2O_3	51.0
Zircon	1 Z = 1/2 ($SiO_2.ZrO_2$)	91.6
Calcite	1 Cc = 1 ($CaO.CO_2$)	100.1
Anhydrite	1 A = 1/2 ($CaO.SO_3$)	68.1
Thenardite	1 Th = 1/3 ($Na_2O.SO_3$)	47.3
Na-carbonate	1 Nc = 1/2 ($Na_2O.CO_2$)	53.0
Halite	1 Hl = 1 NaCl	58.5

TABLE V

EXAMPLE FOR CALCULATING THE NIGGLI VALUES

Rock type: hornblende schist. *Locality*: Höyggelet, Norway (ELLIOTT and MORTON, 1966)

Oxides	*Wt.%*	*Approx. mol. wt.*	*Mol. no. (× 1000)*	*Grouped mol. no.*	*Grouped mol. no.(%)*	*Niggli values*
SiO_2	48.10	60	801	si. 801		al = 22
TiO_2	1.80	79.9	22			fm = 46
Al_2O_3	16.10	102	158	al. 158	al. 22	
Fe_2O_3	1.90	160	24 }			c = 25
FeO	10.40	72	144 }			alk = 7
			}	fm. 327	fm. 46	si = 113
MnO	0.12	71	1 }			
MgO	6.30	40	158 }			k = 0.11
CaO	10.00	56	179	c. 179	c. 25	mg = 0.48
Na_2O	2.60	62	42 }			qz = –15
			}	alk. 47	alk. 7	
K_2O	0.50	94	5 }			
P_2O_5	0.23	142	2			
CO_2	tr					
H_2O	2.10					
Total	100.15		1,536	711	100	

Calculations:

$$si = \frac{\text{mol. no. } SiO_2}{\text{mol. no. } Al_2O_3} \times \frac{al}{\text{total mol.}} = \frac{\frac{48.10}{60}}{\frac{16.10}{102}} \times \frac{158}{711} = 113$$

$$si = 100 + 4(\text{alk}\%) = 100 + 28 = 128$$

$$qz = si - si' = 109 - 124 = -15$$

$$k = \frac{\text{mol. no. } K_2O}{\text{mol. no.}(K_2O + Na_2O)} = \frac{5}{42 + 5} = 0.11$$

$$mg = \frac{\text{mol. no. MgO}}{\text{mol. no.}(FeO + MnO + MgO)} = \frac{158}{327} = 0.48$$

placed with brackets in the equation as is illustrated by the following reaction of calcite with quartz:

$$CaO.CO_2 + SiO_2 = CaO.SiO_2 + CO_2$$
$$1Cc + 1Qu = 2Wo + (CO_2)$$

This form of expression has advantages in portraying metamorphism as the process involves the complication of hydroxyl molecules and gases like CO_2 which are either added or expelled during these reactions under elevated temperatures and pressures.

To obtain the Niggli values the percentage weight of oxide obtained by analysis is divided by the molecular equivalent weight of the compound which is derived from the computable equivalent weights of the principal oxides (see Table IV; after BURRI, 1964). In order to interpret the chemical analysis in terms of its position in relation to other rock types it is first necessary to evaluate the distribution of the oxides in terms of silica (si), ferro-magnesian oxides (fm), alkalis (alk), calcium oxide (c), titania (ti) and phosphate (p). From these are calculated the Niggli values. The example given in Table V will serve to illustrate the calculations involved.

These so-called Niggli values are of great value in comparing the composition of one rock with another. In the majority of usages this is probably the most valuable feature of the Niggli method of using chemical analyses. However, it is a straight forward matter to proceed to an evaluation of these Niggli values in terms of normative molecules using the data shown in Table IV and in this way arrive at a basis of comparison with the norms calculated by the C.I.P.W. method.

HYDROTHERMAL PROCESSES

During the crystallization of a magma the escape of volatiles to form pegmatites and sulphide veins is a topic of major academic and economic importance. Our knowledge of terrestrial gases and volatiles is somewhat limited to the analyses of volcanic eruptions or those gases locked inside fresh igneous rocks. Recently KRAUSKOPF (1959) reviewed this problem by calculating the equilibrium factors affecting gases at high temperatures and pressures and relating these to a theoretical magmatic gas. "Different magmatic gases doubtless have widely different compositions, but the results of these calculations suggest that the most important difference is in the state of oxidation. In defining the oxidation state, the partial pressure of free oxygen serves much the same purpose as redox potential in defining the oxidation state of a solution." (KRAUSKOPF, 1959, p.271.) This is complicated by the important, and highly significant occurrence of metallic compounds as volatile additives to these magmatic gases. These are of considerable importance in the understanding of the genesis of ore deposits. The conventional hydrothermal hypothesis for the formation of ores implies that metals are transported from the crystallizing magma as hot aqueous solutions. Most of these metallic compounds have a low solubility which imposes a weakness in the acceptance of this hypothesis. Consequently, there is a growing belief in the possibility that metals, particularly in the form of chlorides and fluorides, are sufficiently volatile to migrate and concentrate themselves as ore deposits. This aspect of the problem enforces the recognition of the importance of gaseous components in hydrothermal reactions. Geological observations uphold the view that most of the magmatic metallic ores are most stable as sulphides. At 600°C the vapour pressure of most metallic sulphides is low and this would suggest that there is little hope of them migrating as volatiles, with the possible exception of mercury, antimony, bismuth, cadmium and zinc. Therefore, to explain their fugitive nature, in relation to magmatic intrusions, volatility must be combined with the solubility effects induced by magmatic water. All too little is known of the effects of super-heated steam and its effects upon non-volatile metallic compounds. Indirectly, it is to be hoped that the study of deposits formed in high temperature steam generators such as those used in electrical power stations, will throw considerable light on this important aspect of petrogenesis. Many of these deposits show the physical characteristics of natural magmatic ores and information is forthcoming on the solubility of substances in fluids at super-critical pressures in excess of 224 kg / cm^2. MOREY and HESSELGESSER (1951) have deduced values for the solubility of silica, sodium chloride, sodium sulphate and calcium sulphate, and although some of their figures for lower pressures are at variance with those of other workers, it is apparent that these substances are readily soluble in supercritical steam. Moreover, this area of investigation is beginning to provide data which will emphasise the function of water in its gaseous phase as a prime agent in the separation and uplift of large masses of rock to provide the conditions for the intrusion and crystallization of pegmatites and basaltic magmas as sills, dykes, laccoliths and cone sheets. Magmatic water under super-critical conditions possesses far more kinetic energy than even the most basic of magmas so that hydraulic movement of this kind would produce the space conditions for magmas such as dolerite to invade widespread areas to form of sills and swarms of dykes. Moreover, it is significant that such sills of dolerite possess structures which would be difficult to form if the magma was forced to support the superincumbent rocks during its cooling stages. On the other hand, if steam uplifted these rocks and maintained them in this position during the cooling of the magma, it would have created conditions under which structures, such as columnar joints, could be readily developed. Such a mechanism would also account for the absence of collapse structures in the so-called "roof" of these sills. On the other hand, the solvent capacity of these super-critical fluids may account for the widespread occurrence of quartz veins in areas affected by minor intrusions of basaltic magma.

GEOTHERMOMETRY

The temperatures at which minerals crystallize from a magma provide obvious clues to the temperatures which may have existed in the crust of the earth. Thus the study of suitable minerals is beginning to provide geothermal knowledge by using them as geological thermometers. Even so the subject is still in its infancy. BARTH (1962) has shown that felspars can be used as thermometers, provided one accepts the assumption that chemical equilibrium is established in the rock at definite cooling temperatures. In addition to this, one must recognise that the temperatures calculated in this way represent minimum values which may be far below the highest of the pre-existing levels of thermal activity. Introducing

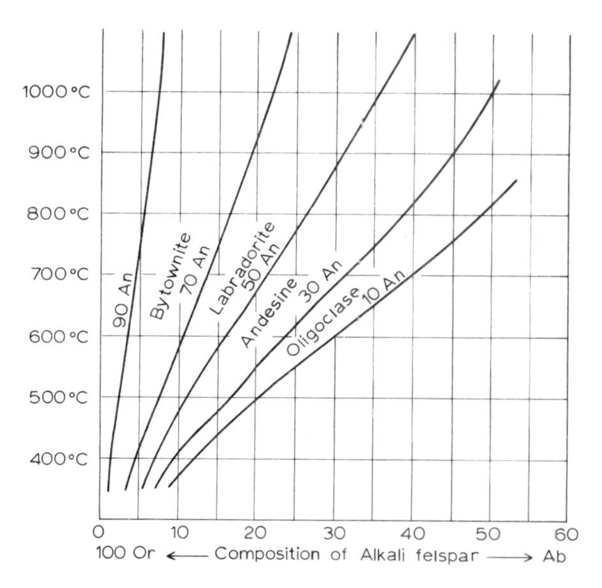

Fig.1. Diagram illustrating the coexistence of alkali felspar and plagioclase showing the equilibrium relationships in terms of anorthite (An) at various temperatures. (After BARTH, 1962, fig.4.)

felspars as geological thermometers, BARTH (1962) used the fact that albite is soluble in orthoclase and anorthite, so that during crystallization the sodium distributes itself between the potassium and the calcium felspars. Assuming that pressure is of less importance than temperature in this phase distribution, Barth has shown that from determinations of the co-existence of alkali felspar and plagioclase it is possible to calculate the temperature at which they co-existed during crystallization. This is illustrated by Fig.1. Others, such as KULLERUD (1959) have carried out systematic studies of synthetic sulphide systems and applied their findings to the stability of mineral assemblages in terms of temperature and vapour pressures. In this way, they have been able to produce a thermodynamic picture of the possible mode of formation and subsequent alteration of ore deposits.

CHEMICAL SOLUBILITY PRODUCTS

Another important area of geochemical studies is that which covers the rate at which reactions proceed at low temperatures and pressures. This is particularly important in the study of sediments. Recent studies (see GARRELS 1959, p.25) have shown that it is unlikely that ionic reactions form precipitates which are initially stable. Continued precipitations under geological conditions enlarges the grain size of the precipitate and this in itself tends to maintain the deposit for some time in its original condition. The addition of other precipitates inevitably leads to interactions in an attempt to achieve chemical stability. Such reactions between successive precipitates probably dominates, for example, the mineralogical composition of evaporites. Associated with precipitation is the relationship of these precipitates to their original solvents. Evaporites, like gypsum, dissolve to equilibrium at the rate of about 10 g/l of pure water. However, such forms of congruent solution rarely occur in nature. By contrast, slight changes in the aqueous environment lead to a complete interplay of solvent action and precipitation. For example, if crystals of sodium sulphate are placed in solutions of calcium chloride the results are not always identical even though the concentrations of each salt are maintained to a uniform degree. Sometimes the sodium-sulphate crystals may develop a coating of gypsum which brings the reaction to an end, or, for no obviuos reason, the reaction may proceed to completion to form a fine-grained gypsum powder. On another occasion, the sodium-sulphate crystals may acquire an incomplete coating of gypsum and from the exposed reaction areas spires of gypsum may develop (see GARRELS, 1959, p.34). This so-called incongruent form of solution is prevalent among most detrital minerals. Felspars react in this way and according to NASH and MARSHALL (1956) the surfaces of the felspar develop disordered K^+ and Na^+ sites upon which a new felspar is formed. This retards the further migration of the K^+ and Na^+ ions outwards through the zone occupied by the hydrogen ions. Such protective mechanisms may well explain many of the characteristics of arkose sediments. Likewise, the development of surface coatings of haematite on grains of magnetite might well account for the preservation of such minerals in placer and foreshore deposits.

The ease with which silica replaces other minerals, but is itself rarely altered into other minerals as pseudomorphs, has not been sufficiently emphasized. It always appears to be the end-product of magmatic reactions and of the genesis of cements and secondary structures in sediments. Silicon itself is unknown in crustal rocks, but its association with other metals and oxygen comprises 95.58% of the composition of the crust. In its affinity to oxygen it sometimes resembles the correlative tetravalent element carbon, but the characteristics of the resultant oxides are quite different, being respectively solids and gases. On the other hand, the nature of the hydrogen derivatives of silicon—the so-called silicones—react and combine with metals and hydrocarbons. Consequently, it might be profitable to visualise the formation of metallo-silicones as intermediate compounds segregating metal sulphides and ferrous iron in the presence of excess oxygen.

KRAUSKOPF (1959) has provided experimental evidence that the solubility of silica ranges from 50–80 p.p.m. at 0°C, 100–140 p.p.m. at 25°C, to 360–420 p.p.m. at 100°C. Of all the crystalline and cryptocrystalline forms of natural silica quartz is the least soluble and yet it is the most common form of silicious cement in rocks. GARDENER (1938) quotes a figure of 6 p.p.m. and SIEVER (1957) a range of between 7 and 14 p.p.m. for the solubility of quartz. On the other hand, its solubility is greatly influenced by physical factors. Freshly fractured quartz is markedly more soluble than well-rounded detrital grains. Conversely, well-rounded grains have a higher adsorption coefficient than angular particles. This is illustrated by the microscopic examination of angular and rounded particles of quartz in contact with hydrofluoric acid (see Fig.2). Freshly fractured quartz was placed in a perspex slide holder and observed microscopically in contact with hydrofluoric acid. The conchoidal edges of the particles should have been the obvious places for solvent action, but it will be seen in Fig.2 that these survived longer than the median parts of the particle. The path of the acid was along a well-defined plane which is now known to be an incipient plane of cleavage in quartz. In addition, the acid revealed that slender microlites of quartz were produced along these crystal surfaces and no trace of them developed upon the conchoidal fracture surfaces. In many ways these microlites of quartz are commonly found in weathered sediments and it would appear from this experiment that they represent the product of chemical action on quartz grains and not a product of secondary crystallization of quartz in soils. By contrast in Fig.2, the rounded grain of detrital quartz was remarkably resistant to solvent action. This serves to indicate the reason why such wide variations occur in the estimation of the solubility of quartz as recorded by various analysts. A reasonable expla-

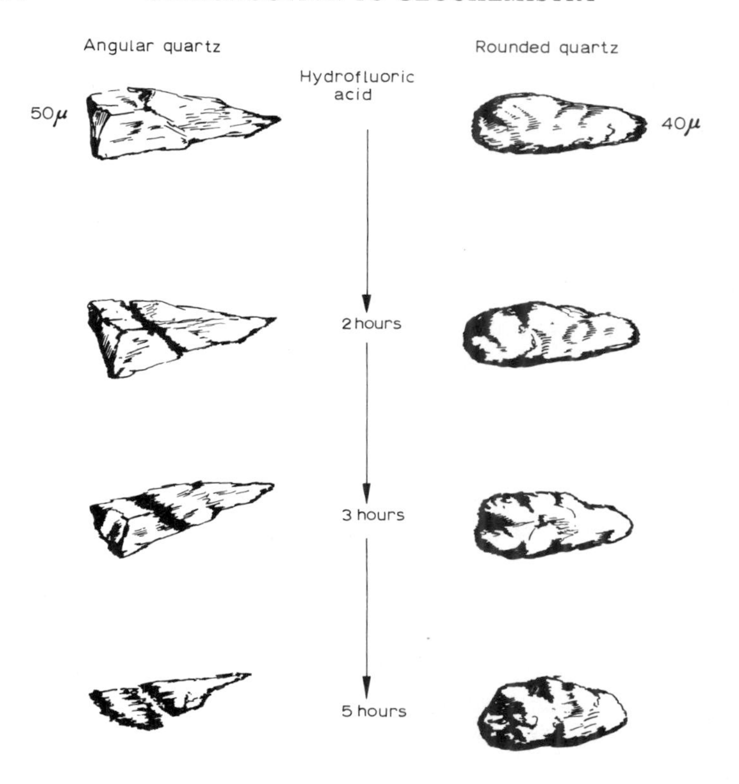

Fig.2. Serial illustrations of quartz particles inmersed in hydrofluoric acid to show the nature of their dissolution relative to each other.

nation of the resistance of rounded grains to solvent action is the existence of adsorbed valency balanced layers of alumina and iron on most detrital grains of quartz. Since quartz grains of this type are the main source of silica to plants and animals in soils this protective coating has to be removed before silica can be released into solution. Consequently, the effect of the sugar acids and amino acids on the dissolution of these layers has been indicated to be the media through which silica is released from soils and subsoils (EVANS, 1955).

Whilst it must be acknowledged that the solubility of quartz is significant at high temperatures it would only account for the formation of quartz veins. Such temperatures are not associated with the more widespread development of quartz as a cement in sediments, or the secondary silicification of rocks and fossils. At high pH, the solubility of silica would reach the levels of concentration which would result in silicification of sediments, but such conditions are rarely attained in natural basins of deposition. It is well-known that marine organisms, particularly diatoms, have the ability to secrete silica and form shells and skeletons of opaline quartz. Moreover, JORGENSON (1953) has shown that diatoms reduce the silica concentrations of sea water from 135 p.p.m. to 0.085 p.p.m., and yet develop opaline shells in the very medium which would seemingly redissolve them, if it was a purely inorganic process. He suggests that this is more probably accompanied by the adsorption of cations to form insoluble silicates or by the formation of an organo–silica complex on the surfaces of the shells and within the skeletons. BIEN et al. (1958) have shown that adsorption on suspended particles of sediments is also an effective means of removing dissolved silica from very dilute solutions. On the other hand, the outstanding characteristics of natural waters is their apparent undersaturation. Most river waters and ground waters contain less than 35 p.p.m. SiO_2 (CLARKE, 1924), so that the widespread development of silica cements and silicates in sediments must reflect an even more subtle interplay of solution and precipitation. This is illustrated by the low concentrations in sea water which, at the surface, appears to range from 0.1 to 4 p.p.m., but in deep water it increases to 5–10 p.p.m. It has been frequently stated (e.g., TARR, 1927; or TWENHOFEL, 1950) that electrolytes coagulate and precipitate the silica. Whilst the average concentration of silica in the surface sea waters of the Gulf of Mexico averages 0.11 p.p.m., that of the Mississippi ranges, from 4 to 7.5 p.p.m. (BIEN et al., 1958). Where the two waters meet, there is no sign of an accumulation of silica in the surface sea water. This depletion is brought about by the mixing of the river and sea waters. It might have been achieved by biological uptake, by the reaction of soluble silica with electrolytes in sea water leading to salt formations, or by adsorption or co-precipitation of soluble silica with suspended solids or colloidal material in river water when it makes contact with saline electrolytes. When the final explanation eventually emerges it will provide the mechanism which leads to the widespread development of siliceous sediments which have accumulated throughout time as estuarine deposits.

Deposits in fissures sometimes resemble the corrosion products of metals. For example, nodules found in limestones, ironstone, and phosphatic deposits closely resemble corrosion deposits. No correlation has yet been established, but in the near future it seems likely that many factors involved in the formation of these industrial secretions will be used to explain a wide variety of sedimentary structures.

FLUIDIZATION

Another field of applied research which has yet to make its full impact in the geochemistry of sediments is that referred to as "fluidized beds" (see DAVIDSON, 1960). The release of air from unconsolidated deposits and the effects of gaseous bubbles passing through sediments bears some relationship to this field of investigation. Powdered solids can be made to behave like a liquid if the gas or liquid passing through it is of such velocity that it will support the particles in isolation. In this condition no particle has to support any other and friction between contiguous particles becomes close to zero. It is well worth applying such considerations to the mechanism involved in the formation of so-called slump deposits. Moreover, the effect of fluidization is to produce a complete distribution of particles in terms of grain size and this may well be the explanation of the occurrence of arenaceous deposits which exhibited no trace of graded bedding. They may have possessed such textures originally but owing to the fluidizing action of water emerging from the sediments it has dispersed the graded particles.

Fluidization might also account for the destruction of bedding planes and an increase in the porosity of deposits found in the anticlinal reservoirs of oilfields. An experimental model of the oil-bearing anticline at Eakring indicated that fluidization does play an important part in the concentration of petroleum in sandstones and such like reservoir rocks. To study this phenomenon (EVANS, 1962) a scale-model of the "Eakring Anticline" was made in an enclosed tank (see Fig.3). The deposits involved were made as accurately as possible to represent the actual deposits of the oilfield. Carboniferous limestone and Millstone grit, which represent the source and reservoir rocks of the oil-bearing anticline, were disintegrated and deposited respectively in actual sea water and brackish waters. These precautions were taken to overcome the difficulty of simulating these deposits by artificial materials and to acknowledge the impossibility of simulating sea water and estuarine waters. Likewise crude petroleum pumped from the Millstone grit at Eakring was added to the slurry of Carboniferous limestone in sea water. As will be seen in Fig.3B the oil disappeared and must have been thoroughly dispersed throughout the limestone deposit which was maintained at a "pore-point" temperature of 40°F throughout the life of the model. Similar geological analogues were incorporated into the construction of the overlying sediments (EVANS, 1962) and, as a result, success was achieved in developing a model oilfield in the anticlinal structure which was elevated pneumatically, by means of an oval bladder situated on the floor of the tank. As time passed the evolution of light hydrocarbons in a gaseous condition fluidized the coarse and fine grades of sandstone representing the Millstone grit. Eventually all traces of graded bedding disappeared and the

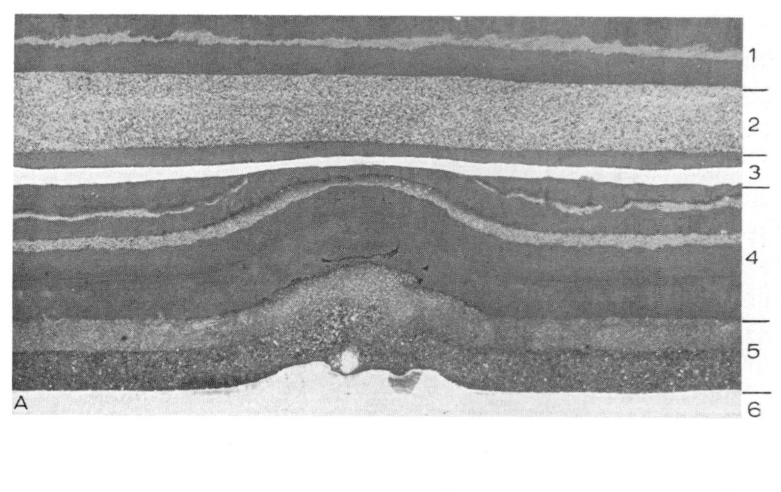

Fig.3. A. The model prior to the migration of oil from the Carboniferous limestone into the coarse and fine sands of the Millstone grit. B. The oil model over 1 year later showing the diffusion of oil into the anticline causing the destruction of the junction of the coarse and fine sandstcne, probably by a process of fluidization. The different layers consist of: *1* = Keuper marl; *2* = Buntsandstein; *3* = magnesian limestone; *4* = Coal Measures; *5* = Millstone Grit; *6* = Carboniferous limestone.

whole area of anticline acquired an increase in the porosity of these reservoir deposits (see Fig.3B). The correlation of this experimental demonstration with observed data in an oil-field is so close as to indicate that the process of fluidization is probably of great importance in the growth of a viable concentration of petroleum in anticlinal structures and also in stratigraphical traps.

ORGANIC GEOCHEMISTRY

The model of the Eakring oilfield also serves to illustrate some of the mechanisms involved in the association of organic matter with sediments. Fugitive hydrocarbons attempting to penetrate the "cap-rocks" are in a highly reactive condition. Evidence is emerging to show that the tensional fissures which were developed during the growth of the anticline were invaded by such hydrocarbons and that they reacted with the clay minerals to form crystalline accumulations of sericitic micas, of which illite is a particularly common variety (EVANS, 1963). In the broader field of sedimentology the impact of organic molecules, and particularly those of biological dimensions, is rapidly being realised by geochemists. The concept of cementation in sedimentary rocks proceeding along conventional inorganic lines over long periods of time and overloading pressures is rapidly being replaced by the metallo-organic reactions which are now known to take place fairly rapidly in marine and estuarine waters. Biogenetic reactions are clearly as important in the growth and decay of rocks as they are in the birth and survival of organisms and it is now difficult to distinguish between results described as the product of "organic geochemistry" and those referred to as the products of "biogeochemistry".

The earliest publications which might be regarded as the origin of this branch of geochemistry emerged from the prolific writings of Carl Neuberg, the acknowledged "father of biochemistry". Of particular interest was his re-interpretation of the origin of stalactites and stalagmites (NEUBERG and MANDL, 1948). This was followed by the discovery of biogenic illite and quartz in pulmonary tissues (EVANS, 1948) and of an antibiotic called "vitricin" which had survived the biogenesis of Carboniferous coals (EVANS, 1951). NEUBERG and MANDL (1948) showed that the transfer of lime derived from limestone was not in the form of a bicarbonate but as a calcium derivative of adenosine triphosphate. They were also able to show that various amino acids could achieve the same end-product more easily than the simple action of so-called carbonic acid derived from rain water. They pointed out that the latter would be immediately neutralized on entering the soils and subsoils overlying limestones, whilst the organic acids would be extracted by rain water to act as solvents capable of maintaining calcium carbonate, as such, under neutral pH conditions.

The conventional approach to the cementation of sediments has been by analogy with established inorganic reactions, even though it is realised that the conditions under which these laboratory reactions proceed could rarely exist in natural basins of deposition. Consequently, it has become customary to assume that long enduring pressure and temperatures have somehow resulted in the dissolution and recrystallization of somewhat insoluble substances like quartz. It must be acknowledged that cumulative and differential pressures developed during earth movements must influence the relative solubilities of compounds like silica and lime, but there are innumerable examples of sediments like quartzites and limestones, which have consolidated somewhat rapidly under negligible superincumbent pressures. Since most sediments have been developed in association with some form or another of biological activity, it is now evident that the function of organic matter must be considered in relation to the dissolution of mineral particles and their subsequent recrystallization. If this be the case, then the source of information concerning the metallo-organic processes involved is more readily obtained from living organisms than from the rocks themselves—unless one is studying the processes of weathering.

It was the mineralogical studies of the effect of dust on human lung processes (pneumokoniosis) which first stimulated these interesting possibilities (EVANS, 1948, 1951). Initially, the possibility of crystallization in lung tissues was regarded as somewhat abnormal, but the conclusions derived from these investigations began to emphasize the possibility of a profound relationship between biogenic processes and secondary mineralization. A significant discovery of a kidney calculus composed of an aggregate of quartz crystals (EVANS, 1955) further emphasized the possibility of solubilization and recrystallization of these commonplace minerals in association with biogenic processes. It is now known that this calculus was by no means unique as three similar specimens have since been recorded (HOLST, 1958). It is noteworthy that these were found in patients who had been treated for peptic ulcers with a synthetic Mg–Na–Al–silicate of zeolitic structure. Similar calculi have been reported as being relatively common in herbivorous animals and particularly in sheep, and this seems to be due to the high silica content of certain types of hay (SWINGLE, 1953; FORMAN et al., 1959).

Likewise, plants assimilate dissolved mineral matter with ease and allow it to recrystallize. It occurs as an encrustation on cell walls or as masses in the interior of cells. An interesting example is the presence of quartz in optical continuity which has grown in the sclerentymatous layers of cardamom seeds (Fig.4). In the same category are the various mineral aggregates associated whith petrified wood. This is particularly well illustrated by the remarkable substitution of plant structures by such minerals as calcite, iron pyrites and even opaline quartz in the Carboniferous flora preserved in coal balls (EVANS and AMOS, 1961). Further references to mine-

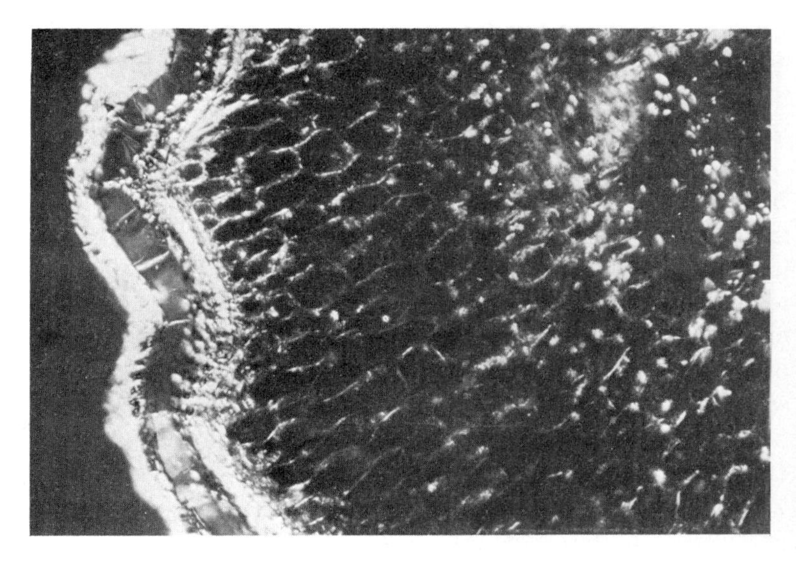

Fig.4. Thin section of a Cardamom seed showing clear areas containing aggregates of quartz crystals (×250).

ral growth and replacements in the botanical field are too numerous to record, but they all resolve themselves into the existence of a process of mineral solubilization which bears little resemblance to the relative solubility of these minerals in inorganic media. Of even greater importance than this is the possibility that the emergence of minerals from an organic solute is controlled crystallographically. If this was not the case then some other explanation must be found from the fact that the cellular tissue of plants remains practically undeformed by the minerals which replace them in petrified wood. In other words the molecular geometry of the organic molecule probably determines the crystalline habit of the emergent mineral. A notable example of this is the dual emergence of trigonal calcite and orthorhombic aragonite found in stalactites.

Probably the first constructive attempt to identify these solubilizing processes resulted from the remarkable experiments of Carl Neuberg and his associates, Ines Mandl, Amelie Grauer and L.S. Roberts. NEUBERG and MANDL (1948) observed a peculiar effect when determining zinc in a solution of L-alanine. Instead of the zinc being precipitated by Na_2S it remained in solution. It was eventually shown that alanine and all the amino acids which were tested exerted a profound effect upon the relatively insoluble sulphides of Mn, Co, Ni, Fe, Cu and most other metals. This enhanced solubility was also comparatively independent of pH. Extending these studies Neuberg and co-workers showed that the function of biogenic exudates extended beyond what takes place when relatively insoluble substances are attacked by acids or bases. "In general the latter reaction comes to a standstill on neutralization. Observations made on a vast number of organic acids and also some inorganic acids have led to the conclusion that exactly at this point of neutralization a special function of the newly formed neutral salts sets in, i.e., the salts themselves show a marked power of solubilization. This unexpected phenomenon which proved impressive in the case of triphosphates as well as salts of nucleic acids and nucleotides, applies to a wide variety of substances including adenosine triphosphates"(MANDL et al., 1952). The choice of ATP salts was obvious from their wide distribution in all cells, but the range of their solvent capacity was unexpectedly wide when applied to commonly occurring minerals in solids and sedimentary rocks. In the biochemical fields of study these observations were of great importance, but in relation to sediments these important studies have been almost ignored.

In order to establish techniques in the use of these nucleic acids, nucleotides, amino acids and phosphates, much of Neuberg's work has been repeated and confirmed (EVANS, 1963). Extraordinary solvent effects were obtained using *M*/5 Na-ATP solutions on known weights of freshly precipitated compounds. Over 200 such specimens were tested, which completely confirmed the findings of Neuberg and his associates. These so-called substrates were prepared by precipitations of the metal salts which, after washing with distilled

TABLE VI

RESULTS OF SOLUTION OF MINERAL POWDERS IN 1/5 *M* Na ATP

Mineral powders	*Observed results*
Calcite $CaCO_3$	Completely soluble between pH 7. 5 and 8.5; finely divided precipitate occurs on standing
Apatite $Ca_5F(PO_4)_3$	Completely soluble between pH 7. 5 and 8.5; minute needles develop on standing for 2 weeks
Augite $MgCa(SIO_3)_2$	Completely soluble between pH 7. 5 and 9.0; gelatinous precipitate develops on standing
Rhodonite $MnSiO_3$	Soluble on warming the suspension; gelatinous black precipitate develops on standing
Cerussite $PbCO_3$	Passes into solution for a short while, but soon forms a finely-divided precipitate
Malachite $(CuOH)_2CO_3$	Completely soluble at pH 7. 5
Rhodochrosite $MNCO_3$	Completely soluble between pH 7. 5 and 8.5, but oxidizes to manganese dioxide on standing
Galena PbS	Completely soluble at pH 7. 5
Siderite $FeCO_3$	Remains in solution at pH 7. 5, but on long continued standing forms a precipitate of ferric oxide.
Dolomite $CaMg(CO_3)_2$	Completely soluble at pH 7. 5
Kaolinite $H_4Mg_3Si_2O_9$	Completely soluble; remains in solution until ATP decomposes to yield a finely-divided white precipitate
Gypsum	Very readily soluble, but forms a precipitate on warming
Precipitated alumina	Insoluble
Gelatinous silica	Insoluble
Hydrated ferric oxides	Slightly soluble

water, were concentrated in a centrifuge and then taken up in solution in *M*/5 Na-ATP. In all cases the pH was maintained around 7.5. When a clear solution had been achieved it was found that the addition of the usual reagents failed to reprecipitate these compounds and they remained in solution until the Na-ATP decomposed. To test the relationship of this remarkable series of results to rocks, natural minerals were ground into powder and then taken up in *M*/5 Na-ATP (Table VI). Very little differences were observed which emphasizes the potentialities of nucleic acids and nucleotides as solubilizing agents in natural sedimentation and cementation under more or less neutral or natural conditions.

From Table VI it will be seen that relatively water-insoluble minerals such as carbonates, phosphates and silicates are easily brought into solution by ATP. In contrast it was found that many oxides, and particularly hydrated oxides of iron and alumina, were practically insoluble, and that they were often formed as precipitates from solutions of minerals during the decomposition of the solubilizer, Na-ATP. This may well be one process by which such hydrated oxides are developed in many types of sediments. On the other hand, such natural solubilizers may well prove able to hold salts in solution within the freshly formed sediment and ultimately allow them to crystallize to yield the ferrous and calcareous cements common in clays and sandstones developed under aerobic conditions.

In addition to enhancing the solubilities of inorganic salts it was shown by MANDL et al. (1952, 1953) that a wide range of so-called "solvents" would also solubilize organic derivates such as certain humates and alginates which are themselves relatively insoluble in water. Beginning with ATP it was natural to extend the work to the use of phosphates. Similar effects were noted using meta-, pyro-, and triphosphate salts. More significantly, similar solubility effects were produced using sodium ribonucleate, sodium galacturonate, sodium gluconate, lactates, glycolates and sodium desoxyribonucleate. As in the case of ATP it seems significant that this wide range of solubilizing agents had little effect upon the solubility of alumina, but exhibited a profound influence over the retention of most metallic salts in solution. Since these experimental solutions were maintained at a pH about 7.5 the ability of relatively small quantities of such compounds as adenosine-triphosphate and ribo- and desoxyribonucleates to maintain potential minerals in solution is of obvious geochemical significance. Moreover, if it is finally established that these minute concentrations of biogenic derivatives are not used up in the process of solubilization, or that they are continuously replenished, their effect upon the accumulation and induration of sediments would be widespread.

In marine sediments, as well as in continental deposits, the variety of "solvents" produced by decaying or living animal or plant tissues would be considerable. On the other hand, such solvents seem incapable of dissolving oxides so that minerals of such type which frequently coat other minerals as adsorbates, or oxidation products thereof, would inhibit such forms of solubilization. Consequently, coated grains of quartz, and the clay minerals in general would remain almost intact to form the bulk of such sediments. Likewise, the secondary effects produced in a sedimentary rock by subsequent associations with organic processes may well leach these rocks of their matrix minerals. A typical example seems to be the seat earths of Carboniferous coal seams. In these deposits, plant growth has leached the original shales and siltstones and re-precipitated nodular growths of limonite and other minerals as secondary structures. Within the coal seams a great variety of crystalline "insolubles" exists. It is not unusual to find cavities containing crystals of millerite enclosed in cubes of galena and these enclosed in quartz. In all cases it is impossible to provide geological evidence involving the requisite temperatures and pressures and inorganic solvents capable of yielding these crystallized sulphides of nickel and lead embedded in quartz.

The geochemistry of limestone formation is still in a similar hypothetical position, but such carbonates have always been recognized as organic derivates. Calcite has been crystallized in the laboratory, but under conditions which would scarcely develop in nature. Consequently, experiments were carried out at pH 7.5 to see whether the up-take of $CaCO_3$ and its subsequent crystallization would be influenced by the presence of ATP, and similar natural "solubilizers". Freshly precipitated $CaCO_3$ was taken up in $M/5$ Na-ATP at pH 7.5 to form a clear solution. This was allowed to crystallize at room temperature under a cover glass. Aggregates of rhombohedra possessing the optical properties of calcite developed within a few days. It is interesting to compare these with crystals of natural calcite (Fig.5). This promising result offers a possibility of prefabricating limestones. Consequently, larger quantities of calcium-carbonate solutions in Na-ATP were allowed to crystallize for several days in the dark to retard the breakdown of the so-called solvent.The breakdown of the ATP resulted in the crystallization of an aggregate of calcite crystals which compare with those of a normal fine-grained limestone (Fig.6). Such a small-scale experiment must not be regarded as proof positive that limestones are always formed in this way. On the other hand, natural substances like calcium alginate from seaweeds might well break down on their own accord, giving a calcareous cement. Consequently, these promising results appear to offer a fresh approach to the possibility of studying limestone formation under laboratory conditions (EVANS, 1963).

Such a process also offered a possible explanation of the origin of the calcite cement of calcareous deposits. The aforementioned solutions of calcium carbonate were allowed to crystallize along with particles of detrital quartz grains. The degree of aggregation was not too convincing, but there are other possibilities of repeating such experiments under more realistic conditions. Nevertheless, it was of interest to compare this residue with a natural calcareous sandstone (Fig.7). The example chosen was a calcareous, oil-bearing sandstone of the Ventersdorp System (Precambrian) of South Africa. As is usual for such deposits the calcite cement not only welds together contiguous particles of quartz and felspar, but also fills re-entrant solution cavities in these particles. The nature of the bituminous material was determined by vapour-phase chromatography which indicated that this residuum was very

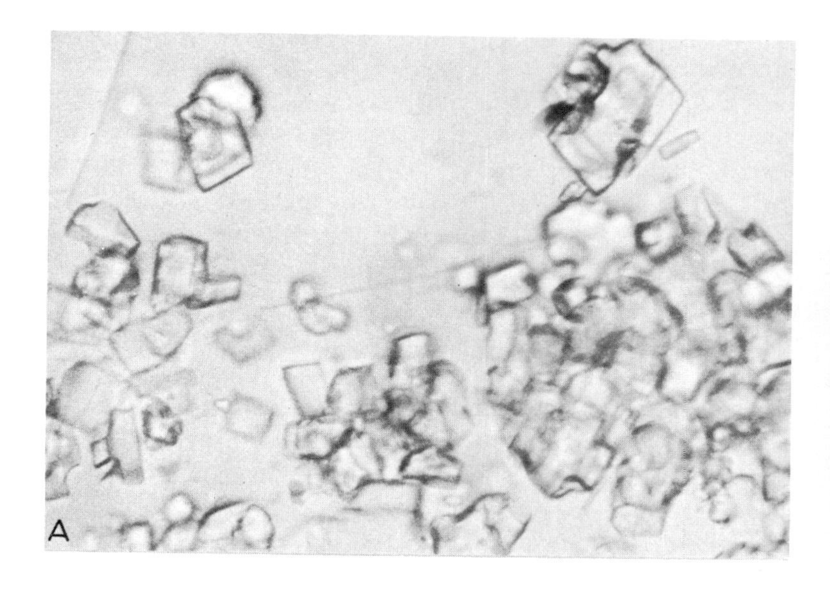

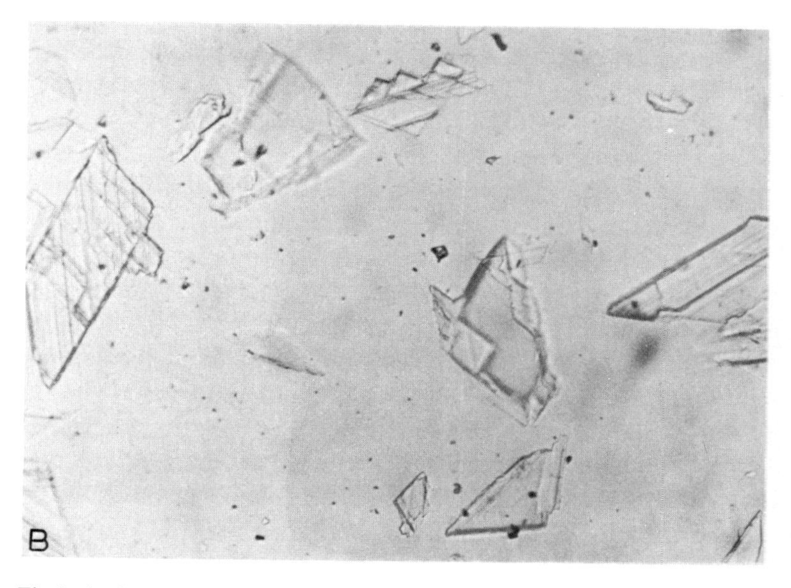

Fig.5. A. Crystals of artificial calcite (× 230). B. Crystals of natural calcite (× 230).

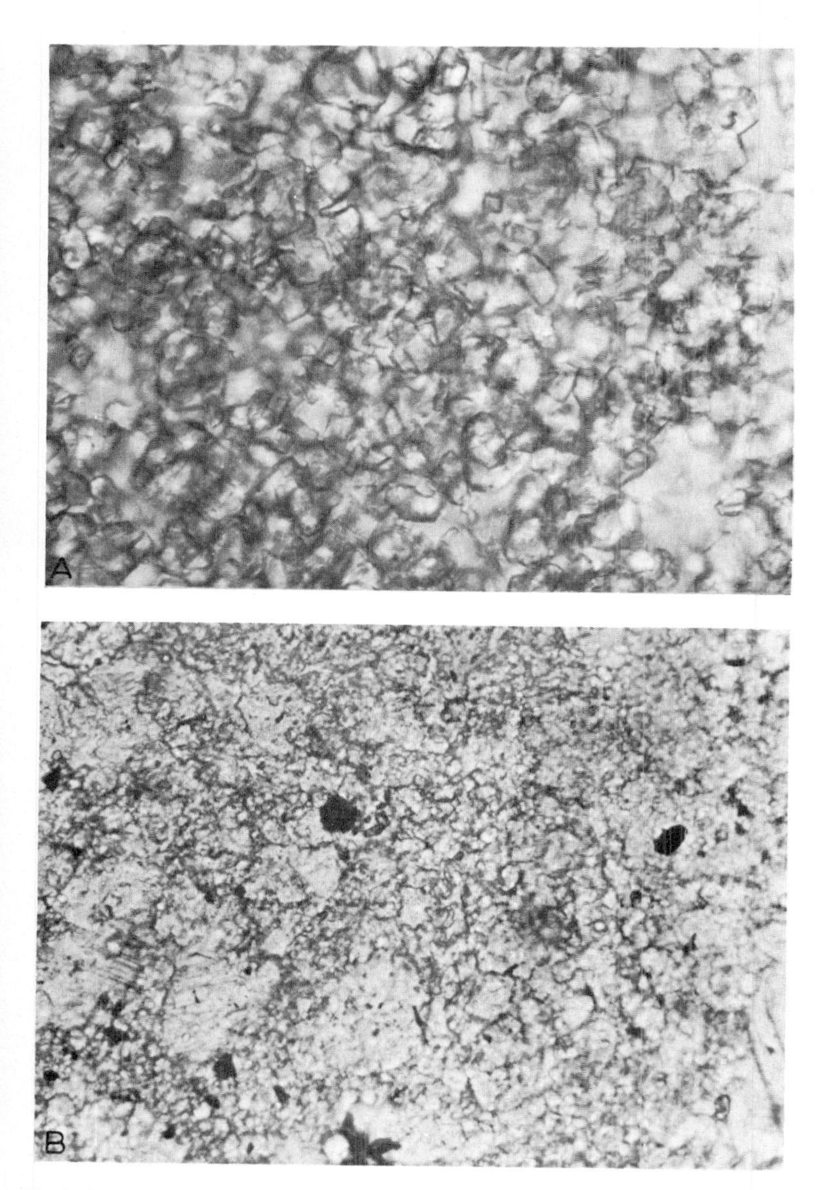

Fig.6. A. A thin section of a limestone formed artificially (× 240). B. A thin section of a typical natural fine-grained limestone (× 240).

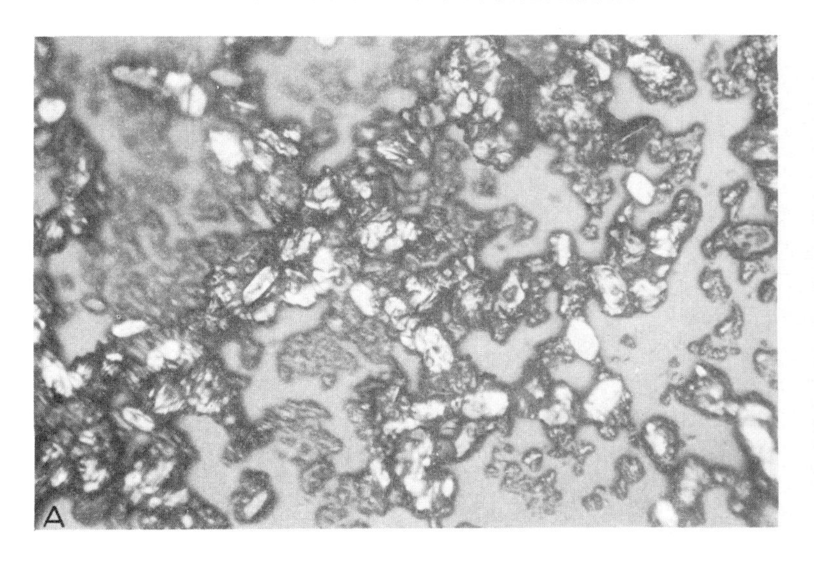

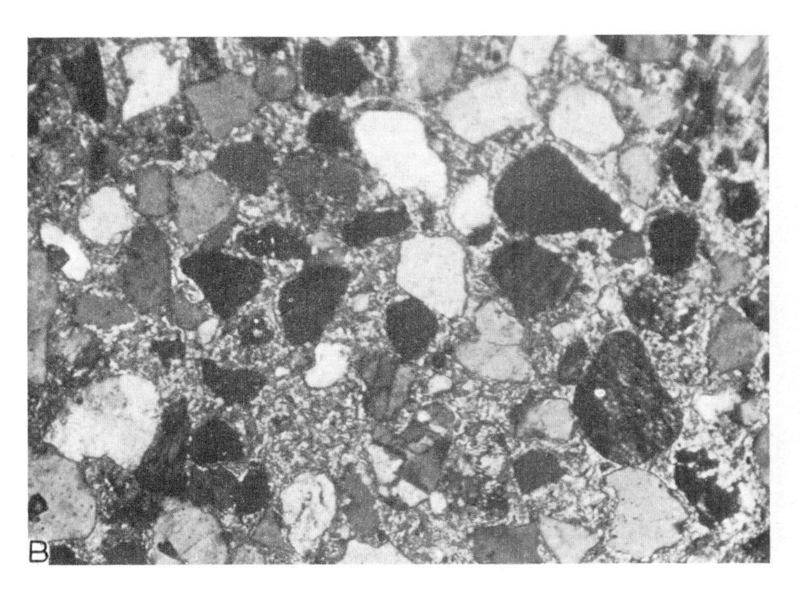

Fig.7. A. A thin section of an artificial calcareous sandstone ($\times$ 230). B. A thin section of a natural calcareous sandstone ($\times$ 230).

similar to a normal petroleum of biogenic origin. Examples of this kind can be quoted from almost all oil-bearing calcareous sandstones. The similarity which exists between the degradation products of Na-ATP and the bitumen of these very ancient sandstones suggests that the latter may have functioned as a "solubilizer" for the cements of these calcareous sediments.

In an additional experiment the Na-salts of alginic acid were added to sands and clays. In the latter they produced compaction and size grading so commonly found in natural argillaceous deposits of carbonaceous composition such as those of the Coal Measures. Apart from the precipitation of prefabricated secondary "minerals" of the composition of calcite and gypsum, these potential solubilizers induced deformative characters closely resembling slump structures (Fig.8). The production of alginates and humates in the environment of argillaceous sediments is a readily acceptable

Fig.8. A nodular slump structure formed from a mixture of Upper Liassic clay mixed with sodium alginate. The specimen is 2.5 inches long.

possibility. Consequently, this initial experiment may well lead to the further prefabrication of such sedimentary structures and a consequent reappraisal of the interpretation of the development of slump sheets found in a wide variety of geological formations.

Algal structures found in a wide variety of rocks have always been regarded as biogenic. Chromatographic techniques now make it possible to investigate such structures in some detail. An interesting example is afforded by the so-called "carbon leaders" of the gold-bearing conglomerates and sandstones of the Witwatersrand System. They are often 14–30 inches thick and are usually situated some 200 ft. beneath the Main Reef. Composed of carbon and gold set in matrix of quartz sand they often present structures resembling algal remains. A vapour-phase chromatogram of these Precambrian deposits has yielded nine organic compounds, which suggests a biogenic origin for the carbon and secretions of gold, since two of these compounds are in excess of C-10. It this proves to be true, then this discovery opens up the possibility of evaluating those "carbon leaders" as deposits developing in slack-water areas impounded between the gravels and sands containing detrital gold. Algal and similar growths would thrive under such conditions and secrete the finely disseminated gold. The consequent geological implications of this would be twofold. Firstly, these "carbon leaders" would delineate the nature of the developing deltaic deposits called reefs, and the location of the shore lines of this great sedimentary basin. Secondly, it would lend support for the possibility of the biogenic secretion of gold from organic solutions. This is by no means unlikely, since Lundberg has demonstrated that horsetails grown on auriferous gravels in northern Indiana have yielded 4 ounces of gold per ton of plants (NEWMAN, 1959).

Special attention has been paid to the behaviour of quartz –the most ubiquitous mineral known. Apparently resisting solution and abrasion, it is the principal constituent of sand-

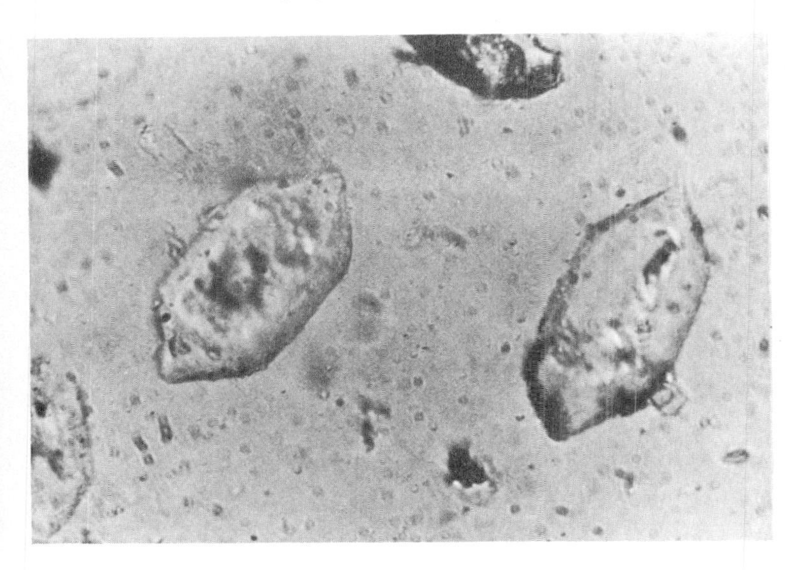

Fig.9. A thin section showing typical authigenic crystals of quartz in limestone (× 160).

stones, and particularly quartzites. Yet it can form quartz cements in these deposits, and develops as holohedral crystals in such rocks as limestones (Fig.9). KING (1945) has shown that 100 ml of blood serum dissolved 7.4 mg of quartz. BRISCOE et al. (1937) demonstrated that freshly-ground minerals yielded more silica than those which had been aged by preservation in bottles for some time. In the presence of alumina and alumino-silicate ash (i.e., the inherent sediment) derived from various types of coal, KING (1945) showed that the solubility of quartz dropped to 0.8 mg in 100 ml acetic fluid. Moreover, using aurine, he demonstrated the existence of protective coatings of alumina on detrital grains of quartz which in themselves inhibited the release of silica—a factor which might well control the ability of many types of quartz-rich sediments to produce cements of secondary silica.

A very pure form of quartz sand was suspended in *M*/5 Na-ATP and allowed to stand under a cover glass for several days. The resultant residue was sufficiently coherent to be sectioned (Fig.10), owing to the development of secondary silica cementing the contiguous particles of quartz. This compares favourably with any natural quartzite. To investigate this phenomenon further a few grains of pure quartz sand were allowed to stand in Na-ATP in the dark for several days. Solution of silica must have occurred to yield the cement welding contiguous particles of quartz shown in Fig.11. As in the case of the calcareous sandstone (Fig.7) the grains of quartz in this prefabricated quartzite are pitted with re-entrant cavities which are evidently due to the solvent effect of the Na-ATP. In contrast, in the case of a specimen of impure quartz sand, in which the particles are coated with hydrated oxides of iron and alumina, the Na-ATP had no obvious effect upon the solubility of the quartz grains and failed to produce a cemented sandstone. As previously indicated this is indirectly supported by the findings of BRISCOE et al. (1937) and KING (1945).

When sodium silicate was dissolved in Na-ATP and allowed to stand under a cover glass, after a few days radiating microlites of a silicate began to crystallize. This unusual result has not been fully investigated as yet, but on examination of the sodium silicate it was found that it contains alumina as an impurity. For the time being it is being presented as a possible route into the investigation of the mode of formation of low-temperature alumino-silicates and, especially, the widespread occurrence of chamosite in sedimentary ironstones. Initially it was anticipated that this type of experiment would yield quartz crystals such as the authigenic quartz crystals in limestones and coal seams.

The ease with which silica replaces many other minerals, but is itself rarely ever altered into other minerals as pseudomorphs, is well established. It evidently always appears to be an end-product of solubilization, or the final limit to which any biogenic silicate or a potential silicone could be oxidized. An interesting example is the silica replacement of the beauti-

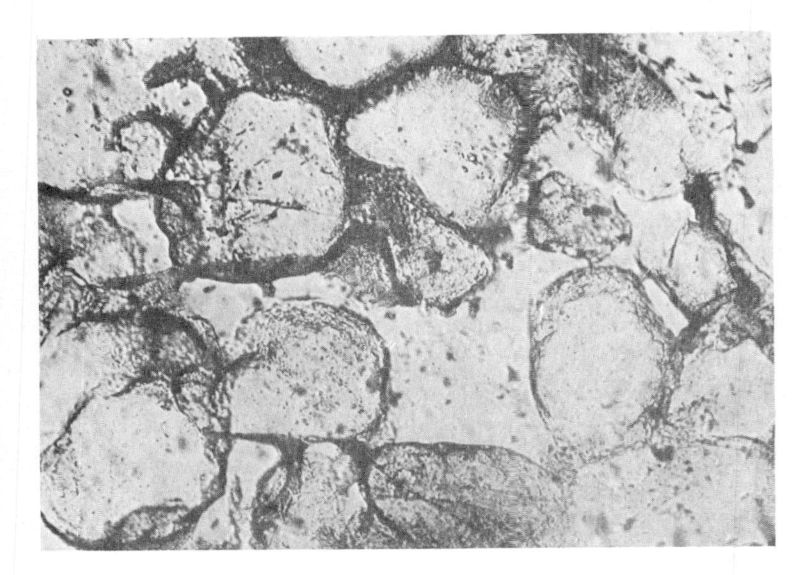

Fig.10. A thin section of an artificial quartzite formed by the cementation of quartz sand with adenosine triphosphate (× 160).

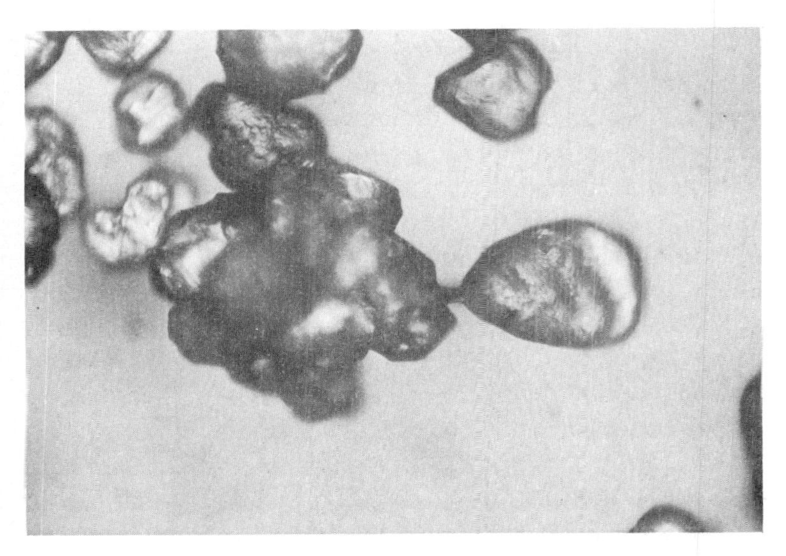

Fig.11. Grains of quartz sand linked by secondary quartz deposited from the solvent action of adenosine triphosphate (× 30).

ful golden asbestos, crocidolite, from Griqualand in South Africa. Recently, HARINGTON (1962) has shown that similar types of crocidolite contain 3,4-benzpyrene plus oil-wax constituents and nine amino acids. This implies that the crocidolite has developed in association with biogenic processes during the Transvaal Period. In one of these asbestos mines the geological evidence presents no trace of high grade dynamic metamorphism and the deposit is essentially a fibrous replacement of black carbonaceous shales. The remarkable evidence produced by HARINGTON (1962) strongly supports the possibility that these sodium-iron silicates have crystallized from, or have at least been influenced by, biogenic solutions embodied in these very ancient sediments.

The more amorphous forms of silica minerals, such as chalcedony, have been deposited from aqueous solutions. One of the most remarkable examples are the aborescent stalactites found in the Trevascus Copper Mine near Camborne, Cornwall. Opal is also worthy of note, and especially the beautiful blue and green precious opals from Barcoo River in Queensland, where it occurs as nodules and pipe-shaped pieces in a brown, jaspery limonite. Such opals, and moss agate, on being subjected to chromatographic analysis have revealed the existence of six hydrocarbons ranging up to C-8, but as yet unidentified. It now seems reasonable to assume that such opaline colours might in part be due to fluorescent films of organic matter, and not entirely to the variation in refractive index of the constituent layers as was formerly supposed. Likewise, it now seems reasonable to assume that the widespread occurrence of "wood opal", which is wood converted cell by cell into opaline quartz, has been formed by silica deposited from solubilizing plant media. It can also be shown that the greyish-brown substance characterizing "menilite" or "liver opal" found in such shales as at Menilmoutant near Paris, is an intergrowth of quartz and humates. Likewise, flint, hornstone and chert are probably derived from one organic source or another.

LOUW and WEBLEY (1959) in an elegant series of experiments have demonstrated the solubilization of calcium carbonate by soil bacteria which they associate with the release of lactic acid and possibly 2-ketogluconic acid. Extending this study, WEBLEY et al. (1960) have shown that micro-organisms are capable of breaking down silicates such as wollastonite, apophyllite and olivine into amorphous residues through the action of 2-ketogluconic acid. Until ABELSON (1957) demonstrated the extent to which alanine would degrade with rising temperature and time, it was popularly believed that such amino acids would be too fragile to exert any long-lived effect in sedimentary environments. The presence of alanine, glycine, glutamic acid, leucine and valine in the Ordovician brachiopod *Plaesiomys subquadrata* in itself demonstrates that these have survived in the shell for at least $430 \cdot 10^6$ years (ABELSON, 1957).

At this stage in the development of researches into the impact of organic matter upon sedimentation and subsequent cementation, it is dangerous to draw firm conclusions from small-scale laboratory experiments (EVANS, 1964). Whether one is dealing with processes of chelation or straightforward solubilization to form true solutions is still questionable. Nevertheless, the production of calcite in association with natural compounds, and the cementation of quartz with secondary silica do, in themselves, make a step forward in this newly emerging aspect of sedimentary petrology.

CHAPTER 2

Qualitative Analysis

INTRODUCTION

A considerable amount of time is usually saved by evaluating the chemical components of minerals and rocks on a qualitative basis. In this way the quantitative characteristics of the geochemical problem in hand are brought into focus. Moreover, the decision to employ either complete or partial forms of analysis for the solution of a problem can be taken in a positive fashion. Furthermore, the existence of systematized reagents used for qualitative tests will improve the facilities of a geochemical laboratory.

For qualitative work a small quantity of the powdered specimen is usually all that is necessary, and the first step is to bring it into solution. Silica and a number of silicates will require the formation of melts using fluxes or the use of hydrofluoric acid. However, a large number of specimens can be dissolved in acids and alkalis in sufficient quantity for qualitative manipulation. Hydrochloric acid can be used for carbonates of Ba, Mn, Fe, Mg, Sn, Ti, Zn and U. The addition of an oxidizing agent such as nitric acid, or potassium chlorate will result in the solution of some silicates and the sulphides of Cu, Co, Pb, Mo, U and Zn. Nitric acid is best used for the ores of Cu, Cd, Mo, Co and Ni, and aqua regia for Au, Hg, V and Pt. Sulphuric acid is employed for the ores of Al, Be, Mn, Pb, Th, Ti, U and the rare earths. Sodium hydroxide (35%) or potassium hydroxide (35%) can be used in some cases. When a sample is not attacked by these acids and alkalis, it is fluxed to a melt and then dissolved in acid.

TABLE VII

HCl solution of sample

Insoluble acid group: Si, Nb, Ta, W, Ag.

H_2S

Precipitated H_2S group: Hg, Pb, Bi, Cu, Cd, Re, As, Sb, Sn, Ge, Mo, Se, Te, Au, Pt.

Oxidize with NH_4OH

Precipitated NH_4OH group: Fe, U, In, Ga, Al, Be, Cr, Th, Sc, Zr, Hf, Ti, Nb, Ta and rare earths.

$(NH_4)_2S$

Precipitated $(NH_4)_2S$ group: Ni, Cu, Zn, Mn, Tl, V.

$(NH_4)_2C_2O_4$

Precipitated $(NH_4)_2C_2O_4$ group: Ca, Sr.

$(NH_4)_2HPO_4$

Precipitated $(NH_4)_2HPO_4$ group: Mg, Ba.

Elements in solution: Li, Na, K, Rb, Cs, V, Br, I.

After the solution has been prepared the elements can be separated in a qualitative fashion into an acid group, hydrogen sulphide group, ammonium hydroxide group, ammonium sulphide group, ammonium carbonate group and an ammonium phosphate group. The schematic procedure for such group analyses is outlined in Table VII since the detailed chemical information is readily available in most standard works on inorganic analysis.

Numerous spot tests have been devised for the identification of the principal metals in ore-deposits (see WEINIG and SCHODER, 1947). Polished surfaces of rocks and minerals respond to etching techniques which sometimes provides additional evidence relating to geochemical problems. The application of reagents such as HCl, HNO_3, KOH and $FeCl_3$ at selected areas of a polished specimen enables one to observe the reaction and the ensuing precipitates under the microscope and these observations often yield valuable information concerning the constituent minerals of a rock or ore deposit. An elegant refinement is the use of microchemical tests carried out on small quantities scratched or gouged from selected areas of a specimen. These methods are of particular value to the economic geologist who wishes to make a rapid but accurate appraisal of the existence of certain valuable metallic elements without the facilities of a chemical laboratory. They are also of value to the geochemist at the stage when an assessment has to be made of a large number of specimens designed to survey the distribution of a limited number of elements.

MICROCHEMICAL TESTS

Initially the etching effects produced by acids and alkalis were used to determine certain opaque minerals, but these methods lacked precision as many minerals react in such a feeble manner that the degree of cleanliness of the surface of the specimen often affected the nature of the results. From this approach, SHORT (1940) developed a series of microchemical tests which were essentially designed to indicate the presence of specific metals in a rock or mineral. The reactions were carried out on glass slides and observed under a polarizing microscope. The reagents employed were designed to precipitate from a solution of a rock, or mineral, specific elements as crystals possessing characteristic forms and colour. This technique possesses considerable advantages over more classical methods of qualitative analysis as it is rapid and requires very little material. In order to ensure that the

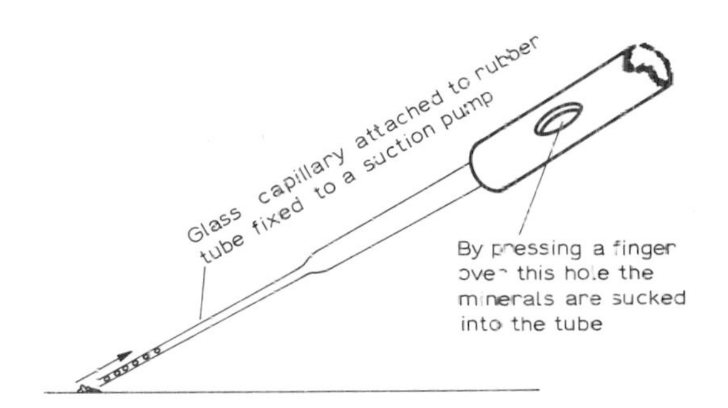

Fig.12. A suction pipette for removing small quantities of finely powdered minerals into reagents for microchemical qualitative tests.

specimen for analysis does not contain extraneous minerals a polished surface is prepared and the specimen is gouged from its surface under a reflecting microscope. The powder so obtained is transferred by means of a wooden needle or by means of the simple suction device illustrated by Fig.12. By either means, the powder is transferred to drops of acid or alkali placed on a glass slide under a microscope. In this way it is possible to note the solubility or otherwise of the minerals in the potential solvent which, in itself, broadly classifies the powder as belonging to one or another group. The solvent is then evaporated on a hot plate and the residue is usually deposited as a finely divided smear on the glass slide. In this state it is in an ideal condition for the application of reagents. These are applied to the smear and the reaction observed under the microscope. The criterion used in the recognition of nearly every element is the character of the crystals formed from the solution or the colour and nature of the precipitate. It is, therefore, essential to have sufficient concentrations of solvent to mineral to produce saturated solutions from which reasonably large crystals can be expected to develop. This is usually achieved by experience.

This method can be applied to any mineral but it is most suited to opaque or nearly opaque minerals, as most of these are reasonably soluble in acids in the small quantities em-

TABLE VIII

REAGENTS REQUIRED FOR MICROCHEMICAL TESTS

Reagent	*Composition*
Nitric acid	1:1 HNO_3, 1:7 HNO_3, 1% HNO_3
Aqua regia	1 volume concentrated HNO_3 in 3 volumes concentrated H
Hydrochloric acid	1:5 HCl, 5% HCl
Sulphuric acid	1:3 H_2SO_4, 20% H_2SO_4
Ammonium hydroxide	concentrated NH_4OH, 20% NH_4OH
Stannous chloride	2% solution in 1:5 HCl
Ammonium molybdate	1.4% solution in 1:7 HNO_3
Sodium sulphide	saturated solution
Potassium thiocyanate	10% solution in H_2O
Pyridene-hydrobromic acid	1 volume pyridene in 9 volumes 40% HBr
Cinchonine reagent	1 g cinchonine in 8 g 1:1 HCl
Diphenyl carbazide	1% solution in alcohol
Benzidine reagent	0.05 g benzidine in 10 ml glacial acetic acid and 90 ml H_2O
Ammonium bichromate reagent	0.1 g $K_2Cr_2O_7$ in 33 ml H_2O
Rhodamine reagent	0.03% dimethylaminobenzole-rhodamine in acetone
Thiourea	10% solution in H_2O
Dimethyl glyoxime	2% solution in alcohol
Potassium mercuric thiocyanate	3% solution in H_2O
Solids	cesium chloride (CsCl), potassium iodide (KI), potassium nitrate (KNO_3), calcium acetate [$Ca(C_2H_3O_2)_2$], metallic zinc (Zn), soda-nitre flux consisting of 10 parts Na_2CO_3 in 1 part KNO_3

ployed for these tests. The quantity of material required is usually no more than that which can be collected from an area of the slide about 0.3 mm diameter. Naturally certain elements will interfere with the precipitation and crystallization of others. Lead, for example, is precipitated simultaneously with silver as a bichromate and this masks the recognition of silver compounds. Nevertheless, it is possible to remove the excess lead before proceeding with the silver test.

With experience this offers no serious obstacle, but it does limit the number of elements which can be swiftly and accurately identified in this way. In Table VIII the tests have been chosen which offer the least likelihood of interference effects.

The reagents required for these microchemical tests are as follows:

Nitric acid: 1 : 1 HNO_3, 1 : 7 HNO_3, 1% HNO_3.
Aqua regia: 1 vol. conc. HNO_3 in 3 vols. conc. HCl.
Hydrochloric acid: 1 : 5 HC1, 5% HCl.
Sulphuric acid: 1 : 3 H_2SO_4, 20% H_2SO_4.
Ammonium hydroxide: conc. NH_4OH, 20% NH_4OH.
Stannous chloride: 2% solution in 1 : 5 HCl.
Ammonium molybdate: 1.4% solution in 1 : 7 HNO_3.
Sodium sulphide: saturated solution.
Potassium thiocyanate: 10% solution in H_2O
Pyridine-hydrobromic acid: 1 vol. pyridine in 9 vols. 40% HBr.
Cinchonine reagent: 1 g cinchonine in 8 g 1 : 1 HCl.
Diphenyl carbazide: 1% solution in alcohol.
Benzidine reagent: 0.05 g benzidine in 10 ml glacial acetic acid and 90 ml H_2O.
Ammonium bichromate reagent: 0.1 g $K_2Cr_2O_7$ in 33 ml H_2O.
Rhodamine reagent: 0.03% dimethylamino benzole-rhodamine in acetone.
Thiourea: 10% solution in H_2O.
Dimethyl glyoxime: 2% solution in alcohol.
Potassium mercuric thiocyanate: 3% solution in H_2O.
Solids: cesium chloride (CsCl), potassium iodide (KI), potassium nitrate (KNO_3), calcium acetate ($CaC_2H_3O_2)_2$, metallic zinc (Zn), soda-nitre flux consisting of 10 parts Na_2CO_3 in 1 part KNO_3.

The principal characteristics of the microchemical tests are summarized in Table VIII and illustrated by Fig.13.

The physical characteristics of minerals have been thoroughly covered by a recent publication by JONES and FLEM-

Fig.13. Micro-drawings of the typical crystals obtained by qualitative tests (after SHORT, 1940). A. Black crystals of zinc mercuric thiocyanate. B. Greenish-yellow crystals of copper mercuric thiocyanate. C. Indigo-blue crystals of cobalt mercuric thiocyanate. D. Light brown and pink spherulites of nickel mercuric thiocyanate. E. Ruby-red crystals of silver bichromate. F. Yellow crystals of arseno-molybdate. G. Orange to red crystals of antimony-cesium iodide. H. Orange crystals of bismuth-cesium chloride. I. Colourless crystals of tin-rubidium chloride. J. Lemon-yellow crystals of lead iodide. K. Indigo-blue crystals of cobalt mercuric thiocyanate. L. Blue crystals of mercuric chloride.

ING (1965). This is mentioned as the reason for the omission of such techniques. Nevertheless, they form an essential facet of the work of a geochemist as in the final analysis of his results he must resolve his quantitative determination of oxides and elements in terms of assemblages of modal or normative minerals.

MEMBRANE COLORIMETRY

This is a technique which has originated from a variety of methods employing the transfer of chemical images from flat surfaces of metals, rocks and plants. Such images were then processed by the use of colorimetric reagents designed to reveal the distribution of metals throughout the surface of the specimen. With the replacement of opaque transfer papers by transparent membraneous films, these several techniques have been enhanced both qualitatively and quantitatively. Moreover, membranes can be used which are extremely thin and, in themselves, take part in the transfer and capture of the elements contained in the surface of the specimen being analysed. Consequently, it was deeemed advisable to collectivize these several techniques under the title of "membrane colorimetry" (EVANS, 1966).

The idea of using electrolysis to transfer metal ions from a metallic specimen and testing the solution by colour test originated independently with FRITZ (1929) and GLAZUNOV (1929). Fritz transferred small quantities of solubilized metals into filter paper by electrolysis under standardized conditions. Glazunov employed electrolytic transfer of metal ions to bring out the macro-structure of specimens of metal. Combined, these techniques are known to metallurgists as "electrographic analysis" (see LINGANE, 1958, p.446). Minerals and rocks were first analysed in this way by GUTZEIT et al. (1933), and later their work was developed to a high degree of accuracy by WILLIAMS and NAKHLA (1950). Until recently only specimens which were ionizable in an electrical field were analysed in this way. With the introduction of ion exchange transparent membranes the process of making contact prints has widened in its application to metallurgical, geological and biological specimens (EVANS, 1966).

WILLIAMS and NAKHLA (1950) proposed the term "chromographic contact printing" to describe the way in which they produced specifically coloured images of minerals and ore deposits. They used gelatine coated papers similar to those employed for "electrographic analysis" (LINGANE, 1958, p.446). The results they obtained were of great qualitative value in the study of the distribution of cations and anions in mineralogical specimens. In fact it has been the only means available of studying on a chemical basis the qualitative nature of the distribution of metals in ore deposits. With the use of transparent membranes these chemical images can now be produced with greater accuracy and magnified to reveal the fine structure of the specimen in terms of its cations and anions (see frontispiece and Fig.14).

In its simplest form, a contact print is achieved by causing a transfer of an element from the surface of a specimen into an adsorbent film either under pressure or by means of an electrical direct-current field. The gelatine surface of photographic paper provides a convenient surface film which is capable of trapping the element in situ. The paper backing serves to carry the reagents involved in the reaction between the surface of the specimen and the gelatine film. Transparent membranes of material like cellophane produce precision images. The reagents involved can either be inserted into the membrane or applied to the upper side of it by means of a pad of porous paper.

The reagents which are employed may be classified as "attacking reagents" and "specific reagents". Attacking reagents are those wich render the desired element into a soluble condition. The specific reagents are those which function as colorimetric agents by forming coloured compounds or complexes with the ionized element.

Reactions can offtimes be effected by the straightforward application of the attacking reagent under pressure to the surface of the rock or mineral. This involves the use of a press such as that illustrated by Fig.15A, but for the production of refined images it is necessary to use an hydraulic press as it sometimes takes ca. 1,000 lb./sq. inch to force the reagents into the irregularities and pinholes of the surface of the specimen. Direct printing can be achieved for certain groups of ions when the attacking reagents can be used with the specific reagent. Excellent single transfer prints can be achieved, for example, by saturating the gelatine or membrane with a solution of sodium carbonate and sodium nitrate. For specimens which are electrically conductive such a print will register iron, chromium, copper, nickel, silver, lead, tin, cadmium and zinc in a definitive fashion. Interference by other groups of ions often limits this attractive aspect of the technique so that the contact paper or membrane must be processed after contact has been made with the specimen.

Using the simple devices illustrated by Fig.15 contact prints can be achieved in the following manner:

(*1*) A flat, or preferably polished surface, is developed in

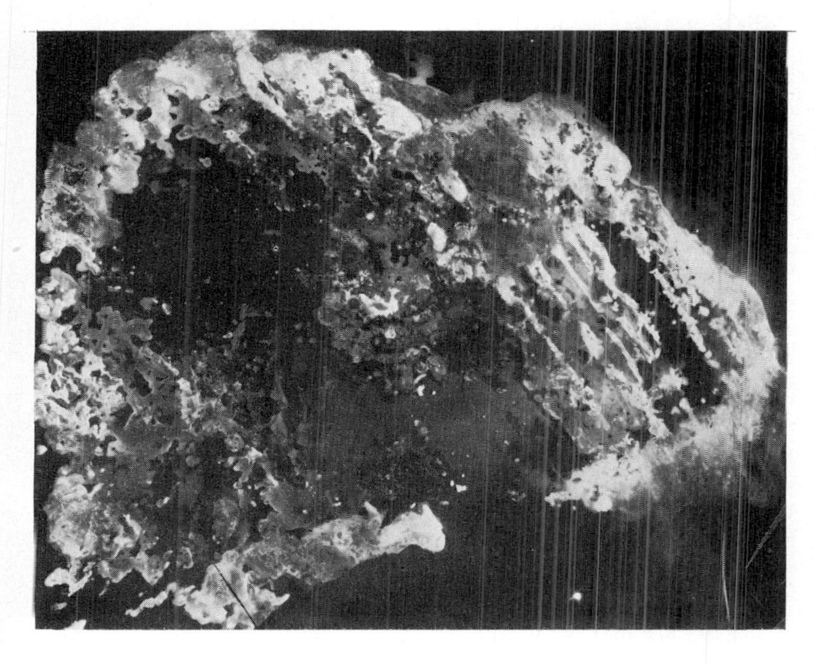

Fig. **14A.** Texture and mineral intergrowths in a vein of pyrite, chalcopyrite, galena and quartz (× 1.5).

B1

14B. Nickeliferous breccia, **B1** Texture and mineral intergrowth (x 2). **B2** Intergrowth of millerite and pyrite (x 57.5). **B3** Zonal areas of pyrite and millerite (x 136). **B4** Quartz with fractures filled with pyrite and millerite (x 85).

B2
B3
B4

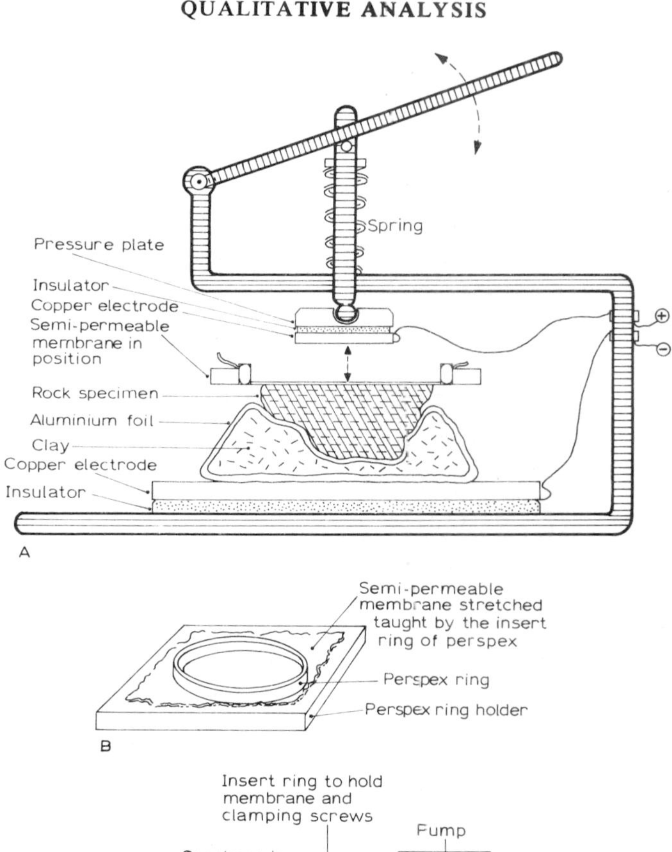
Spring
Pressure plate
Insulator
Copper electrode
Semi-permeable membrane in position
Rock specimen
Aluminium foil
Clay
Copper electrode
Insulator
A
Semi-permeable membrane stretched taught by the insert ring of perspex
Perspex ring
Perspex ring holder
B

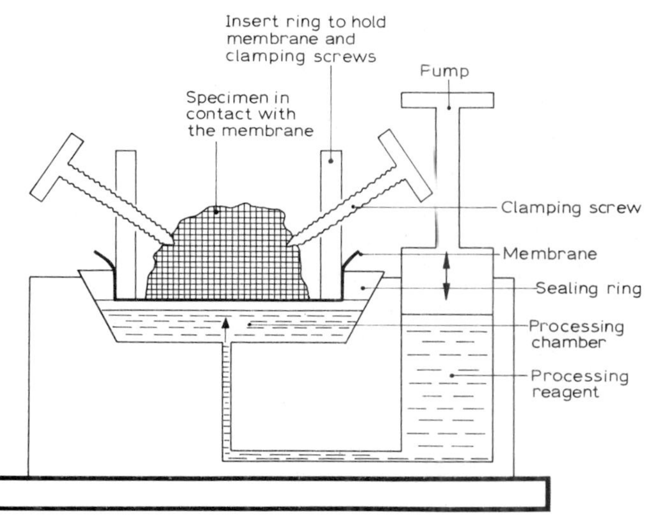
Insert ring to hold membrane and clamping screws
Fump
Specimen in contact with the membrane
Clamping screw
Membrane
Sealing ring
Processing chamber
Processing reagent

C

the specimen. Crystal faces are suitable without further preparation. The nature of the surface will determine the sharpness of the contact print. Mineral grains embedded in a bakelite-carbon mount and then polished will produce contact prints of the larger grains.

(*2*) The specimen is pressed firmly into a plastic bed forming the base of the press. This is a convenient method of orientating rough-sided specimens so that the cut surface is brought into a position parallel to the pressure plate of the press.

(*3*) The gelatine-coated paper can be purchased without the photographic salts it normally contains. Otherwise, it can be preparared by dissolving and removing the silver salts from ordinary photographic glossy bromide paper. This is achieved by soaking the paper under a safe-light for not less than 15 min in a standard acid-fixing solution commonly known as "hypo", which consists of 400 g of sodium thiosulphate and 25 g of potassium metabisulphite dissolved in 1 l of water. The paper must be washed thoroughly in running water for at least 1 h, and then dried. The membrane is produced by stretching it between two rings, one inserted into the other. A strong taut membrane is obtained in this way and it can be used in a wet or dry condition depending upon circumstances.

(*4*) The gelatine-coated paper or membrane is firmly applied to the dry surface of the specimen after it has been impregnated with the attacking reagents. Excessive quantities of the attacking reagent are usually removed from the gelatine-coated paper or membrane by pressing it between filter papers prior to its application to the surface of the specimen.

(*5*) The reaction is allowed to proceed for specific periods of time. Similar to a photographic print, the image becomes

---

Fig.15. Devices for making contact prints. A. The press for making contact prints. B. Plate and ring for holding the semi-permeable membrane. C. Device using the semi-permeable nature of cellophane membranes and their cationic properties.

increasingly more diffuse with an increase in the period of exposure owing to the sideways diffusion of the elements entering the gelatine film. This occurs to a lesser degree in membranes than it does in the gelatine-coated papers.

(*6*) When the attacking reaction has reached the requisite degree of completeness, the paper or membrane is removed and washed. It is then immersed in the specific reagent which develops the coloured image.

If the specimen is electrically conductive the application of a direct current to the specimen and the pressure plate (see Fig.15A) produces prints rapidly and accurately. Moreover, by making the specimen respectively the cathode or the anode contact print will show the distribution of cations and anions in the specimen. The simple equipment used for direct contact prints can be modified for this method by simply inserting electrode plates into the base and pressure plate of the press. To accommodate irregularities in the specimen and its surface and to achieve parallelism between it and the pressure plate, WILLIAMS and NAKHLA (1950) devised simple, but effective, plastic electrodes by wrapping a pad of plasticine in aluminium foil. The foil deforms to give excellent electrical contact between the irregular surfaces of the underside of the specimen and the copperplate electrode in the base of the press. The plastic electrode inserted between the copper electrode of the pressure plate and the print is employed to take up imperfections in the surface of the specimen. This ensures reasonable control over the contact paper or film. Both the voltage and the intensity of the current must be carefully regulated, but for most specimens the applied voltage varies from 4 to 30 V. The intensity of the current is a function of the specific resistance of the rock of minerals involved, the size of the specimen, the humidity of the reaction paper or membrane and the concentration and composition of the attacking reagents. Consequently, it is essential to control the intensity of the electrical field to keep it as low as possible in order to prevent undue lateral diffusion of the ions into the print. About 20–40 mA is usually sufficient, which means that dry batteries can be used as a convenient source of current. This is particularly valuable for the use of this method under somewhat primitive conditions in the field or as portable equipment at remote mining centres.

The replacement of contact papers by the use of membranes has achieved two advantages. For purely qualitative purposes the transparent coloured images can be magnified to almost any extent (see Fig.14) and this reveals details of the chemical distribution of elements which were not visible in the gelatine films of contact papers. Secondly, the ease with which these images can be scanned photometrically provides a straightforward method of obtaining a semi-quantitative analysis of the surface of the specimen. Additional to these advantages is the ease with which permanent records can be maintained since the membrane can be preserved between plates of glass or mounted on paper, like a coloured photograph. At the present stage in the development of membrane colorimetry the physico-chemical properties of thin films of transparent media have only been exploited to a limited degree. Films possessing enhanced ion-exchange properties will be developed as the technique is extended and it is reasonable to visualize the discovery of membraneous materials which will respond effectively under pressure. At present the most convenient materials to employ for the manufacture of the membrane are thin films of commercial cellophane or the cellulosic materials used for dialysis or micro-filtration purposes. In short, the future development of membrane colorimetry will depend upon the application of various ways and means of ionizing the surface of a specimen in contact with the membrane. An electrical potential is the obvious means of dealing with specimens of low resistance but this eliminates all trace of the silicates, alumino silicates and carbonates in rock specimens. For such insulating substances ion bombardment has been exploited but this involves the use of high energy sources or the use of fuming acids. The latter have been used to a limited extent with promising results.

For normal rock specimens use is made of the semi-permeable nature of the cellophane membranes and their cationic properties. As shown in Fig. 15C, the specimen is placed in one half of an acid proof vessel divided into two sections by a cellophane membrane. Acid is poured into the other half of the vessel and allowed to make contact with the specimen by diffusion through the membrane. Alternatively, the fumes from such acids are allowed to saturate the membrane and so induce cationic exchange between it and the rock specimen. This produces a much more definitive image for colorimetric processing than the use of liquids brought into direct contact with the membrane. When hydrofluoric acid is used for such ion exchange processes the vessel is made of either P.T.F.E. gutta percha or from metals coated with paraffin wax. The vessel shown in Fig. 15C may be made out of any of these materials and is designed so that the vessel is easily taken apart for the processing of the membrane. As the technique develops it is probable that other forms of equipment will be designed to overcome many of the practical difficulties which obstruct the more widespread use of membrane colorimetry to the analysis of rock surfaces. The examples illustrated in Fig.14 serve to show the potentialities of the technique especially in the location of trace metals in relation to the texture and mineralogy of igneous and metamorphosed rocks. Rocks like limestones and evaporites present no problem and the application of membranes to the study of corals and such like fossils have certain advantages over the use of thin sections or the so-called "cellulose peels" used by palaeontologists. In fact, the latter, when produced by coating etched surfaces with cellulose preparations are strictly speaking in themselves membranes capable of being processed by colorimetric reagents (cf. DICKSON, 1966).

To translate membrane colorimetry into a semi-quantitative technique the image can be scanned at various wavelengths photometrically. From standardized specimen membranes it is possible to obtain semi-quantitative values for the various elements adsorbed by the membrane. In addition to this, it has proved possible to obtain quantitative estimates of cation transfer in ore deposits by recording, potentiometrically, the release of ions from the specimen into the membrane by mechanically controlling the increase in the electrical field. Each element, as it reaches its excitation potential, causes a sharp increase in the current flow.

As mentioned previously, it has been found convenient to refer to the reagents involved as "attacking reagents" and "specific reagents". The attacking reagents include hydrochloric acid, nitric acid, various organic acids, ammonia, cyanides and alkalis. Their concentration depends upon the minerals and rocks under investigation, but within limits it is determined by experience. Some attacking reagents yield diagnostic prints without the aid of an additional specific reagent. For example, native silver gives a grey to black print with ammonia and native bismuth yelds a bright yellow image when treated with a 20% solution of potassium thiocyanate, whilst smalite—$(Co,Ni)As_2$—and safflorite—$(CO,Fe)As_2$—produce deep yellow prints, signifying cobalt, with a 2–5% solution of potassium cyanide. One of the striking features of the electrographic method is that the polished surfaces of minerals are more or less easily attacked by certain reagents which do not dissolve or even etch, these minerals under ordinary conditions. For instance, pentlandite—$(Fe,Ni)S$—and nickel-bearing pyrrhotite are normally insoluble in acetic acid or ammonia but under electrolytic dissolution with ammonia as the attacking reagent nickel ions are liberated and respond to the characteristic dimethylglyoxime test for nickel. Similarly, using acetic acid as the attacking medium iron ions released from these minerals are trapped in the gelatine layer, where they give the diagnostic prussian blue reaction of iron with potassium ferricyanide. Galena and chalcopyrite are two other minerals which are dissolved by ammonia under electrolytic attack, but not under simple contact print conditions (WILLIAMS and NAKHLA, 1950, p.265).

The specific reagents are really those colorimetric reagents often used in titrimetric analysis. As such, they are usually quite stable and in many cases develop specific colours. They are also extremely sensitive to certain metal ions. GUTZEIT (1942) illustrates this by examples, such as the detection of 0.00005 mg of cobalt in a specimen using α-nitroso-β-naphthol; of 0.0001 mg of copper using α-benzoinoxime; of 0.0002 mg of manganese using benzidine; and of 0.00004 mg of lead by using diphenyldithiocarbazone. On the other hand, it must be recognized that the identification of an element by these specific reagents is limited by the interference of other elements. Fortunately, in many cases, steps can be taken to inhibit interferences by producing prints under modified conditions. Complexes which are mostly colourless can be made of the interfering elements by using suitable reagents in conjunction whith the attacking reagent. For example, iron can be masked from nickel or copper by adding sodium potassium tartrate to the attacking reagent. Similarly, interfering elements can be made non-reactive to the specific reagent by prior oxidation or reduction. Advantage may also be taken of the instability of some metallo-organic compounds in the presence of certain media. A typical example would be the instability of rubeanates of copper and nickel in a dilute solution of potassium cyanide, whilst the corresponding rubeanate of cobalt is relatively stable in that medium. Some interfering elements can be removed by washing in acids. In practice, however, the suppression of interfering elements only becomes really necessary when they are present in sufficient quantities to obscure the recognition of the element in question. The ideal situation is that in which the reaction paper carries both the attacking reagent and the specific reagent. In this way contact prints of great clarity and precision are obtained which provides the geochemist with a convenient means of recording the distribution of a particular element in a mineral of a rock. Unfortunately, most combinations of attacking and specific reagents deteriorate rapidly on mixing and it is advisable to prepare them in small quantities for immediate use. However, the most serious limiting factor is the investigation of those minerals which are resistant to chemical attack and are also electrically non-conducting. As a preliminary approach to this chromographic technique, the procedures used for the determination of some of the principal elements are summarized in Table IX. As the technique develops, there will be future additions to this list of recipes.

The following account of the specific reagents used for the detection of elements has been quoted by kind permission of Professor Williams from the classic paper on this subject (WILLIAMS and NAKHLA, 1950).

Antimony

(*1*) 9-methyl-2, 3, 7-trioxy-6-fluorone: a saturated alcoholic solution of the reagent gives with antimony ions a red precipitate in weak acid solutions. At pH4 the test is specific for antimony, as under this condition only ferric ions interfere and these can be masked as phosphates. In neutral solutions arsenic and titanium interfere with the test. The addition of tartaric and phosphoric acids to the attacking reagent is recommended.

(*2*) Rhodamine B: antimony ions in the presence of an aqueous solution of rhodamine B change the colour of the dyestuff, which is usually red, to violet or blue, with the formation of finely-divided precipitates. The test must be carried out in the presence of hydrochloric acid and if necessary the antimony is oxidized to the pentavalent state. Small amounts of iron do not interfere, but mercury, gold, bismuth, and tungstates in acid solutions give similar reactions with rhodamine B to those given by antimony salts (FEIGL, 1946).

(*3*) 2, 5-dimercapto- 1, 3, 4-thiodiazole: with antimony ions this reagent gives a yellow precipitate. The reaction is

TABLE IX

MEMBRANE COLORIMETRY

Element	*Attacking reagent*	*Voltage (V)*	*Time (sec)*	*Specific reagent*	*Print colour*
Antimony (Sb)	1:1 HNO_3 add tartaric and phosphoric acid to suppress ferric iron, arsenic and titanium	4	30	alcoholic solution of 9-methyl,-2,3,7-trioxy-6-fluorone; ferric ions masked as phosphates	red
				or: aqueous solution of Rhodamine B with trace of HCl	red to violet or blue
				or: 2.5-dimercapto-1, 3, 4,-thiodiazole; similar colours yielded by Ag, Pb, Hg, Cu, and Bi	yellow
				carvacrol in chloroform	red to purple
Arsenic (As)	1:1 HNO_3 or aqua regia or:	10	30	1.5% ammonium molybdate in 1.7% HNO_3	yellow
	ammonia or caustic soda with hydrogen peroxide	10	30	or: 0.5% aqueous solution of N-ethyl-8-hydroxytetrahydroquinoline hydrochloride; add HCl to suppress Cu, Pb, Hg.	reddish-brown
				or: 1% solution of silver nitrate after acidifying paper with acetic acid	brown
				or: stannous chloride in 35% HCl	brown-black metallic arsenic

Bismuth (Bi)	20% KCNS for native bismuth	5	30	direct print	yellow
	or: 1:15 HNO_3	10	25	1 g cinchonine in 100 cm^3 warm H_2O containing trace of HNO_3, after cooling dissolve 2g potassium iodide; Cu interference removed by washing print in a ION solution of sodium thiosulphate; Pb and Hg suppressed by addition of HCl to reagent	orange-red
				or: ammoniacal alcoholic solution of dimethylglyoxime; nickel when rarely present in bismuth ores suppressed by cyanide	yellow
Cadmium (Cd)	1:7 HNO_3	12	10	Cadion 2B (nitronaphthalene-diazoaminobenzene-4'-azobenzene in alcoholic solution in presence of sodium potassium tartrate)	pink
Chromium (Cr)	20% H_2SO_4	12	10	1% alcoholic solution of diphenyl carbazide	orange-yellow violet
	or: 1:7 HNO_3	12	10	2% silver nitrate	red
	caustic soda + hydrogen peroxide				
Cobalt (Co)	tartaric acid + acetic acid + 0.5% nitroso-R-salt buffered with sodium acetate	25	15	direct print—Cu, Fe, Pb, Ni, and Ag removed from print by 2:1 HNO_3	red
	or: 1:7 HNO_3	25	10	3% potassium mercuric thiocyanate	dark blue

Continued on p. 70

TABLE IX (continued)

MEMBRANE COLORIMETRY

Copper (Cu)	NH_4OH + α-benzoinoxime in 1–5% alcoholic solution	5	15	direct print	green
	or: 1:7 HNO_3	5	15	3% potassium mercuric thiocyanate	yellow to green
Gold (Au)	aqua regia	5	10	1 volume pyridine in 9 volumes 40% HBr	orange-maroon
				or: benzidine in acetic acid	blue
Iron (Fe)	1:15 HNO_3 or glacial acetic acid + $K_4Fe(CN)_6$	8	25	direct print	blue
	or: 1:5 HCl + $K_4Fe(CN)_6$	20	25	direct print	blue
	or glacial acetic acid	8	30	direct print	blue
	or: 1:20 HNO_3 + 2% KCNS	12	25	direct print	red
	or: 1:10 HNO_3 + 5% sulpho salycilic acid	10	25	direct print	violet
	or: 1:15 HNO_3 + iodo-hydroxy-quinoline sulphonic acid	8	35	direct print	green

Lead (Pb)	1:20 HNO_3	20	15	$SnCl_2$ + KI	yellow
	or: 2% KCN + 0.04% ammoniacal solution of dithiozone	20	40	direct print	red-violet
	or: 1:1 HNO_3	100	10	10% ammonium sulphide	brown
Manganese (Mn)	1:5 HCl + H_3PO_4	5	20	acetic benzidine solution after exposure to ammonia	blue
	or: 1:10 HNO_3		5-10	ammoniacal $AgNO_3$	brown-black
Molybdenum (Mo)	1:1 HNO_3	20	20	potassium ethyl xanthate	red-violet
				or: phenylhydrazine in acetic acid	reddish-brown
Nickel (Ni)	NH_4OH + rubeanic acid	7	20	direct print	red to violet
	or: NH_4OH + dimethylglyoxime	10	20	direct print	red
Silver (Ag)	NH_4OH for native silver	8	35	direct print	grey-black
	or: 3 to 5% KCN + acetone solution of paradimethylamino-benzylidenerhodamine	12	30-180	1:15 HNO_3	red-violet
Sulphur (S)	5% NaOH-electrodes reversed	5	25	$SbCl_3$ + HCl	orange

Continued on p. 72

TABLE IX (continued)

MEMBRANE COLORIMETRY

Tin (Sn)	concentrated HCl or concentrated HNO_3	20	90	saturated aqueous solution of cacotheline	red-violet
				or: 2-benzylpyridine-specific test	green to red
Titanium (Ti)	$H_2SO_4 + H_3PO_4$	22	90	chromotropic acid	brown-red
	or: H_2SO_4 + phosphoric acid + hydrogen peroxide			direct print	yellow
Tungsten (W)	1:1 HCl			Rhodamine B (dilute solution)	red to violet
Vanadium (V)	1:5 HCl + 1:5 HNO_3 + 1 % acetic acid solution of diphenylamine		90	direct print	green
	or: 1:5 HCl + 10% tannic solution		60	expose to ammonia	black
Zinc (Zn)	1:5 HCl	20	18	acridine hydrochloride + KCNS	yellow-green
	or: HCl + 20% cobalt nitrate	20	30	potassium mercuric thiocyanate followed by ammonium fluoride to mask iron	blue

very sensitive, but not specific, as coloured precipitates are also obtained with silver, lead, mercury, copper, and bismuth. Bismuth cannot be masked and in its presence another reagent must be used (WELCHER, 1948).

Arsenic

(*1*) N-ethyl-8-hydroxytetrahydroquinoline hydrochloride: a 0.5% aqueous solution of the reagent gives a reddish-brown coloration with arsenites in the presence of a 1% solution of ferric chloride. Copper, lead, and mercury interfere, but can be eliminated by using an excess of hydrochloric acid.

(*2*) Silver nitrate: arsenates yield a brown precipitate of silver arsenate with a 1% solution of silver nitrate. Oxidation to arsenates is effected by attacking the mineral with ammonia or caustic soda to which hydrogen peroxide has been added. The reaction paper is then acidified with acetic acid and developed in the silver nitrate solution. The print must be thoroughly washed with distilled water after development.

(*3*) Stannous chloride (Bettendorf's reaction): arsenic compounds are convenient and satisfactory in most cases, particularly when the electrographic contact print method is applied.

Bismuth

(1) Cinchonine and potassium iodide: cinchonine potassium iodide solution (prepared by dissolving 1 g of cinchonine in 100 cm^3 of warm water containing a little nitric acid; after cooling, 2 g of potassium iodide are added) gives with bismuth ions in a weak acid medium an orange red precipitate. Copper, lead, and mercury interfere owing to the presence of potassium iodide. The lead and mercury can be suppressed by hydrochloric acid, while copper can be removed

by washing the print in a 10 *N* solution of sodium thiosulphate after developing in the specific reagent.

(*2*) Dimethylglyoxime: an alcoholic solution of dimethylglyoxime in ammoniacal solution gives a yellow print with bismuth. Nickel interferes, but can be masked by cyanide. Fortunately, for the purpose of this test, nickel is seldom associated with bismuth ores.

Cadmium

(*1*) Nitronaphthalenediazoaminobenzene-4 -azobenzene (Cadion 2B): in an alcoholic solution and in the presence of sodium potassium tartrate, cadmium ions give a pink colour with Cadion 2B. This reagent is more sensitive than p-nitrodiazoaminobenzene (Cadion) and in the presence of excess potassium cyanide the test is specific, since magnesium is the only metal which interferes under this condition (DWYER, 1938).

(*2*) Dinitrodiphenylcarbazide: in an alkaline medium and in the presence of a 10% potassium cyanide and 40% formaldehyde solution cadmium ions yield a brown precipitate with an alcoholic solution of dinitrodiphenylcarbazide, which changes to greenish blue on standing. An alkaline solution of the reagent, which is red, is coloured violet by formaldehyde, so that with minor amounts of cadmium the colour should be compared with that of a blank test.

Chromium

(*1*) Acid alizarin yellow R.C.: this dyestuff gives an orange-yellow colour with chromium ions. After developing, the reaction paper should be washed in 2*N* sulphuric acid to destroy the effect of interfering elements and neutralized with dilute ammonium hydroxide. In the presence of chromium an orange-yellow print remains on the paper, otherwise the dyestuff turns orange and dissolves in the ammonia.

(*2*) Silver nitrate after conversion into chromate: a 2% silver nitrate solution gives a red print with chromate ions $(CrO_4)_6{}^{2-}$, which is insoluble in acetic acid. The conversion into chromate can be achieved by oxidation in an alkaline medium–such as caustic soda and hydrogen peroxide.

(*3*) α-naphthylamine after conversion into chromate: an alcoholic solution of α-naphthylamine in a hydrochloric acid medium gives a violet-brown colour with chromate ions.

(*4*) Phenylazodiaminopyridine hydrochloride after conversion into chromate: an aqueous solution of the reagent forms a yellow precipitate with chromate ions (WELCHER, 1948).

Cobalt

(*1*) α-nitroso-β-naphthol: a 0.1% acetic solution of α-nitroso-β-naphthol yields an insoluble brown complex with cobalt ions. Iron and uranium interfere but can be masked by converting them into non-reactive phosphates by adding phosphoric acid to the attacking reagent. Cupric salts also interfere, but by adding potassium iodide cuprous iodide is formed with the liberation of iodine, which is removed by treating with sodium sulphite; the white insoluble cuprous iodide does not react with nitrosonaphthol.

(*2*) Nitroso-R-salt: a 0.5% aqueous solution of the reagent reacts with cobalt ions in a solution buffered with sodium acetate, giving a red print. Copper, iron, lead, nickel, and silver interfere, but are eliminated by washing the paper in 2 : 1 nitric acid, followed by rinsing in water. The reagent is extremely sensitive.

(*3*) Rubeanic acid: cobalt yields a brownish print in a weak ammoniacal or acetic medium with a 1% alcoholic solution of rubeanic acid. Nickel under similar conditions gives a blue colour, while copper yields a black to olive-green precipitate. Both nickel and copper are removed by washing the reaction paper in a 1% solution of potassium cyanide for a very short time in order to dissolve the nickel or copper rubeanates and leave a clear cobalt print.

(*4*) Piperidine: a 1% aqueous solution of the piperidine derivative $Na_3Fe(CN)_5C_6H_{11}N.6H_2O$ is a very sensitive reagent for cobalt, with which it yields a green colour. With the same reagent ferric iron produces a prussian blue test, while copper yields a brown precipitate. In the presence of these other elements it is advisable to use another reagent (WELCHER, 1947).

(*5*) Benzimidazole: if an electrolytic attack is used a 5% alcoholic solution of benzimidazole gives a violet-blue precipitate with cobalt in an ammoniacal solution containing ammonium chloride. The reagent is very convenient for detecting cobalt in such minerals as cobaltite, smaltite, safflorite, and skutterudite. Copper interferes by giving a brown print (WELCHER, 1947).

(6) Potassium cyanide: a 2–5% solution of potassium cyanide alone gives a direct yellow-orange print with some cobalt-bearing minerals, notably smaltite and safflorite, when applying the electrographic contact method. The test is due to the formation of a cobalticyanide, which is yellow in colour, the reaction being specific but not sensitive.

(*7*) Antipyrine and potassium thiocyanate: a 10% aqueous solution of antipyrine in the presence of 2% potassium thiocyanate, gives a green complex with cobalt ions. Under the same conditions ferric ions yield a red or reddish-brown precipitate, which masks the presence of cobalt; in such circumstances it is advisable to employ another reagent.

For detecting cobalt in various mineral species it is more convenient to use benzimidazole in the cases of cobaltite, safflorite-loellingite and skutterudite. This reagent is very stable and does not deteriorate on standing for a long time. In the case of carrolite and linnaeite, freshly-prepared solutions of nitroso-R-salt or antipyrine-thiocyanate reagents yields very satisfactory results.

Copper

(*1*) α-benzoinoxime: in an ammoniacal medium a 1–5% alcoholic solution of the reagent yields an insoluble green precipitate with copper ions. The test is specific in ammoniacal tartrate solution.

(*2*) Rubeanic acid: in neutral, acetic, and ammoniacal solutions a 0.5% alcoholic solution of rubeanic acid gives a black to olive-green precipitate with copper ions, the depth of colour being proportional to the amount of copper present. Cobalt and nickel interfere with the test, owing to the formation of brown cobalt rubeanate and blue nickel rubeanate. In the presence of only small quantities of these interfering elements a more selective copper reaction is obtained by using free acetic acid and ammonium acetate. The reagent must be freshly prepared.

(*3*) Salicylaldoxime: an alcoholic solution of the reagent reacts with copper ions in an acetic acid solution yielding a yellow-green print. In acid solution only gold, palladium, and vanadate ions interfere, but many metallic ions interfere in neutral or alkaline media.

(*4*) Sodium diethyl dithio-carbamate: in neutral, slightly alkaline, or slightly acid solutions, copper gives a brown precipitate with an aqueous solution of the reagent. Iron interferes, as it also produces a brown colour, but it can be masked by phosphoric acid (WELCHER, 1948).

(*5*) Potassium ferrocyanide: a 1% aqueous solution of potassium ferrocyanide gives a brown precipitate with copper ions in a neutral or acidic medium. Iron interferes by yielding the prussian blue test, but can be suppressed by developing the paper in ammonia.

If the mineral is attacked by ammonia, α-benzoinoxime in the presence of sodium potassium tartrate is the most convenient specific reagent for copper. In the absence of nickel and cobalt ions, the test with rubeanic acid is preferable, as it is very sensitive and delicate. If the mineral is attacked by strong acids, as in the cases, for example, of bournonite and tetrahedrite, then either rubeanic acid or potassium ferrocyanide is a more suitable reagent for copper. In the case of the rubeanic acid test the reaction paper must be neutralized with ammonia before developing in the specific reagent.

Gold

Benzidine: auric salts give a blue precipitate with an acetic acid solution of benzidine. Oxidizing agents and auto-oxidizable substances produce the same reaction; ferric iron can be eliminated by reduction.

Iron

Since ferrous ions are quickly oxidized to the ferric state on the reaction paper, the reagents employed for detecting iron are limited to those which indicate the presence of ferric ions.

(*1*) Sulphosalicylic acid: a 5% aqueous solution of the reagent yields a violet colour with ferric ions in an acid medium.

(*2*) Chromotropic acid: in an acid medium an aqueous solution of the reagent gives a deep-green print with ferric salts. This reaction is quite effective in determining iron in the presence of copper. Titanium interferes by yielding a brown coloration.

(*3*) Potassium thiocyanate and antipyrine: a mixture in equal proportions of a 2% solution of potassium thiocyanate and a 10% aqueous solution of antipyrine reacts with ferric ions, giving a red to brown precipitate. Cobalt interferes by giving a greenish-blue colour that cannot be masked.

(*4*) Potassium ferrocyanide: a 1% aqueous solution of the reagent gives the well-known prussian blue test with ferric ions in an acid medium. Copper interferes by yielding a brown precipitate and in the presence of this element it is advisable to develop the paper in ammonia gas, or, better still, to use another reagent, preferably chromotropic acid.

(*5*) 7-iodo-8-hydroxyquinoline-5-sulphonic acid: in weak acid solution ferric ions yield a green colour or precipitate with a 0.5% aqueous solution of the reagent. The intensity of the colour is proportional to the concentration of ferric ions. The reaction is sensitive and highly selective (YOE and ROBERT, 1937).

The detection of iron is often carried out by using either potassium ferrocyanide or sulphosalicylic acid. Both reagents are readily available, do not deteriorate, and their solutions can be kept for a considerable length of time. Moreover, they yield results which are very characteristic, conclusive, and sensitive enough to be suitable for detecting iron in most ore minerals, even if it is present only as a minor constituent.

Lead

(*1*) Stannous chloride and potassium iodide: a complex yellow to orange salt of $2PbI_2.SnI_2$ is formed with the reagent in the presence of lead ions. Since lead sulphate can also react, this test can be used in the presence of interfering elements such as copper, iron, and bismuth. The attack should be carried out in a mixture of acetic and nitric acids, but if the electrographic method is used acetic acid alone should be adequate. The excess of acid should be neutralized by ammonia gas. After the attack, the reaction paper is immersed in dilute sulphuric acid, then thoroughly washed in water, neutralized, and developed in the specific reagent. The reagent must be freshly prepared.

(*2*) Gallocyanine: lead salts in neutral solutions or lead hydroxide in the presence of ammonia give a blue precipitate with gallocyanine. The test is sensitive but not specific since many other elements interfere, though they can be removed as soluble sulphates. If disturbing elements are present the test can be conducted in two ways: (*a*) after the attacking process the gelatine paper is treated with dilute sulphuric acid, washed in alcohol and pyridine, and developed in the specific reagent; (*b*) after attack, the reaction paper is developed in gallocyanine, exposed to ammonia gas and then thoroughly washed. The presence of lead is indicated by a persistent violet-blue stain on the paper. Gallocyanine solution must be freshly prepared.

(*3*) Dithiozone: an ammoniacal solution of dithiozone in the presence of potassium cyanide yields a red precipitate with traces of lead ions. Under this condition the test is specific, since most of the interfering heavy metals are masked. The reagent must be freshly prepared (FEIGL, 1946).

(*4*) Sodium rhodizonate: in a neutral of weakly-acid medium an aqueous solution of the reagent yields a violet to rose-red colour in the presence of lead ions. Strontium and barium react similarly (WELCHER, 1947).

(5) Dioxydiquinone or tetrahydroxyquinone: both reagents give a red precipitate with lead ions. Barium and strontium behave in the same way.

Manganese

Sulphurous acid is the most suitable attacking reagent for dissolving selectively the higher oxides of manganese, such as pyrolusite.

(*1*) Benzidine: manganous ions in the presence of a base will oxidize benzidine to the blue meriquinoid compound. According to Gutzeit, the best way to obtain a definite print is to develop the paper in ammonia gas and allow it to oxidize in the air for 5 minutes; then to treat it with an acetic benzidine solution, which will produce a dark-blue print. If the electrographic method is used the paper is moistened with an acetic solution of benzidine to which hydrochloric acid has been added. After the attack the print is greenish-yellow and turns blue when the paper is developed in a 1% caustic soda solution. The test is sensitive, but not specific, since it is also given by elements which yield auto-oxidizable compounds as well as by oxidizing agents.

(*2*) Tetramethyl-p-diaminodiphenylmethane: in the presence of manganese dioxide a solution of the reagent is oxidized to an intensely blue-coloured compound. The test is more sensitive than that preceding, but is non-specific. The reagent must be freshly prepared.

(*3*) Ammoniacal silver nitrate: manganese ions in the presence of dilute solutions of ammoniacal silver nitrate give a brown to black precipitate of manganese dioxide and black metallic silver. The test is very sensitive and is useful for detecting manganese in most of its higher oxides. Its sensitivity is decreased in the presence of coloured metallic hydroxides, particularly that of iron, which gives a brown print.

Mercury

(*1*) p-dimethylaminobenzylidene rhodanine: mercury gives a red-violet precipitate in neutral or weakly acid solutions with a saturated alcoholic solution of the reagent. Large amounts of chloride interfere, since they reduce the ionization of the already weakly ionized mercuric chloride. Buffered sodium-acetate solutions suppress the interference of chloride. The mineral preferably should be attacked by either nitric or sulphuric acid. As silver ions react similarly they should be removed by precipitation with hydrochloric acid and the test carried out in a buffered acetate solution. Copper also interferes with the test, but can be masked by using sodium phosphate or phosphoric acid. The violet reaction of the mercury test is visible even in the presence of light-green copper phosphate.

(*2*) Stannous chloride and aniline: mercury ions are reduced to metallic mercury by freshly-prepared stannous chloride solution in the presence of aniline. Large amounts of silver produce a similar reaction.

Molybdenum

(1) Phenylhydrazine: molybdates react with an acetic acid solution of the reagent to give a reddish-brown precipitate. Interfering elements are gold, silver, and palladium, which, however, are rarely associated with molybdenum minerals.

(*2*) Potassium ethyl xanthate: in acid solution molybdates react with the reagent to produce a red coloration. The reaction is sensitive and specific for molybdates, but has the disadvantage that diffusion quickly takes place after development of the print. Nevertheless, the test serves readily to distinguish molybdenite from graphite.

Nickel

(*1*) Dimethylglyoxime: a 1% alcoholic solution of dimethylglyoxime gives a red precipitate with nickel ions. Reaction takes place in neutral, acetic, or ammoniacal solutions. Under electrolytic attack a direct print is obtained by mixing ammonia with the specific reagent. For minerals containing much iron—such as pentlandite, bravoite, or nickeliferous pyrrhotite—a 10% solution of sodium potassium tartrate (rochelle salt) is added to the attacking reagent in order to suppress interference by iron ions. In the case of cobalt minerals that contain minor quantities of nickel, the attack is carried out by a mixture of potassium cyanide and hydrogen peroxide, as a result of which the corresponding nickel and cobalt complex cyanides are formed. The reaction paper is then developed in a mixture of 40% formaldehyde solution and 1% alcoholic solution of dimethylglyoxime. In consequence, the nickel cyanide complex is decomposed and the liberated nickel ions react with the specific reagent, while the cobalticyanide complex remains stable and unaffected. Iron is masked as ferricyanide by the same treatment, and the nickel test can therefore be conducted in the presence of cobalt alone or of both cobalt and iron.

(*2*) α- furildioxime: a saturated alcoholic solution of the reagent gives a reddish-brown precipitate with nickel ions in an ammoniacal medium. The test is more sensitive than that preceding, and, like dimethylglyoxime, is specific in ammoniacal tartrate solution (BYRON, 1925).

(*3*) Rubeanic acid: a freshly-prepared 0.5% alcoholic solution of the reagent gives a blue-violet insoluble precipitate with nickel salts. The reaction is very sensitive, but less specific than the dimethylglyoxime test, since both copper and cobalt interfere.

Of the reagents mentioned dimethylglyoxime is recommended, because its alcoholic solutions can be kept indefinitely without deterioration and it is sensitive enough to detect nickel even if it occurs as a minor constituent of pyrrhotite.

Silver

(*1*) p-dimethylaminobenzylidene rhodanine: in an acid medium (dilute nitric acid), an alcoholic or acetone solution of the reagent yields a red-violet colour with silver ions. The test is not specific, as cuprous, mercury, gold, palladium, and platinum ions interfere.

(*2*) Orthotolidine: at a pH of 4 an alcoholic solution of the reagent gives an intense blue colour with silver ions. Nitric acid should be used as an attacking reagent and the paper dried before the print is developed. The reaction is very sensitive, but not specific, as many other metals and oxidizing agents interfere.

(3) Deposition of silver: two separate solutions are prepared: (*a*) dissolve 10 g of p-methylaminophenol (metol) and 50 g of citric acid in 500 cm^3 of distilled water, to 50 cm^3 of this solution add 2 cm^3 of 0.1 *N* silver nitrate solution immediately before use; (*b*) 0.02 *N* potassium bromide solution.

When using the electrographic method the mineral is attacked with 1 : 10 nitric acid. The paper is then immersed in a dish containing potassium bromide solution for about half a minute, in order to fix the silver as silver bromide in the gelatine layer. The paper is thoroughly washed, first with nitric acid, and then several times with distilled water, to remove all traces of potassium bromide. It is then developed in the specific reagent (a freshly-prepared mixture of metol, citric acid, and silver nitrate). After a short time a grey to black print appears, the intensity of the colour depending on the concentration of silver ions. The developer at first becomes violet, due to the separation of colloidal silver, and finally turns black. In this test all the reagents and equipment must be scrupulously clean and it is important that all the potassium bromide should be washed out of the gelatine layer before development in the specific reagent.

(*4*) Reduction: silver can be detected by reducing an alkaline solution of silver salts with formaldehyde. A freshly-prepared mixture of 20% formaldehyde solution and 10% potassium hydroxide in the proportion of 2 : 5 is suitable as a reducing agent (WELCHER, 1947).

The test with p-dimethylaminobenzylidene rhodanine is very satisfactory for detecting silver in sulphides, sulpharsenides and sulphantimonides. As an attacking reagent potassium cyanide (2–5% solution) is selective in its effect. When applying the electrographic process native silver gives a direct grey print with ammonia alone. The test by deposition of silver on a silver bromide surface in the presence of a reducing agent is a very sensitive and specific one but is rather elaborate. It should be used preferably for detecting minute amounts of silver in sulphides such as galena, pyrite, and chalcopyrite.

Sulphur

(*1*) Antimony trichloride: a weak acid solution of the reagent gives an orange print of antimony sulphide with sulphur ions.

(*2*) Lead acetate: in an acetic acid medium, a dilute solution of lead acetate gives a brown to black print with sulphur ions.

(*3*) Sodium nitroprusside: in an ammoniacal medium the reagent yields an intense violet colour with sulphur ions. An excess of the alkali should be avoided, as it retards the reaction, owing to the formation of a reddish-yellow compound. The main disadvantage of this test is that the violet-coloured compound is unstable and hence does not produce a permanent print.

(*4*) Catalytic effect of sulphides on iodine-azide solution: the reagent is prepared as follows: in 100 cm^3 of 0.1 *N* iodine solution (12. 69 g iodine and 22 g potassium iodide in 1,000 cm^3 of water) dissolve 1. 3 g of sodium azide. This reagent is very stable, but in the presence of sulphur ions it is immediately decomposed into sodium iodide and free nitrogen. The liberated bubbles of nitrogen are clearly visible and after a few minutes larger bubbles form and remain on the gelatine paper.

When the electrographic contact print method is applied the tests with antimony trichloride and lead acetate are quite satisfactory. In this case the mineral should be attacked with a dilute alkaline solution (1–5% sodium hydroxide) and the reaction paper connected to the anode. The sodium azide-iodine test is very sensitive and specific, although it does not give a permanent print.

Tin

(*1*) Cacotheline: a saturated aqueous solution of the reagent in the presence of excess hydrochloric acid yields a red-violet colour with stannous ions. Antimonious ions and reducing substances, such as ferrous salts or sulphites, give similar reactions under the same conditions.

(*2*) Anthrapurpurin: an alcoholic solution of the reagent yields an orange-yellow colour with stannic ions. Iron, chromium, molybdates, and titanium interfere (WELCHER, 1948).

(*3*) Anthraquinone-1-azo-dimethylaniline: stannic ions in the presence of an alcoholic solution of the reagent and hydrochloric acid yeld a bluish-violet precipitate. The test is sensitive, but non-specific, as iron, cadmium, zinc, lead, gold, and molybdenum interfere (WELCHER, 1948).

(*4*) 2-benzylpyridine: the reagent is prepared by dissolving 0.2–0.5 ml of 2-benzylpyridine in 100 ml of 96% ethyl alcohol in a round-bottom flask, 100 ml of water is then added and the mixture shaken for 2 min. It is then exposed to ultraviolet rays (300–350 mμ) for at least 1 h in order to produce the desired photo-product. The green reagent-impregnated paper turns red in the presence of stannous chloride. The test is very specific, as the only interference is caused by sulphur dioxide.

Titanium

(*1*) Chromotropic acid: a 5% aqueous solution of the sodium salt of chromotropic acid in the presence of hydrochloric or sulphuric acid gives a reddish-brown precipitate with titanium salts. Ferric iron and uranyl salts interfere, but can be eliminated by reduction with stannous chloride. In addition, iron can be masked as phosphates by the use of phosphoric acid.

(2)- Hydrogen peroxide: on applying the electrographic contact print procedure, a direct yellow print is obtained by using a mixture of sulphuric acid, phosphoric acid, and hydrogen peroxide as an attacking and specific reagent.

(3) 3-(ortho-arsenophenylazo)-4, 5 dihydroxy-2, 7-naphthalene disulphonic acid: titanic ions in an acid medium

yield a violet colour with the reagent. The test is specific, as the only interfering elements in an acid medium are thorium, zirconium, niobium, tantalum, and uranium (WELCHER, 1948).

Tungsten

Rhodamine B: in hydrochloric acid solutions tungstates give a red to violet reaction product with a dilute solution of rhodamine B. Antimony, mercury, gold and titanium, if present in excess, interfere with the test, but this seldom occurs in tungsten ores.

Vanadium

(*1*) 5, 7-dibromo-8-hydroxyquinoline: in a solution containing 20% nitric acid and an alcoholic solution of the reagent vanadium ions yield a brown precipitate. Ferric ions give a green colour, but can be masked as phosphates. Molybdenum and titanium also interfere by yielding a yellow colour.

(*2*) Tannin: metavanadates give a black precipitate with tannin. The test is very delicate and a direct print can be obtained by using a mixture of 10% aqueous solution of tannin and 1 : 5 hydrochloric acid, and then exposing the reaction paper to ammonia after the attacking process. Iron interferes by giving a similar test, but it can be masked by phosphoric acid (WELCHER, 1947).

(*3*) Diphenylamine: a freshly-prepared 1% acetic solution of the reagent in the presence of 1 : 5 hydrochloric acid gives a direct green print with some vanadium minerals. The test is not specific, as manganese, zinc, and oxidizing agents interfere, but it is nevertheless useful in certain cases (WELCHER, 1947).

Zinc

(*1*) Co-precipitation with cobalt mercury thiocyanate: very dilute cobalt salts react with potassium mercury thiocyanate, Hg $(CNS)_4K_2$, to form a blue complex compound. The reaction is very slow, but in the presence of a small amount of zinc the blue precipitate is formed at once, owing to catalytic action. Iron and copper interfere; the former gives a red colour which can be masked by a rinse in ammonium fluoride, while the latter yields a green precipitate.

(*2*) Dithiozone: in an alkaline medium dithiozone reacts with zinc ions to form a purplish-red colour. Copper, cobalt, and mercury interfere with the test, but can be removed by the addition of tartrate and potassium cyanide to the solution.

(*3*) Acridine hydrochloride: an alcoholic solution of acridine hydrochloride in the presence of potassium thiocyanate gives a yellow-green complex with zinc ions. Iron gives a red colour and cobalt yields a green reaction (WELCHER, 1947).

(*4*) p-dimethylaminostyryl-β-naphthothiazole methiodide: in a neutral or slightly acid solution an alcoholic solution of the reagent in the presence of potassium thiocyanate gives a rose to violet colour with zinc ions. The reaction is sensitive, but not specific, as under the same conditions silver, copper, mercury, and iron interfere (WELCHER, 1947).

CHAPTER 3

Quantitative Chemical Analysis

INTRODUCTION

The need for large numbers of chemical analyses of rocks and minerals has resulted in considerable advances over the classical methods initiated by HILLEBRAND (1900) and eventually by WASHINGTON (1932). For example, KRALHEIM (1947) and HEGEMANN and ZOELLNER (1952) devised rapid spectrochemical methods for the determination of the common elements in silicate rocks. Likewise, SHAPIRO and BRANNOCK (1952, 1956) combined various techniques of their own with those of many other analysts and thus established a widely accepted scheme for the analysis of silicate rocks. RILEY (1958) modified certain parts of this procedure which resulted in increased accuracy and the rate of output of analytical results. Consequently, it is now possible to assemble these several forms of analytical procedure for the convenience of analysts concerned with geochemical problems.

In all forms of chemical analysis the production of a representative sample is essential. With mixtures of chemicals this can be obtained with reasonable ease. On the other hand, geological material is exceedingly difficult to sample and one is never absolutely sure of obtaining representative samples of natural processes under investigation. If a specimen has been obtained from a natural outcrop, as distinct from a deep excavation, a borehole or a mine, one must acknowledge that it will not be entirely free from the effects of weathering. Consequently, no effort must be spared in extracting a sample of rock from the innermost parts of an exposure. Wherever possible, a number of samples should be taken to cover the possibilities of atmospheric oxidation which are not always evident to the naked eye. Moreover, the specimens should never be collected in an haphazard fashion. All too frequently, the exposures of rocks are scattered and deeply eroded, so that extreme care must be exercised in the replicate collection and logging of samples for analysis.

Whilst no precise rules can be given for the assembly of specimens for geochemical analysis, some generalizations might be made. Among large intrusions of igneous rocks one might assume that rapid changes in composition will be somewhat exceptional so that a few samples can often reveal the geochemical characteristics. On the other hand, the margins of such intrusions usually contain rapid variations in the composition of the rocks, so that considerable numbers of specimens will be required to reflect the geochemical changes involved. Dykes and sills of basic rocks usually remain fairly uniform in composition, but more acid intrusions tend to be characterized by mineral segregations which require special treatment in the collection of samples. Volcanic rocks of low viscosity such as basalts, but to a lesser extent andesites, can be sampled at widely spaced intervals. In contrast, viscous lavas such as rhyolites, dacites and trachytes are characterized by rapid variations in composition which require rather specialized attention to obtain a geochemical picture of their origin and post-volcanic history.

Sedimentary rocks present great difficulties. From the unconsolidated condition of a sediment to its completely indurated condition there are ofttimes no well-defined limits. Consequently, a large number of specimens will be required to reveal the geochemical processes involved in the formation of individual deposits. In many sedimentary studies, partial analyses suffice to illustrate the processes involved. Metamorphic rocks are exceedingly difficult to deal with, and it has been found from experience that geochemical analyses

TABLE X

MINIMAL SIZES OF SAMPLES FOR ANALYSIS

Grain size	*Texture*	*Min. sample weight (g)*
3 cm	conglomeratic or pegmatitic	5,000
1–3 cm	coarse grained or phenocrystic	2,000
1–10 mm	medium grained	1,000
0–1 mm	fine grained	1,500

mainly assist in the clarification of physico-chemical processes revealed by petrographic studies. Magmatic rocks present considerable sampling problems which usually consist of the collection of closely spaced samples which might cover the qualitative passage of one rock-type into another. The effects of weathering cannot be too strongly impressed upon the geologists collecting samples for geochemical studies. Such near-surface changes by no means consist of a one-way process of oxidation. Secondary mineralization of the products of weathering often leads to the formation of new rock-types which are different from the parent deposits. At this end of what might be termed the metasomatic alteration of near-surface rocks one approaches the geochemistry of sub-soils and eventually soil. Finally, it must be remembered that the size of a sample is of considerable importance. Too frequently, collections have been made from remote places which have provided the analyst with too little material for a full-scale programme of analysis to be completed. From experience it has been found necessary to obtain samples of minimal sizes as given in Table X. When obtaining such samples care should be taken to label and pack the specimens carefully. The low cost of polythene bags, and their astonishing strength, make them ideal for the collection of rock samples, especially under difficult climatic conditions. Such bags can be serially numbered before embarking for the field and this leads to a more accurate description in a field note book of the location and geological significance of each specimen. All these several points may seem trivial and obvious at first sight, but the experienced worker will agree that any suggestion which results in an improvement in the collection and location of specimens and field data is more than welcome. All too frequently has it proved impossible to repeat collections from certain areas and the application of modern techniques to the study of such collections has proved impossible.

Prior to the preparation of the sample for chemical analysis, a full description of its colour, texture and all other macroscopic characteristics should be fully recorded. In certain cases it is advisable to have a photographic record of the hand specimen as part of the analytical data. With the exception of very soft rocks such as clays, shales, mudstones and soft sandstones all surface contamination must be removed by washing the surface of the specimen with running water and finally rinsing them with distilled water. The specimen is then dried in the atmosphere or in a stream of compressed air.

The initial stage in the preparation of the sample for chemical analysis consists of the fragmentation of the specimen into pieces approximately 1–2 cm in size either with a hardened steel hammer or one of the standard rock-breakers. Each fragment is then hand-picked into a pile to ensure as far as possible that no weathered particles enter the sample at this stage of its preparation. Some of these fragments can also be selected for rock sectioning if at a previous stage the sample has not been covered by a microscopic examination. In many laboratories a few grams of such fragments are stored in bottles for future reference. The fines produced at this stage of the fragmentation of the sample are discarded as they are almost certain to be contaminated by the hammer or the blades of the crusher. From this stage onwards the further preparation of the sample is performed with great care. The reduction of rock fragments to powder requires the use of grinding processes which might contaminate the resultant powder and also oxidize some of the ingredients, especially minerals containing ferrous iron. These hazards have long been recognized as being almost impossible to eliminate, but by rigorous attention to detail they are usually reduced to recognizable and reproducible limits. At first, relatively coarse powders are produced by crushing the rock-fragments, without selection on a tungsten carbide plate with a pestle of the same alloy. The powder so produced is passed through a 12-mesh sieve and the emergent material poured through an automatic sample spitter to yield about 40–50 g. This is then ground down in stages in an agate pestle and mortar and passed through a 100-mesh bolting-silk sieve contained in a polythene former. The reduction of the rock sample to this powdered condition is performed in stages to avoid excessive oxidation of minerals such as ferrous iron minerals that might be present. In practice it is advisable to grind the coarse powder for not longer than 3 min at a time and sieving off the powder produced into a perfectly clean glass bottle. Prior to analysis the powdered specimen must be dried at 100°C for at least 2 h.

PREPARATION OF SOLUTIONS

It has now become common practice to prepare sample solutions for silicate analyses in two separate ways. Such solutions are now usually referred to as "solution A" and "solution B". Solution A is derived from dissolving a fusion cake formed by heating the sample with a flux and is mainly used for the determination of silica and alumina. Solution B is obtained by digesting the sample with hydrofluoric and perchloric acids and is mainly employed for the determination of metals as oxides. Certain mineral oxides such as corundum, rutile and spinel, and silicate like staurolite and tourmaline are resistant to these forms of attack. Consequently, they must be separated mechanically from the source of solution B and digested separately with potassium pyrosulphate. Likewise, some resistant silicates require the use of hydrofluoric acid and perchloric acid under pressure in a "Teflon" bomb.

Solution A is usually prepared as follows:

(*1*) Either 0.1 g or 0.05 g of the powdered sample is weighed accurately in a covered silver or nickel crucible.

(*2*) Add 1.5 g of sodium hydroxide which has been carefully stored in a firmly stoppered polyethylene bottle. About seventeen pellets of sodium hydroxide is approximately the right weight and as this is not critical it is common practice to add this number of pellets to avoid contamination of the alkali.

(*3*) The covered crucible and contents are placed in a muffle furnace at a temperature of between 800 and 850°C. At this temperature complete fusion is attained in about 5 min. Experience of the efficiency of the furnace employed will determine the precise duration of complete fusion, but this should rarely take more than 10 min, even when using 0.1 g of powdered sample.

(*4*) The crucible is removed to a safe cooling plate and allowed to cool. Some workers swirl the sodium hydroxide melt around the sides of the crucible during transfer to the cooling plate as this tends to assist eventual removal of the melt into solution.

(*5*) The covered crucible and contents are warmed on a water bath after adding 20 ml of distilled water to the solidified melt. This accelerates the solution of the solidified melt.

(*6*) A solution of 20 ml of 2.5 *N* sulphuric acid in about 200 ml of distilled water is made up in a 1 l graduated flask fitted with a stopper. A solution of 20 ml of 1 : 1 hydrochloric acid is used by many workers in place of sulphuric acid as this reduces the number of acid radicals in the eventual solutions and facilitates the use of solution A for the determination of alumina.

(7) Great care has to be taken to remove all traces of the melt into solution. Consequently, the crucible and lid are carefully washed through the polyethylene funnel into the flask. The scouring of the funnel and the crucible is done with a rubber policeman.

(8) The washings and the solution of the hydroxide melt are made up to 1 l in the flask by washing distilled water through the polyethylene funnel to ensure that no traces are allowed to remain upon it. Before finally making up the solution to the graduation mark, a small crystal of ferrous sulphate should be added if the solution has a pink colouration. This is usually due to the presence of excessive quantities of manganese. The ferrous sulphate reduces the permanganates to a colourless solution which is more suitable for colorimetric determinations.

This completes the formation of solution A which should be visibly free from suspended matter and visually colourless.

DETERMINATION OF SILICA

The most widespread method for the determination of the total silica in solution A is based upon the ability of silica to form a yellow-coloured complex with ammonium molybdate which in turn can be reduced to a blue colouration by means of a metol-sulphite reagent. The transmission of light through such solutions provides a direct estimate of the complexed silica. Known solutions of silica-complexes are calibrated at convenient wavelengths of transmitted light using a photometer. Most laboratories possess reasonably accurate spectrophotometers which provide convenient ranges of wavelength of transmitted light. Some instruments are calibrated to work at 650 mμ (SHAPIRO and BRANNOCK, 1956) whilst other instruments have been used at 390 mμ. The choice of wavelength is often controlled by the type of spectrophotometer which is available. When this choice has been made the analyst prepares standard solutions of silica in order to calibrate the quantitative values of light transmitted at the accepted wavelength. Since the ultimate accuracy of silica determinations will depend upon the validity of the photometric readings obtained from these standard solutions of silica considerable care has to be exercised in the preparation and preservation of these solutions in the laboratory. Since these solutions will be in constant use, it has been found undesirable to store them in colourless glass bottles. Carefully cleaned, opaque polyethylene stoppered bottles have superceded glass for this purpose.

To prepare standard solutions of silica, it is absolutely necessary to use pure solids such as quartz crystals. Some workers have found that commercially prepared transparent fused silica rods are convenient sources of reasonably pure silica. On the whole it is desirable to obtain small transparent crystals of quartz for this purpose as when these are powdered in a scrupulously clean percussion mortar they yield as pure a reference sample as it is possible to obtain. When powdered to pass a 12-mesh steel sieve the particle size is sufficiently small to produce a complete melt by fusion with sodium hydroxide. This powder is then boiled with concentrated hydrochloric acid to remove iron. Filter and wash the powder with hot distilled water until all traces of the hydrochloric acid have been removed. Transfer the moist powder to a clean silica crucible and raise the temperature to red heat. At this stage it is necessary to establish the degree of purity of this silica powder. This is done by weighing out a small quantity (about 0.5 g) into a platinum crucible to which is added 15 ml of 40% hydrofluoric acid and 0.25 ml of concentrated sulphuric acid. Evaporate the whole by heating up to a temperature of 900°C, allow to cool and reweigh the crucible. Likewise evaporate the same quantities of hydrofluoric acid and sulphuric acid in a platinum crucible in order to determine the quantity of impurities in these reagents. The weight of residue in the evaporation of the silica specimen should not exceed 0.25 mg after deducting the weight of the residue obtained from the reagents. Up to this limit, the purity of the silica powder can be accepted for the preparation of standard solutions of different concentrations of silica. The standard solution is prepared in the same way as for solution A, using 0.1000 g of pure silica fused in 1.5 g of sodium hydroxide and the ultimate solution made up to 1 l. This solution can be accepted as containing 0.1 mg SiO_2/ml. In practice a number of such standardized concentrations should be prepared in order to construct a standard curve for the photometric determination of silica.

When silica is the predominant constituent of the vast majority of rocks and minerals it must be determined with greater precision than is necessary for the constituents which are present in lower concentrations. Consequently, all sample weighings and volumetric measurements must be carried out in a meticulous fashion. Unless the greatest care is exercised in using scrupulously clean glass and plastic ware the ultimate readings on the photometer can be misleading and lacking in reproducibility. This is particularly necessary in the preparation of the following reagents which are required for the determination of the total silica in "solution A".

(*a*) Ammonium molybdate reagent is prepared by dissolving 2 g of ammonium molybdate in 75 ml of distilled water, to which 6 ml of concentrated hydrochloric acid is added and the whole diluted to 250 ml. This reagent must be stored in a plastic bottle or a pyrex glass bottle. In the latter, the colourless solution may turn blue, at which stage of decomposition it should be discarded. It is advisable to make up fresh molybdate reagent at fairly regular intervals.

(*b*) Metol-sulphite reagent is prepared by shaking 5 g of metol (p-methylaminophenol sulphate) with approximately 230 ml of distilled water in which 3 g of anhydrous sodium sulphite has been previously dissolved. The solution is then diluted to 250 ml and filtered through a Whatmann No.1 paper into a plastic bottle, or one made of brown glass.

(*c*) The reducing reagent is made by adding 85 ml of the metol-sulphite reagent to 50 ml of 10% oxalic acid. Place this solution in an ice-bath and add 100 ml of 25% sulphuric acid. When quite cold make up the solution to 250 ml with distilled water. This reducing reagent should be made up at frequent intervals as it loses its powers of reduction on standing. SHAPIRO and BRANNOCK (1956) used a reducing solution composed of 0.7 g of sodium sulphite dissolved in 10 ml of water, to which is added 0.15 g of 1-amino-2-naphthol-4-sulphonic acid and stirred until dissolved. To this was added a solution of 9 g of sodium sulphite in 90 ml of water and after thoroughly mixing stored in a plastic bottle.

Having prepared these necessary reagents the procedure for the determination of the total silica in the standard solutions and in "solution A" of samples to be analysed, is as follows:

(*1*) By means of a pipette place accurately known quantities of the silica solutions into 100 ml graduated flasks. The quantities chosen will depend upon the sensitivity of the spectrophotometer to be employed. Some workers (RILEY, 1958 b) use 2 ml of solution whilst others employ as much as 10 ml for the molybdate reaction. Assuming the use of 2 ml of solution, add about 10–15 ml of distilled water and then 10 ml of the ammonium molybdate reagent with a pipette swirling the flasks during the additions. It is important to allow the reagent to mix well with the silicate solution and then allow them to stand for at least 15 min in order to ensure that the molybdate complex has been completely formed.

(*2*) After this interval of time, add 15 ml of reducing reagent by means of a pipette whilst swirling the flasks. Dilute to 100 ml with distilled water and thoroughly mix the contents of the flask. The solution is now ready for optical determination in the spectrophotometer.

In order to zero the spectrophotometer at the wavelength

selected, a blank solution of the reagents must be prepared in the same manner as the aliquots designed for standardization and analysis. In a 1 cm cell, a standard solution of 0.2 mg of silica measured at a wavelength of 812 mμ gives an optical density of about 0.745, after deducting the optical density of the reagent blank which should have yielded a value of about 0.004. If the rock sample is known to contain less than about 40% silica it is necessary to increase the volume of solution A, and as a consequence the volume of the blank should be increased to the same value. It must also be appreciated that sources of technical error are almost inevitable, so it is always advisable to carry out in quadruplicate the spectrophotometer measurements of the blank solutions and of those being analyzed. Moreover, each of these optical readings should be determined at least three times. These values are then averaged and the percentage of silica calculated from the following formula:

$$\frac{\mathrm{OD} - \mathrm{ODB}}{(\mathrm{ODS} - \mathrm{ODSB}) \times \mathrm{V} \times \mathrm{W}}$$

of which: OD = optical density of the unknown sample, ODB = optical density of the corresponding reagent blank, ODS = optical density of the standard silica solution, ODSB = optical density of the standard reagent blank, V = volume of the unk nown sample, and W = weight of sample.

SHAPIRO and BRANNOCK (1962) suggest that the transmission readings are converted to values of absorbance and quote a table of values from which these may be readily evaluated. They then calculate all the absorbance values to a 50 mg sample-weight basis using the original sample weights as follows:

$$\text{absorbance} \times \frac{50}{\text{sample weight}} = \text{absorbance on a 50 mg sample-weight basis}$$

They then compute a factor for each of the two comparative standards as follows:

$$\frac{\text{percent SiO}_2\text{ of comparison standard}}{\text{absorbance of comparison standard (on 50 mg basis)}} = \text{factor}$$

The average of the two factors then enables them to calculate the silica percentage of the specimen as follows:

$$(\text{av. factor}) \times (\text{absorbance of sample solution on 50 mg basis}) = \text{percent SiO}_2$$

DETERMINATION OF ALUMINA FROM SOLUTION A

The same standard solutions of the alkali melts and of the solution A can be used for the determination of total alumina. As in the case of silica, evaluation is obtained by the use of a spectrophotometer, using complexes of aluminium. SHAPIRO and BRANNOCK (1962, p.34) recommend calcium and alizarin red-S as described by PARKER and GODDARD (1950) as this has a much greater absorption for a given amount of aluminium than that formed by simply using alizarin red-S. Using this method the following reagents are required:

(*a*) Hydroxylamine hydrochloride as a 10% solution in distilled water. It is usual to prepare 500 ml quantities.

(*b*) Calcium chloride solution made by transferring about 7 g of calcium carbonate to a 250 ml beaker containing about 50 ml distilled water. To this, dilute hydrochloric acid is added drop by drop whilst constantly stirring the suspension of calcium carbonate. The solution is then boiled for a few minutes and allowed to cool, when it is diluted to about 500 ml.

(*c*) Potassium ferrycyanide solution must be made up just prior to use to a strength of 1%.

(*d*) A buffer solution, consisting of 70 g of sodium acetate and 30 ml of glacial acetic acid diluted to 500 ml with distilled water.

(*e*) Alizarin red S solution made up to 500 ml as a 1% solution.

With these reagents the determination of alumina is obtained as follows:

(*1*) Transfer to 100 ml flasks, which have been acid cleaned, 15 ml aliquots of "solution A" samples, and of each reference solution. To these add 1 ml of the calcium-chloride solution and mix thoroughly.

(*2*) To this add 1 ml of the hydroxylamine hydrochloride reagent, thoroughly mix and then add 1 ml of the potassium ferricyanide solution, mix and allow to stand for at least 5 min to complete the reaction.

(*3*) Add 10 ml of the buffer solution to each flask and again allow them to stand for at least 10 min after they have been thoroughly mixed.

(*4*) The solution is now ready to receive the 5 ml of alizarin red S reagent. After mixing this thoroughly the solution is allowed to stand for about 2 h before carrying out the readings on the spectrophotometer in the same way as described for the determination of total silica. Complexes, additional to alumina, which absorb light at 475 mμ are also formed by iron and titanium, but the interference from iron is eliminated by the formation of ferrous ferricyanide prior to the addition of the alizarin red S solution. On the other hand, this does not eliminate titanium. Consequently, the quantity of titanium, which is determined otherwise, is subtracted from the apparent value of alumina obtained by the amount of absorbance at a specified wavelength. Empirically, it has been found that for each percent of TiO_2 in the sample there is an apparent increase of 0.5% in the values of alumina. This provides a correction in the alumina results which is achieved by subtracting one half of the percentage amount of TiO_2 from the apparent percentage values for alumina.

Other reagents such as ammonium aurintricarboxylate (aluminon) and solochrome cyanine R (also termed eriochrome cyanine R) have been used as colorimetric reagents for the determination of aluminium. The results obtained with solochrome cyanine R have proved to be somewhat more reliable and offers considerable advantages in controlling interferences from zinc, nickel, manganese and cadmium. On the other hand, iron and copper seriously affect the readings. These can be removed by passing the sample solutions through a cellulose column. Iron, and many other elements, are separated by elution with a mixture of concentrated hydrochloric acid and freshly distilled methyl-ethyl ketone (8 : 192 v/v). The aluminium and nickel retained on the cellulose are recovered by eluting the column with dilute hydrochloric acid (1 : 5 v/v) and then determined spectrophotometrically using solochrome cyanine R. For this method the following reagents are required:

(*a*) A concentrated buffer solution composed of 320 g of ammonium acetate dissolved in sufficient distilled water to which 5 ml of glacial acetic is added, and the whole made up to 1 l. VOGEL (1962, p.793) recommends the use of 27.5 g of ammonium acetate plus 11.0 g of hydrated sodium acetate and 1.0 ml of glacial acetic acid in 100 ml distilled water for the concentrated buffer solution.

(*b*) A solution of 1.5 g of sodium mercaptoacetate (Kodak) with 50 ml of 95% ethyl alcohol diluted in distilled water to 500 ml. This solution dispenses with the use of hydrated sodium acetate in the concentrated buffer. If this is employed, then a dilute buffer solution is made by taking up one volume of concentrated buffer solution in five volumes of distilled water and adjusting the pH to 6.1 by adding acetic acid or sodium hydroxide solutions.

(*c*) Solochrome cyanine R solution, which is made up by dissolving 0.75 g of the solid reagent in 200 ml distilled water and then adding 25 g of sodium chloride, 25 g of ammonium nitrate, and 2 ml of concentrated nitric acid. After vigorous stirring this mixture, dilute this solution to 1 l.

These reagents are somewhat unstable, and they should be prepared as frequently as possible. In batch analysis they are best prepared daily. Using these reagents, the procedure for the determination of alumina is as follows:

(*1*) Prepare a standard solution of alumina. This may be done in either of the two following ways: (*a*) Dissolve 1.3192 g of analar reagent aluminium potassium sulphate in distilled water and dilute to 1 l in a volumetric flask. This represents 1 ml = 75 g of aluminium in solution. (*b*) Dissolve a known weight of a standard sample of a soda feldspar. This is usually obtained as a standard sample from the Bureau of Standards, Washington, U.S.A. When dried at 105°C, Standard Sample No.99 has the following composition:

SiO_2	68.66 %
$R_2O_3 + RO_2 + R_2O_5$	19.28 %
Al_2O_3	19.06 %
Fe_2O_3 (total Fe as Fe_2O_3)	0.067 %
MgO	0.053 %
CaO	0.360 %
Na_2O	10.730 %
K_2O	0.410 %
TiO_2	0.017 %
P_2O_5	0.142 %
MnO	$<$ 0.001 %
BaO	0.010 %
Loss on ignition	0.520 %

A known weight of this felspar sample is taken up as a sodium hydroxide melt and prepared in the same way as solution A.

(*2*) Take 10 ml aliquots of the sample solutions, the reagent blank, and the standard solution of alumina, in 100 ml flasks and add to each 50 ml of the sodium mercaptoacetate solution.

(*3*) After thoroughly mixing these, add 10 ml of the solochrome cyanine R solution to each flask and mix thoroughly.

(*4*) Add 10 ml of ammonium acetate buffer solution and

dilute the whole to 100 ml. Mix thoroughly and after eight minutes determine the absorbance of the solutions at an established wavelength such as 475 or 535 mμ in the spectrophotometer.

(*5*) Compute the factor as follows:

$$\frac{(\%\ Al_2O_3 \text{ in standard}) \times (\text{average wt. of standards})}{\text{average absorbance of standards}} = \text{factor}$$

(*6*) Compute apparent weight of Al_2O_3, as follows:

$$\text{factor} \times \frac{\text{absorbance of sample}}{\text{weight of sample}} = \text{apparent } \%\ Al_2O_3$$

(*7*) Correct for TiO_2 from Fig.17.

(*8*) Percentage of alumina = {apparent % Al_2O_3 – (TiO_2 correction × % TiO_2 in sample)}.

Other analysts (e.g., RILEY, 1958 b) prefer to determine alumina from fluoride solutions along with iron, titanium, manganese, magnesium, phosphorus and the alkalis. This practice is to be commended for those rocks composed of melts. This will be apparent when the melt is observed to contain gritty particles after the cake has been dissolved from the crucible.

PREPARATION OF SOLUTION B

These solutions must be made up with extreme care. Platinum or silver crucibles must be used as the solvents employed include hydrofluoric acid and perchloric acid. Both these acids present laboratory hazards owing to the dangerous nature of the hydrofluoric acid fumes and the explosive nature of the mixture with perchloric acid. The hydrofluoric acid is used as an analar reagent of 48% strength. Perchloric acid is used as a 60% analar reagent. SHAPIRO and BRAN-

NOCK (1962) and previous workers preferred the use of concentrated sulphuric acid in place of perchloric acid or a mixture of equal portions of 72% perchloric acid with concentrated nitric acid. Used with extreme caution the mixture with perchloric acid effects a more rapid solution of the specimen.

Using hydrofluoric acid and perchloric acid solution B is prepared as follows:

(*1*) Weigh out exactly 0.5 g of the sample in a thoroughly clean, dry platinum or silver crucible of 25 ml capacity with a lid.

(*2*) With a safety pipette, add 4 ml of perchloric acid.

(*3*) From a polyethylene measuring cylinder add 15 ml of hydrofluoric acid.

(*4*) Heat the covered crucible on a water bath overnight. Care must be taken to see that the bath does not boil dry as a serious accident could occur.

(*5*) The following morning remove the lid from the crucible and evaporate the contents to dryness on the hot plate.

(*6*) When no fumes are visible place the crucible either under a vitreosil infra-red heater, or upon a hot plate at 200-220°C, until most of the perchloric acid has evaporated, and allowe it to cool. Do not allow the contents to become perfectly dry.

(*7*) Add 4 ml of perchloric acid from a pipette and follow this with 15 ml of distilled water.

(*8*) Heat the covered crucible in the water bath. Using a platinum rod, stir the contents occasionally until the contents have been completely dissolved. If the cake proves difficult to dissolve remove it with the solution to a 150 ml transparent silica flask and make it up to about 100 ml. Boil until all the cake has been dissolved.

(*9*) When cold, transfer to a 500 ml graduated flask and make it up to this volume with distilled water. This completes the formation of solution B.

In the same way prepare blanks of the hydrofluoric acid and perchloric acid in order to balance out the impurities in these reagents from estimates to be carried out with solution B. These blank solutions will also be required for calibration of colorimetric and other forms of instrumental techniques employed for the determination of the various oxides.

Occasionally rocks contain minerals which are not decomposed by this acid treatment. Under these circumstances it helps to use powders which pass a 140 mesh sieve. The principal resistant minerals are either oxides such as corundum, rutile and spinel, or silicates such as tourmaline and staurolite. Rock sections, which are an essential feature of geochemical work, should give a prior indication of their presence in the rock. If the rock contains such resistant oxides, transfer the potential solution B with its undissolved ingredients from the crucible, at stage (*8*) of the foregoing procedure, to a centrifuge tube and thus precipitate these oxides. Decant the supernatant liquid into a 250 ml graduated flask. Wash the residue several times with distilled water and add this to the contents of the flask. Make this up to 250 ml and regard this as "solution B1". The residue is then washed back into the platinum crucible with a jet of water and evaporated to dryness on a hot plate, or under an infra-red heater. When dry, add 0.50 g of fused potassium pyrosulphate A.R. Cover the crucible and heat over a low gas flame until the pyrosulphate becomes molten. Maintain it in this condition for at least 10 min before raising the temperature to a dull-red heat for a few minutes. Allow the crucible to cool and then add 1 ml of perchloric acid and 15 ml of distilled water. Warm on a water bath until a clear solution is obtained by occasionally stirring with a platinum rod. Transfer and make up this solution to 250 ml in a graduated flask. This is termed "solution B2". At this stage the two solutions B1 and B2 are combined by transferring 200 ml of each solution to a dry stoppered flask in which they are thoroughly mixed. This combined solution is used for the determination of all those elements which are normally derived from solution B with the exception of the alkalis, which are obtained from solution B1. With respect to the alkalis it must be noted that the solution is made up to 250 ml and not to 500 ml.

Resistant silicate minerals, such as tourmaline and staurolite, may be dealt with in one of the following ways:

(*a*) If the alkali content is known to be negligible the insoluble residue precipitated in the centrifuge is fused in a 10 ml silver crucible with 1 g of sodium hydroxide in a muffle furnace at 750°C for 15 min. When cool, add 5 ml of water to the cake and heat on a water bath until the contents are thoroughly disintegrated. Transfer the solution and residue to a platinum basin and evaporate down to a concentrated volume. To this, add 4 ml of perchloric acid and 10 ml of hydrofluoric acid. Then slowly evaporate the hydrofluoric acid on a water bath. When this is complete transfer the basin to an infra-red heater or a hot plate, and at about 200°C carefully evaporate most of the perchloric acid. When nearly dry, dissolve the residue in 30 ml of distilled water containing 1 ml of perchloric acid. Transfer, and make up the solution to 250 ml in a graduated flask with distilled water. This corresponds to solution B2.

(*b*) If alkali metals are present in significant concentrations, transfer the insoluble residue precipitated in the centrifuge to a reaction bomb composed of polytetrafluoroethylene (Teflon). To this is added 5 ml of hydrofluoric acid and 1 ml of perchloric acid. Securely screw down the lid to the bomb and place it in a steel vessel fitted with safety valves. The steel vessel should be quarter-filled with water. The safety jacket and bomb are heated in a thermostatically controlled oil-bath for about 4 h at 150°C. When quite cold unscrew the lid of the pressure jacket and remove the Teflon bomb. Open the bomb and rinse the contents and lid into a platinum dish by means of a jet of distilled water. To this, add 1 ml of perchloric acid and then evaporate on a water bath or a hot plate until nearly dry. When sufficiently concentrated, add a further 1 ml of perchloric acid and 20 ml of water. Warm on a water bath until solution is complete. This solution is then added to that decanted from the centrifuge tube and made up to 500 ml in a graduated flask. This solution is equivalent to solution B1 and may be used for all determinations, with the exception of silica and ferrous iron.

From the solutions collectively referred to as "solution B" it is usually found convenient to process the analyses within the following pattern:

(*a*) By spectrophotometry: alumina if not previously determined from solution A, titanium dioxide, total iron oxide, manganese, and phosphorus pentoxide.

(*b*) By titration: calcium oxides and magnesium oxide.

(*c*) By flame photometry: sodium oxide and potassium oxide.

DETERMINATION OF ALUMINA FROM SOLUTION B

For this determination the following reagents are required:

(*a*) Hydroxylamine hydrochloride 25% w/v.

(*b*) Sodium acetate (0.5*N*) prepared by dissolving 17.0 g of hydrated sodium acetate in distilled water and diluting to 250 ml

(*c*) Dipyridyl solution prepared by dissolving 2,2′–dipyridyl in 100 ml of 0.2 *N* hydrochloric acid.

(*d*) Beryllium sulphate tetrahydrate (4.0 g) dissolved and diluted to 100 ml in distilled water.

(*e*) Oxine solution, prepared by dissolving 1.25 g of 8-hydroxyquinoline A.R. in 250 ml of chloroform B.P. This solution, which is stored in an amber glass botlle in a refrigerator, should be discarded when it becomes coloured.

(*f*) A complexing reagent composed of 4 ml of hydroxylamine hydrochloride in 50 ml of 0.5 *N* sodium acetate, 20 ml dipyridyl solution, and 10 ml of the beryllium sulphate reagent; dilute to 100 ml.

A standard solution of aluminium is required to calculate

the quantity of alumina in solution B and it is prepared by dissolving 0.0529 g of "Specpure" aluminium (which is 99.99% Al) in a sligthly excessive volume of dilute hydrochloric acid and then diluting this solution to 1 l. This standard solution contains the equivalent of 100 g of alumina per ml of solution.

For the ultimate photometric determination of alumina take 1 ml of solution B and add it to 10 ml of distilled water in a 50 ml stoppered separating funnel. If the rock is known to contain more than 20% alumina, only 0.5 ml of solution B will be required. Add 10 ml of the complexing reagent and allow to stand for about 5 min. From a measuring cylinder, add 20 ml of the oxine solution, tightly stopper the separating funnel and shake mechanically, or by hand, for 5–8 min. Allow the chloroform layer to settle out at the bottom of the separating flask. To break any emulsion which might form stir with a very fine stainless steel wire or add a few drops of alcohol. Run off the chloroform layer into a dry 25 ml graduated flask through a 2.5 cm filter funnel whose stem has been tightly plugged with filter paper. Add a further 2–3 ml of chloroform to the separating funnel and rotate it gently before allowing it to settle to the bottom of the separating funnel. Run off this volume of chloroform through the filter paper into the 25 ml graduated flask. Make up the eluate in the flask to 25 ml by pouring chloroform through the filter paper plug. In this way the total alumina content of solution B is contained in the chloroform. By following this procedure, the iron is complexed as a ferrous dipyridyl, whilst the beryllium sulphate complexes any traces of the fluorides which might still exist in solution B. It is essential that the pH of the solution should be maintained between 4.9 and 5.0 prior to extraction. Whilst manganese does not present any interference effects, titanium does and it is therefore necessary to correct this by using the values obtained later for this metal dioxide. The extraction of the alumina complex from the chloroform by means of the oxine solution has to be done with great care in order to avoid dispersing the chloroform as a colloidal suspension in the separating funnel. The vibration of the chloroform and oxine solution with a stainless steel wire assists in breaking these colloidal suspensions. Moreover, since the aluminium oxinate solutions are light-sensitive they should be kept in the dark until they are measured in the spectrophotometer. For like reason the extraction should not be carried out in strong sunlight. The optical density of the extract is measured at 410 mμ in a covered 1 cm cell. The colour of the oxine extract is pure yellow. Traces of green will indicate that iron is being extracted. This will occur if the pH of the aqueous solution is too low for the reduction of the iron to a ferrous state by the hydroxylamine. When this occurs sufficient buffer solution should be used to bring the pH to 4.9–5.0.

The method is standardized by carrying out this procedure with blank solutions of the reagents to which 1 ml of the standard aluminium solution has been added. At a wavelength of 410 mμ, 100 μg Al_2O_3 gives an optical density of 0.394, after deduction of the readings obtained from the reagent blank which usually amounts to about 0.014.

To calculate the percentage of alumina apply the readings to the following formula evaluated by RILEY (1958, p.419):

$$\% \ Al_2O_3 \text{ in solution B} = \frac{10 \times D}{\text{opt. d. of } 100\ \mu g\ Al_2O_3 \text{ standard} \times \text{vol. of solution B}}$$

in which D = corrected optical density of oxinate solution. The correction for D is obtained by subtracting from the observed optical density of the solution the following factor:

$$(\text{vol. of solution B} \times \%\ TiO_2 \times 0.0106)$$

DETERMINATION OF TITANIUM DIOXIDE IN SOLUTION B

The photometric determination of titanium depends upon the formation of yellow complexes in acidified solutions of the samples. One of the commonly used complexing agents is called "tiron" (YOE and ARMSTRONG, 1947). It is powder composed of disodium-1,2-dihydroxybenzene-3,5-disulphonate, and is used with a buffer solution consisting of 40 g of ammonium acetate and 15 ml of glacial acetic acid made up to 1 l with distilled water. Another dry powder, sodium dithionite (sometimes called sodium hydrosulphite) is also employed in this determination. SHAPIRO and BRANNOCK (1956, p.37) recommend the use of the standard sample of titanium dioxide, No.154 of the National Bureau of Standards. RILEY (1958b, p.418) uses either "Specpure" titanium dioxide or an analyzed sample of anatase for the preparation of sodium bisulphate. When cool, place the crucible in a drogen peroxide reagent in acid solution for the formation of a yellow titanium compound for photometric purposes. Consequently, the preparation of the standard solutions will depend upon the colorimetric reagent employed.

By using "tiron" (SHAPIRO and BRANNOCK, 1956, p.37) the standard solution of titanium dioxide No.154, N.B.S. is obtained by fusing 0.1013 g in a platinum crucible with 2 g of sodium bisulphate. When cool, place the crucible in a beaker containing 50 ml of 1 + 1 H_2SO_4. Heat and stir until the melt has been completely dissolved. When cool, dilute the solution to 250 ml by washing the contents of the crucible and beaker into a volumetric flask. Take 50 ml of this solution and further dilute to 1 l in another volumetric flask. This standard solution will contain 0.02 mg/ml.

To determine the concentration of titanium using "tiron" the following procedure is used by SHAPIRO and BRANNOCK (1956, p.37):

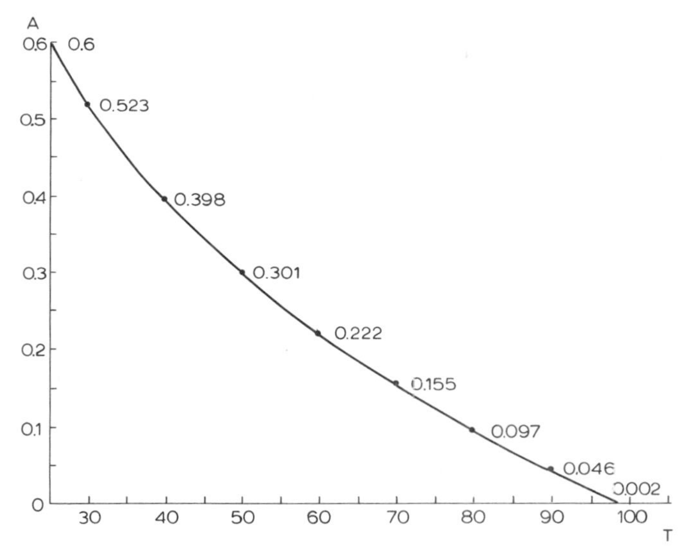

Fig.16. Conversion curve for transmission (*T*) and absorbance (*A*).

(*1*) To a series of dry 100 ml beakers add 5 ml of distilled water.

(*2*) To another series of dry 100 ml beakers add 5 ml of the standard solution of TiO_2.

(*3*) To a third series of beakers add 5 ml of the sample solution B.

(*4*) Add about 130 g of dry tiron powder to each beaker.

(*5*) With a pipette add 50 ml of the buffer solution of ammonium acetate and glacial acetic acid and mix thoroughly.

(*6*) Add 10–20 g of sodium dithionite powder to the reagent blank solution and gently mix by rotating the beaker two or three times. Do not overmix. Allow this to stand for about 1 min before pouring the solution into the adsorption cell of the spectrophotometer.

(*7*) Prior to this, the spectrophotometer is set to 430 mμ and on inserting the cell with the reagent blank the slit is opened to give maximum transmission.

(*8*) Repeat these last three operations with the standard solution of TiO_2 and the samples of solution B.

(*9*) Convert each of the values obtained for percentage transmission to absorbance by means of Fig. 16.

(*10*) Compute the factor for the standard solution:

$$\frac{1}{\text{absorbance of standard solution}} = \text{factor}$$

Compute the percentage of TiO_2: factor × (absorbance of sample solution) = percentage of TiO_2.

When using the hydrogen peroxide reagent as described by RILEY (1958b, p. 418) the following reagent is required: hydrogen peroxide reagent composed of 200 ml of 50% v/v sulphuric acid, 200 ml of syrupy phosphoric acid and 200 ml of 30% (100 volume) hydrogen peroxide. When mixed in this order the solution is made up to 1 l with distilled water and stored in an amber glass bottle with a loosely fitting glass stopper. Under these conditions it is a stable reagent which will last indefinitely.

The standard solution of titanium is prepared by using either "Specpure" titanium dioxide or an analysed sample of anatase. With either of these, 0.1000 g of the powdered sample is weighed in a platinum crucible to which 2 g of fused potassium pyrosulphate is added and heated gently, covered by a lid. The temperature is raised, after a while, to a dull red heat for about 15 min. The initial gentle heating removes all traces of water which is necessary before raising the temperature to finalize the pyrosulphate fusion. To remove the melt, place the crucible, when cool, into a 100 ml beaker containing 50 ml of 50% v/v sulphuric acid and allow it to simmer on a hot plate until the fused cake dissolves. Wash the contents of the beaker into a graduated flask and make up to 1 l with distilled water. This solution will contain 100 g TiO_2 per ml.

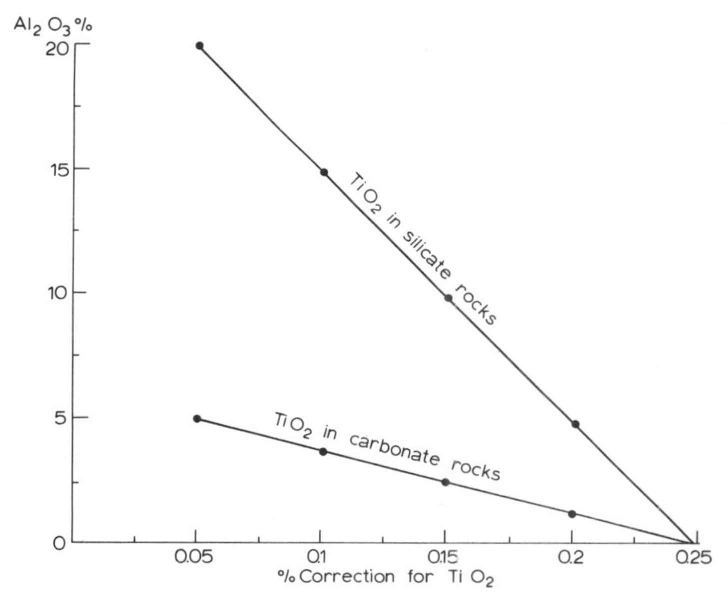

Fig.17. Correction graph for Al_2O_3/TiO_2 in silicate and carbonate rocks.

To analyse the samples of the reagent blank, the standard solution and the solutions B, take 40 ml of each and make them up to 50 ml with hydrogen peroxide reagent in a graduated flask. This will produce the yellow coloured solution necessary for spectrophotometric analysis. In a 4 cm cell the optical density of the yellow coloration is determined at a wavelength of 400 mμ. If one wishes to compute the absorbance of these solutions by means of Fig.17, it will be found that 800 μg of TiO_2 gives an optical density of about 0.565 at 400 mμ in a 4 cm cell, after the value for the reagent blank has been deducted. The computation of absorbance is not really necessary as the percentage of TiO_2 can be calculated as follows:

$$\frac{\text{opt. d. of unknown} - \text{opt. d. of reagent blank}}{\text{opt. d. of 800 g } TiO_2 - \text{opt. d. reagent blank}} = \%\ TiO_2$$

DETERMINATION OF TOTAL IRON AS Fe_2O_3 FROM SOLUTION B

In solution B, the total iron is determined spectrophotometrically by measuring the transmission of light through a complex formed either with orthophenanthroline (SHAPIRO and BRANNOCK, 1956, p.36) or with dipyridyl (RILEY, 1958b, p.420).

The procedure with orthophenanthroline is as follows:

(*1*) Prepare 500 ml of 10 % hydroxylamine hydrochloride which is to be used to reduce the iron at pH 4.8–5.0.

(*2*) Make up 500 ml of 10 % sodium citrate solution in distilled water.

(*3*) Prepare 500 ml of a 0.1% solution of orthophenanthroline.

(*4*) Make up a standard iron solution composed of either ferrous ammonium sulphate, or "Specpure" iron sponge. To prepare the former dissolve 0.491 g $FeSO_4.(NH_4)_2SO_4.6H_2O$ in 3 ml of 1 + 1 H_2SO_4 and about 300 ml of distilled water in a 500 ml graduated flask. Stir until completely dissolved and make up to 500 ml with distilled water. The Fe_2O_3 concentration of this solution is 0.2 mg / ml. When using "Specpure" iron sponge, weight out 0.1398 g and dissolve it in 5 ml of 2 *N* hydrochloric acid and dilute to 2 l. This solution contains the equivalent of 100 g Fe_2O_3 per ml and is preferable to that produced by the use of ferrous ammonium sulphate.

(*5*) To a set of 100 ml volumetric flasks, add 5 ml of the standard iron solution and 5 ml of the samples of solution B. Use one flask for the production of the reagent blank.

(*6*) Add 5 ml of the solution of hydroxylamine solution to each flask and allow them to stand for at least 10 min.

(*7*) Add 10 ml of the orthophenanthroline solution to each flask followed by 10 ml of the sodium citrate solution.

(*8*) Dilute each flask to 100 ml with distilled water and mix.

(*9*) Allow the flasks to stand for at least 1 h to ensure that the complexing reaction is complete.

(*10*) Determine the optical density of reagent blank, standard iron solution, and the samples of solution B at 560 mμ.

The procedure using dipyridyl solution follows the foregoing steps (*1*) to (*7*). This is followed by:

(*1*) Add 10 ml of the developing dipyridyl reagent. This is prepared by dissolving 0.2 g of 2, 2′-dipyridyl in 100 ml of 0.2*N* hydrochloric acid. To 20 ml of this dipyridyl solution add 4 ml of 25% w/v hydroxylamine hydrochloride and 50 ml of 0.5*N* sodium acetate. Mix thoroughly and make up to 100 ml with distilled water. This reagent is very stable, and can be used for at least one month.

(*2*) To 2 ml of the reagent blank solution, the standard iron solution, and the samples of solution B, add 5 ml of the developing dipyridyl reagent and dilute to 50 ml. Measure the optical density of the red coloured ferrous dipyridyl in a 1 cm cell at 522 mμ. This red colour is extremely stable, so long as the solutions are removed from direct sunlight, so that the measurements may be made up to 30 hours after adding the reagent. RILEY (1958b, p.420) reports that "Beer's law" is obeyed up to at least 500 μg Fe_2O_3, and that 200 μg Fe_2O_3 give an optical density (less reagent blank of about 0.010) of about 0.430 at 522 mμ in a 1 cm cell.

The total iron is obtained from the following calculation:

$$\frac{10}{\text{absorbance of standard soln.}} = \text{factor for the standard soln.}$$

$$\text{percentage of total iron as } Fe_2O_3 = \text{factor} \times (\text{absorbance of sample solution})$$

In each case the absorbance of the reagent blank is deducted from the measurements of the standard solution and the samples of solution B.

DETERMINATION OF MANGANESE IN SOLUTION B

Manganese oxide is determined spectrophotometrically at 525 mμ from samples in which the manganese has been oxidized to permanganate by means of ammonium persulphate in the presence of phosphoric acid and a catalytic quantity of silver ion (RILEY, 1958 b,p.420). Previously MnO was determined by measuring the light transmitted at 525 mμ by a sample solution in which the manganese was oxidized to permanganate by KIO_4. This method is still widely used and recommended by SHAPIRO and BRANNOCK (1956, p.38) to be used as follows:

(*1*) Prepare an acid mixture composed of 10% H_3PO_4 and 25% H_2SO_4 in 2 l of distilled water.

(*2*) Prepare a standard solution of manganese, either by using the manganese ore standardized by the National Bureau of Standards as "Sample No.25b" or by using "Specpure" manganese. Transfer 0.531 g of the manganese ore to a 250 ml beaker, and add 25 ml 1 + 1 HNO_3 and 2-3 ml of 3% H_2O_2. Digest this on a steam bath until no black residue remains. When cold, add 50 ml of H_2SO_4 and then evaporate until fumes of SO_3 are evolved. Allow to cool and dilute the contents of the flask to approximately 200 ml with distilled water. If complete solution of the manganese ore is not achieved at this stage, gently reheat the flask until it has been dissolved completely. Allow the solution to cool and dilute it with distilled water to 2 l. Store in a glass-stoppered bottle, as a standard solution containing 0.02 mg / ml. Using "Specpure" manganese the standard solution is prepared by dissolving 0.0774 g in 20 ml of 0.5*N* sulphuric acid and then diluting it to 2 l with distilled water. This solution contains 50 μ g of manganese per ml.

(*3*) In a series of 250 ml beakers, transfer 20 ml of water to one of them, 20 ml of standard manganese solution to another and 20 ml of samples of solution B to the others.

(*4*) To each beaker, add 50 ml of the acid mixture and then 0.2 g of KIO_4.

(*5*) Cover the beakers with watch glasses, and bring to the boil for 5 min and then allow them to cool.

(*6*) Transfer these solutions to 100 ml graduated flasks. Make up the solutions to this volume with distilled water and with these measure the percentage transmission at 525 mμ, in order to compute the percentage of MnO.

A more rapid method, described by RILEY (1958b, p.420), consists of the following procedure:

(*1*) Prepare an acid reagent composed of 37.5 g of mercuric sulphate dissolved in 200 ml of concentrated nitric acid and 100 ml of distilled water, to which is added 100 ml of 85 % phosphoric acid and 0.017 g of silver nitrate. When cold dilute the solution to 500 ml with distilled water.

(*2*) Into separate 50 ml conical flasks pipette 20 ml of the standard manganese solution, and of the samples of solution B. To each of these, add 1.5 ml of the acid reagent and about 0.5 g of ammonium persulphate. Place the flasks on a hot plate with a surface temperature of 400°C and boil for not longer than 1 min. Cool the flasks under the tap and dilute the contents to 50 ml with distilled water. In this way the manganese is completely oxidized to permanganate as a result of the catalytic action of the ionized silver nitrate.

(*3*) Determine the optical density of these permanganate solutions at 525 mμ, or at 575 mμ if there are excessive quantities of chromium present in the samples of solution B.

The percentage of manganese is given by:

$$\%\text{MnO} = \frac{(\text{opt. d. sample} - \text{opt. d. reagent blank}) \times 0.5}{\text{opt. d. } 100\ \mu\text{g MnO std.} - \text{opt. d. reagent blank}}$$

DETERMINATION OF PHOSPHORUS PENTOXIDE IN SOLUTION B

The phosphate content of the samples of solution B are determined spectrophotometrically by means of the optical density of a solution containing yellow molybdovanado-phosphoric acid at 430 mμ (KITSON and MELLON, 1944). RILEY (1958a, p.425) uses a solution of molybdenum blue with ascorbic acid as the reducing agent.

The procedure using molybdovanado-phosphoric acid is as follows:

(*1*) Dissolve 1.25 g of ammonium metavanadate in 400 ml of 1 + 1 HNO_3, to which add 50 g of ammonium molybdate in 400 ml of distilled water. Mix thoroughly and make up to 1 l with distilled water. This becomes the molybdovanado-phosphoric acid solution.

(*2*) Prepare a standard solution of phosphorus pentoxide by adding to 0.0849 g of National Bureau of Standards phosphate rock (sample No.120) to 25 ml of 1 + 1 HNO_3 in a 150 ml beaker. Heat on a steam bath, until the solution is essentially complete. When cold, transfer to a 2 l graduated flask, and make up to this volume with distilled water. Store in a glass-stoppered bottle, as a standard solution containing 0.015 g of P_2O_5 per ml. A standard solution may also be prepared by dissolving 0.959 g of potassium dihydrogen phosphate A.R. in 1 l of distilled water. This solution will contain 50 μ g of P_2O_5 per ml.

(*3*) Transfer 15 ml of the standard solution, and of samples of solution B, to separate 100 ml beakers. Use 15 ml of distilled water in another beaker for the preparation of a reagent blank.

(*4*) To each beaker, add 10 ml of the molybdivanodate solution, swirling the solutions as it is released from the pipette.

(*5*) Add 25 ml of distilled water, mix and allow to stand for at least 5 min.

(*6*) Determine the optical density of each solution at 430 mμ, using the reagent blank solution as the blank reference solution.

The method described by RILEY (1958a, p.425) proceeds as follows:

(*1*) Dissolve 5 g of ammonium molybdate A.R. in distilled water, and dilute to 250 ml.

(*2*) To 38 ml of this ammonium molybdate solution, add 125 ml of 3 *N* sulphuric acid with 60 ml of ascorbic acid, mix well, and dilute to 250 ml with distilled water. This reducing reagent, which is faintly green in colour, should be made up immediately before use.

(*3*) In separate 50 ml graduated flasks, take 25 ml of a blank reagent solution, the standard P_2O_5 solution, and samples of solution B. To each of these add 20 ml of the reducing reagent and dilute to 50 ml with distilled water. Allow to stand overnight, before measuring the optical density at 827 mμ in cells of appropiate length (1 or 4 cm). The molybdenum blue colour is stable for at least a week.

(*4*) The quantity of phosphorus pentoxide in each of the samples is given by:

$$\% \ P_2O_5 = \frac{(\text{opt. d. sample} - \text{opt. d. reagent blank}) \times 0.2 \times \text{vol. std.}}{\text{opt. d. of standard} - \text{opt. d. of reagent blank}}$$

To obtain good results for such phosphate determinations, the graduated flasks should be allowed to stand for several hours filled with concentrated sulphuric acid. They should then be thoroughly washed with distilled water. If it can be afforded, a set of graduated 50 ml flasks should be retained for exclusive use in the determination of phosphorus pentoxide. This practice relieves the analyst from the tedious task of cleaning this glassware prior to each phosphate determination, as these flasks will not require attention for several weeks.

DETERMINATION OF CALCIUM OXIDE IN SOLUTION B

Using solution B, calcium oxide is determined by titration with disodiumethylenediaminetetra-acetic acid (also known as versene), with ammonium purpurate (called murexide) as an indicator (BETZ and NOLL, 1950). Interference from iron and aluminium is removed, either by complexing them with sodium potassium tartrate, or by extracting their oxinates at pH 4.9–5.1 with chloroform in a continuous extraction apparatus (see Fig.18). An ultrasonic probe can also be used to speed up this chloroform extraction. Manganese causes slight interference. Since it is not removed by one of these two

Fig.18. Continuous extraction apparatus for the removal of interfering elements in the determination of calcium and magnesium (RILEY, 1958, fig.1).

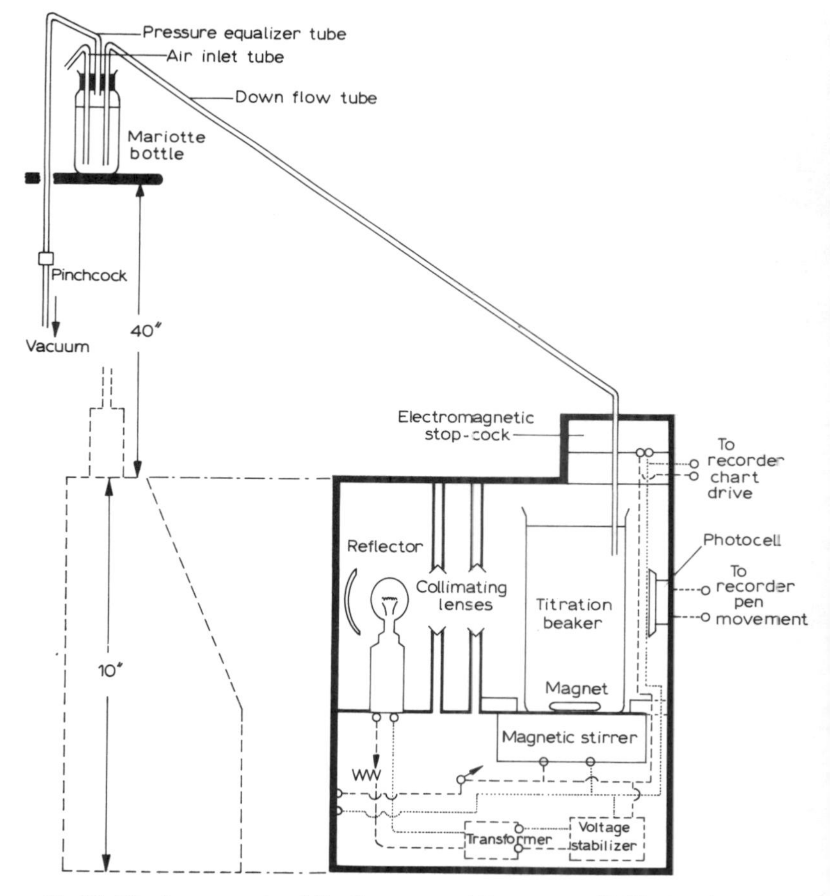

Fig. 19. Titration apparatus (after SHAPIRO and BRANNOCK, 1962), showing complete assembly and the electrical circuit.

methods, a correction must be carried out in the calculation of the concentration of calcium oxide.

Many analysts have difficulty in observing the end-points of subtle colour changes, particularly those involved in the use of murexide as an indicator for the determination of calcium oxide and magnesium oxide. Consequently, it has been found necessary to resort to the use of an automatic titration apparatus. BARREDS and TAYLOR (1947) suggested the design for an inexpensive titrator which was built and employed by SHAPIRO and BRANNOCK (1956, p.27, fig. 7 and 8). This is illustrated by Fig.19. Unfortunately, it includes the use of a pen recorder, the quality of which determines the accuracy of the calcium and magnesium end-points. Moreover, the pen recorder must be calibrated in terms of flow-rate of versene. For obvious reasons this must be kept as near constant as possible and this often causes considerable operational difficulties. Fig.20 is a typical titration curve obtained with this equipment. From this it will be seen how difficult it is to extrapolate the maximum optical density of the indicator on to the curve. In addition to this, one must calibrate the end-point in terms of the volume of indicator delivered from the Mariotte bottle reservoir.

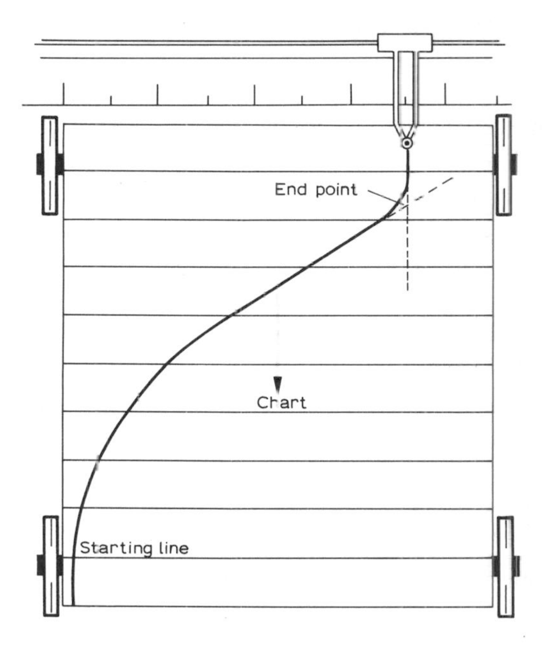

Fig.20. Titration curve for Ca/Mg determinations on recorder chart.

To overcome this difficulty, and to automate most other forms of titration, a new automatic titration apparatus has been designed and described for the first time in this book (see p. 125). This equipment removes the use of expensive pen recorders, the necessity for having one piece of equipment exclusively used for two or three elements, and the need to convert a mechanically drawn curve into the volumetric intake of the processing reagents. In the following accounts of the determinations of calcium and magnesium oxides it has been assumed that the equipment of Shapiro and Brannock is being used. The method of using the automatic titrator is given on p.128.

The procedure for the determination of CaO, as described by SHAPIRO and BRANNOCK (1956, p.39), is as follows:

(*1*) Prepare the versene solution, which is composed of 20 g of disodiumethylenediamine tetra-acetate in 20 l of distilled water. Transfer this either to the Mariotte bottle of the pen-recording titration apparatus (Fig.19).

(*2*) Prepare a standard solution of CaO with 1.000 g of Bureau of Standards sample No.88 (dolomite) dissolved in 20 ml of boiling 1 + 1 HCl in a 250 ml beaker. Cool to room temperature, and dilute to 1 l in a graduated flask. This solution contains 0.305 g of CaO per ml. Store the solution in a plastic bottle for use with successive determinations.

(*3*) Transfer 1 ml of the standard solution of CaO to a graduated 400 ml beaker, suitable for use in this automatic titration apparatus.

(*4*) Fill the beaker to the graduation mark with distilled water, then add a plastic coated magnet for stirring.

(*5*) Start the magnetic stirrer and add 3 ml of 20% potassium tartrate.

(*6*) Add 10 ml of 15% NaOH solution.

(*7*) Add 1 ml of an approximately 0.2% solution of freshly made murexide indicator.

(*8*) Insert the top of the delivery tube from the Mariotte

bottle, check that the orange and green filters are in front of the photocell and close the cover of the apparatus.

(*9*) Adjust the intensity of the light to maximum and then reduce it slightly so that the last movement of the pen is from right to left.

(*10*) Throw the toggle switch so that the electromagnetic hosecock opens and the recorder pen begins to form the trace of its base line.

(*11*) As the titration proceeds the pen traces the colour changes until sufficient versene solution has flowed into the

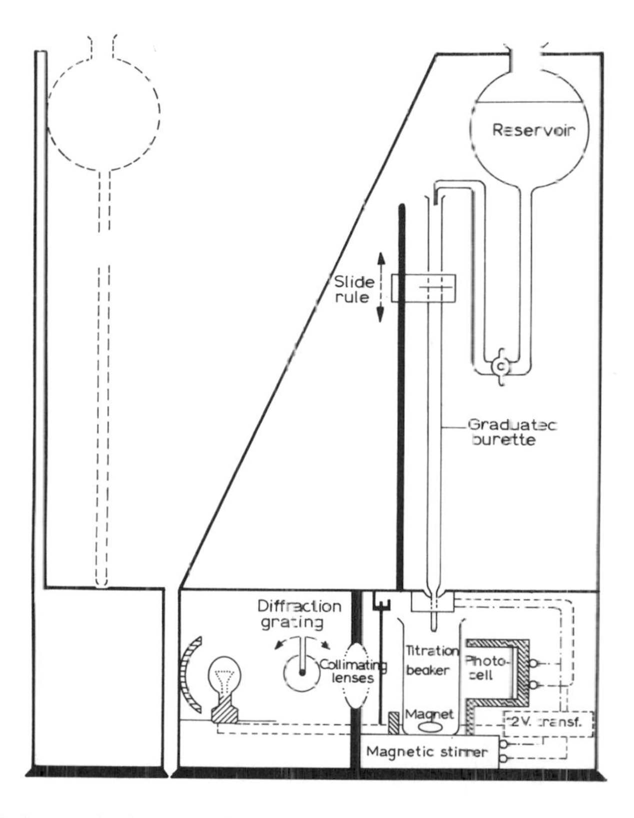

Fig.21. Automatic titrator, designed by W.D. Evans and constructed by W. Wilson of the University of Nottingham.

beaker to complete the reaction. At this stage the pen will begin to trace a line parallel to the movement of the chart. The end-point is determined as shown in Fig.20.

(*12*) Using 10 ml of samples of solution B the percentage of CaO is determined in relation to the curve obtained for the standard solution by following exactly the same operational procedure. It is also necessary to determine the CaO content of a blank solution of reagents in this way.

As mentioned previously (p.121) RILEY (1958b, p.421) carried out the determination of CaO along with MgO after the removal of interfering elements (apart from manganese) in a continuous extractor. This is a very effective method but it involves the use of calein as an indicator and this is not readily available outside the U.S.A. Moreover the effective removal of the interfering elements is rather time consuming and requires a considerable amount of manipulative skill (RILEY, 1958b). The use of the new automatic titrator (Fig.21) speeds the work immensely and it retains the use of versene as an indicator without the associated problems of extrapolating its end-point from curves recorded on a potentiometric recorder.

Using the equipment of Shapiro and Brannock (Fig.19, 20) the calculation of calcium oxide is carried out as follows:

(*1*) Obtain a correction for the blank solution from:

$$\text{(sec for 1 ml std.)} - \frac{\text{(sec for 10 ml std.)}}{10} \times 1.1$$

(*2*) Compute the "factor" as follows:

$$\frac{3.05 \times 100}{\text{(sec for 10 ml std.} - \text{blank correction)} \times 20}$$

(*3*) Calculate the percentage of CaO from:

(factor) × (sec for sample – blank correction)

DETERMINATION OF MAGNESIA

The estimation of magnesia (MgO) is usually based on the method established by BETZ and NOLL (1950) in which they used eriochrome black T as an indicator. This is affected by the metals of the R_2O_3 group which are removed along with calcium as tungstates. The procedure must be controlled by the use of accurately prepared standard solutions of magnesia and of reagent blanks. Standard solutions are conveniently prepared from clean magnesium ribbon or wire. By transferring 0.752 mg of either of these to a 250 ml covered beaker, the solution is made by adding 10 ml of 1+ 3 HCl and boiling it gently for a few minutes. The concentrated solution obtained in this way is allowed to cool and the watch-glass cover washed into the beaker with 200 ml of distilled water. Of this diluted solution 50 ml is transferred by pipette to a 1 l volumetric flask and diluted to volume with distilled water. This solution will contain 0.250 mg MgO per ml and it should be stored in a plastic bottle and not exposed to the atmosphere for long periods at a time.

The determination of magnesia in the reagent blanks, and in samples of solution B, is obtained as follows:

(*1*) Transfer 25 ml aliquots of samples of these solutions to a series of 250 ml volumetric flasks.

(*2*) Add about 200 ml of distilled water and 15 ml of ammoniated methyl red solution, which is prepared by adding 10 ml of 0.05% methyl red solution to 1 l of 15% NH_4Cl.

(*3*) From a burette add 1 + 1 NH_4OH solution, dropwise, until the methyl red indicator turns yellow.

(*4*) Dilute the contents of the volumetric flasks to 250 ml and allow them to stand for about 15 min.

(*5*) Pour the contents of the flasks through dry fluted filter papers held in dry funnels, catching 20 ml of the filtrates in the graduated flasks. These filtrates of 200 ml will contain the equivalent of 40 mg of the original samples.

(*6*) Transfer the contents of the flasks to 400 ml beakers and rinse the flasks with distilled water.

(*7*) To each beaker add 5 ml of calcium chloride solution, prepared by dissolving 500 mg of $CaCO_3$ in a few millilitres of HCl. Then dilute to 500 ml with distilled water.

(*8*) Add 10 ml of a buffer solution, prepared by dissolving 66 g of NH_4Cl in 1 l of 1 + 1 NH_4OH. Store this buffer solution in a plastic bottle.

(*9*) Add 10 ml of a 20% solution of sodium tungstate.

(*10*) Bring these solutions to boiling point for a few minutes, and allow them to cool to room temperature in a water bath. This ensures complete formation of the tungstates, the excess of which precipitates and adheres to the sides of the beakers.

(*11*) Decant these solutions carefully into 400 ml titration beakers, so as to remove as little as possible of the precipitates. The amount of solution left on the walls and bottom of the beaker is inconsequential. It is important to avoid transfer of precipitated tungstates, as these cloud the solutions and interfere with the quality of the end-point of titration.

(*12*) Add a further 5 ml of the buffer solution, together with 1 ml of eriochrome black T indicator solution. This is prepared as an approximate 0.2% solution.

(*13*) Insert a magnetic stirrer in the titration beaker and allow the automatic recording to proceed as in the case of calcium determinations (p.129).

(*14*) The calculation for the determination of magnesium is carried out in the same way as for CaO as follows:

The factor is obtained from:

$$\frac{2.5 \times 100}{(\text{sec for 10 ml std.} - \text{blank correction}) \times 40}$$

The percentage of magnesium oxide is obtained from:

(factor) × (sec. for sample – blank correction) = % MgO

DETERMINATION OF SODIUM AND POTASSIUM OXIDES IN SOLUTION B

These determinations are now carried out by means of a flame photometer. The degree of accuracy obtained in this way depends upon the efficiency of the instrument. In recent years the original simple, but extremely efficient flame photometers have been developed to a sophisticated degree and the whole subject is dealt with in Chapter 5. However, in using solution B, the employment of a simple flame photometer demands the prior removal of the major sources of interference, namely, iron, aluminium and titanium. RILEY (1958b, p.424) removed these oxides by passing solution B through an anion exchange resin in its citrate form. This method has proved to be highly satisfactory and within the range of accuracy of the flame photometer, the estimates of sodium and potassium oxides have proved to be reasonably consistent and valuable. It must be remembered that even with a simple flame photometer the degree of detection is sufficient to record traces of alkali removed by the solution from the glass ware employed. Consequently, the use of resistant plastic containers is recommended. Otherwise the determinations should be carried out as soon as possible after the preparation of solution B in order to minimize this source of contamination.

BRANNOCK and BERTHOLD (1953) used lithium as an internal standard. In this way, they reduced the effects caused by the differences in the concentration of the dissolved salts, as well as to variations in the viscosity which occurred between the standard and sample solutions. This internal standard also minimized the deviations in the readings from the photometer due to changes in the character of the flame caused by small fluctuations in the pressures of the gas and compressed air. They used a Perkin Elmer (model 52c) in their original work, whilst RILEY (1958 b) used the simple but equally accurate flame photometer made by Eel. For more sophisticated equipment see Chapter 5.

Using the internal standard employed by SHAPIRO and BRANNOCK (1962) the following procedure should be adopted:

(*1*) Adjust the flame and air pressure to achieve a steady flame and allow the instrument to warm up for about 30 min.

(*2*) Prepare an internal standard solution composed of 200 ml of a stock solution of lithium sulphate diluted to 2 l with distilled water. A convenient stock solution of lithium sulphate consists of 184.4 g of $Li_2SO_4.H_2O$ dissolved in 10 l of distilled water stored in a pyrex or plastic bottle.

(*3*) Prepare a stock solution of KCl composed of 1.584 g of dry KCl in 2 l of distilled water. This yields a concentration of 0.5 mg of K_2O per ml of solution.

(*4*) Prepare a stock solution of sodium chloride composed of 1.886 g of dry NaCl in 2 l of distilled water. This yields a concentration of 0.5 mg of Na_2O per ml of solution.

(*5*) Prepare standard solutions of K_2O, each containing 50 ml of the stock solution of $Li_2 SO_4$ by diluting 5, 10, 20, 40, 60 and 100 ml of the stock solution of KCl to 1 l with distilled water. This yields standard solutions containing 0.5, 1, 2, 4, 6, 8 and 10% of K_2O respectively.

(*6*) Prepare standard solutions of Na_2O containing Li_2SO_4 in the same way as for K_2O.

(*7*) To 25 ml of samples of solution B add 25 ml of distilled water and 50 ml of the internal standard which contains 200 p.p.m. lithium. Mix well in a pyrex or plastic beaker.

(*8*) When the flame has achieved a steady zero reading, insert a sample of K_2O or Na_2O standard solution and adjust the galvanometer reading to read 50.

(*9*) Remove the beaker containing the standard solution

and, after wiping the atomizer tube, introduce a sample of solution B which also contains its lithium internal standard.

(*10*) The galvanometer response to solution B is directly proportional to the concentration of K_2O or Na_2O in their respective standard solutions.

It is advisable to take several readings for each sample solution at various levels of concentration of the standard solution. The averages of these readings will yield the percentage of K_2O or Na_2O in the sample from the following two-point equation:

$$X = \frac{Y - Y_1}{Y_2 - Y_1} (X_2 - X_1) + X_1$$

in which X = concentration of K_2O or Na_2O, X_1 = concentration of K_2O or Na_2O in the most dilute standard solution, X_2 = concentration of K_2O or Na_2O in the most concentrated standard solution, Y = response reading for the sample solution, Y_1 = response reading for the dilute standard solution, and Y_2= response reading for the concentrated standard solution.

DETERMINATION OF FERROUS OXIDE IN SILICATE ROCKS

Solution B cannot be used for this purpose as the ferrous compounds have been completely oxidized during its preparation. A solution of the ferrous minerals must be derived from the rock in such a way as to inhibit oxidation. In the determination of ferrous iron in solution there has been no significant advance made over the method established by SARVER (1927). In essence this consists of titrating the solution with a standard dichromate solution using diphenylamine as the indicator. The speed of the reaction which yields the purple end-point during titration is somewhat slow and intervals of decolourisation take place before a permanent colour change can be established. This frequently leads to error in the addition of dichromate. With the automatic titrator (Fig.21) the photocell response is so rapid and sensitive that this factor is completely obliterated. If a sample contains a very small amount of FeO, the end-point is made more definitive by a fixed but small amount of ferrous ammonium sulphate solution.

The standard procedure for the determination of FeO in most rock-types is as follows:

(*1*) Weigh out 0.500 g of the powdered rock sample in a platinum crucible of approximately 100 ml capacity.

(*2*) Add 20 ml of 1 + 3 H_2SO_4, cover the crucible and carefully raise the temperature to boiling point.

(*3*) Allow this to cool and then carefully add about 10 ml of hydrofluoric acid, cover and again bring to the boil for about 15 min.

(*4*) While this solution is gently boiling prepare a solution of 10 g of boric acid in 400 ml of distilled water in an 800 ml beaker. Stir the solution thoroughly to ensure complete solution of the boric acid.

(*5*) Take the boiling crucible and transfer it to the beaker, holding the lid firmly over the top with a stirring rod. Immerse it completely into the boric acid solution. When totally immersed release the lid and allow it to stand so that the contents are thoroughly mixed in the boric acid.

(*6*) After allowing to stand for a while add 10 ml of indicator which consists of 0.2 g of sodium diphenylamine sulfonate in 1 l of distilled water, to which is added 1 l of 85% H_3PO_4.

(*7*) Titrate the solution with standard dichromate solution which is prepared by dissolving 2.728 g of dry A.R. potassium dichromate in 2 l of distilled water. This solution contains the equivalent of 2.000 mg FeO per ml.

(*8*) The purple end-point must persist for at least 20 sec

to ensure complete resolution of the FeO in the sample solution.

The percentage of FeO is computed as follows:

$$\text{vol. dichromate (ml)} \times 0.4 = \%\ \text{FeO}$$

DETERMINATION OF CARBON DIOXIDE AND WATER IN SILICATE ROCKS

A considerable amount of work still needs to be done on the determination of carbon dioxide in all types of rock as in some cases it can represent as much as 37% of the specimen. Any of the following methods can be used, but they are open to criticism as little control is exercised over the addition of occluded volumes of this gas derived from air spaces which even exist within the particles of powdered samples. Moreover, excessive pulverisation of the rock specimens can lead to the oxidation of carbonate minerals which might lead to a significant loss of CO_2.

(*A*) SHAPIRO and BRANNOCK (1956, p.46) use a simple device called a "carbonate tube" (see Fig.22) in which the volume of gas emitted from a specimen on treatment with acid is calibrated as a direct reading of the CO_2 content of the sample. The calibration is performed by using a standard sample of fluorspar (No.79 of the National Bureau of Standards), which contains 0.97% CO_2. The procedure is as follows:

(*1*) Insert 1.03 g of the fluorspar sample, by means of a dry funnel, into the bottom of the carbonate tube.

(*2*) Add 2 ml of a 3% mercuric chloride ($HgCl_2$) solution to the powder to prevent the evolution of hydrogen, which might form as a result of iron impurities in the sample reacting with the hydrochloric acid.

(*3*) Fill the carbonate tube to the graduation mark with a moderately light motor oil.

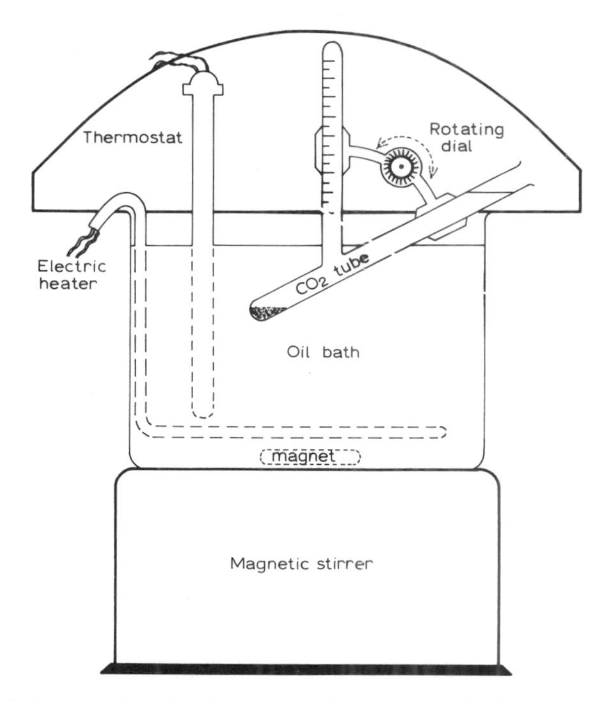

Fig.22. Carbonate tube (after SHAPIRO and BRANNOCK, 1962) for measuring the volume of CO_2 liberated on attacking specimens with acids.

(*4*) Tilt the tube to expell the air from the side arm, and then re-position it so that the main part of the tube is vertical.

(*5*) Add 2 ml of 1 + 1 hydrochloric acid and tilt the tube so that the side arm is now vertical. This allows any CO_2 produced to enter the side arm.

(*6*) Heat the sample and acid and allow to boil for 2 or 3 min.

(*7*) Rapidly cool the tube with a water jacket and mark the position of the meniscus on the side arm. This becomes the mark on the scale representing 1%.

(*8*) Repeat this procedure without inserting a sample, and then with 0.206, 0.412, 1.54, 1.85 and 2.06 g of the standard sample. These will yield calibration marks for 0, 0.2, 0.4, 1.5,

1.8 and 2.0% CO_2 respectively. The side arm can then be marked off at 0.1% intervals, which will give a fairly accurate measure of the gas evolved from specimens.

(*9*) From 1.0 g of the powdered rock sample, direct reading of the CO_2 content can be obtained from this carbonate tube. In many ways this tube is satisfactory but it involves a certain amount of skill in its manipulation. If the sample contains more than 6% of CO_2, it is advisable to add 1 g of mercuric chloride as a solid to the sample and cover with 2 ml of a saturated solution of $HgCl_2$. The sample usually decomposes on heating to between 115°C and 120°C (see Fig.22).

(*B*) Described for the first time is a method initially devised for the rapid determination of CO_2 in limestones. It has the advantage over the carbonate tube in being able to weigh accurate quantities of standard and sample specimens. The details are described in Fig.23. The procedure is identical to that used for the carbonate tubes, and the volume of CO_2 is measured direct from a calibrated scale on the manometer.

Neither of these techniques is free from the error associated with the evolution of moisture along with CO_2 carried over with CO_2. Consequently, several attempts have been made to estimate CO_2 and H_2O as a combined operation. For the determination of H_2O, SHAPIRO and BRANNOCK (1962) still use the technique of evaporating the water from the specimen and causing it to become absorbed by a piece of dried filter paper. This is a reasonably satisfactory method for a rapid, but approximate determination of the total water content of a specimen. The character of the results depends upon the accuracy with which the piece of filter paper can be weighed before and after its contact with the water expelled from the specimen. GROVES (1951, p.97) developed a much more realistic method for the determination of water in silicate rocks. Realizing that certain minerals fail to yield their water, even at temperatures of 1,000°C, he suggested the use of fluxes. Even so, a great deal of time elapses before the

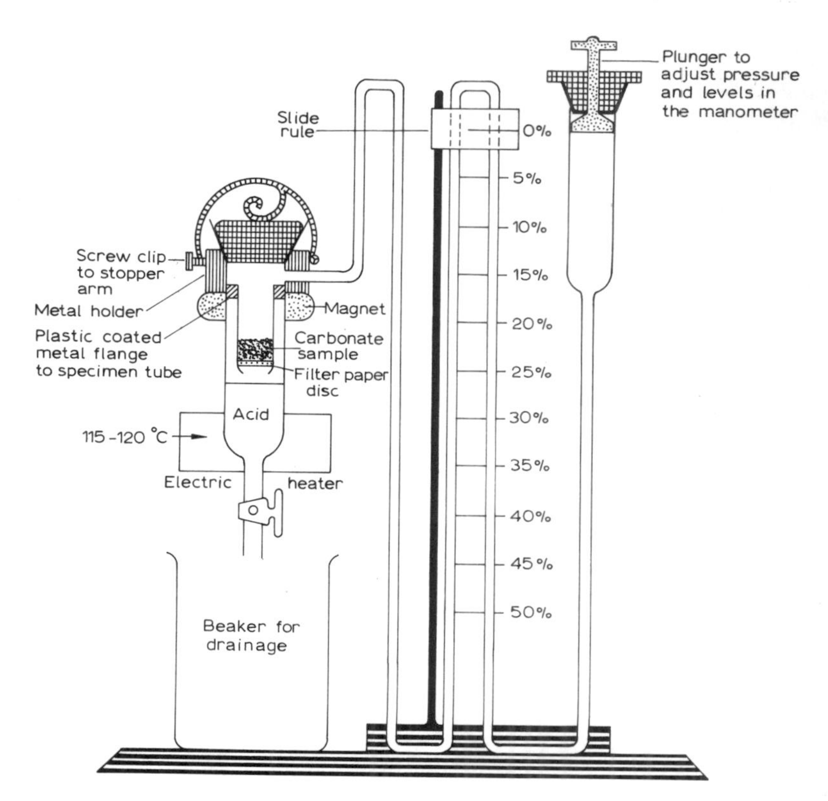

Fig.23. Carbonate determination apparatus which may be calibrated for the CO_2 content of most rocks and minerals, or for the $CaCO_3$ and $MgCO_3$ content of limestones. (Designed by W. D. Evans and constructed by F. Bancroft and W. Wilson, University of Nottingham.)

combustion tube is cool enough to take readings. In essence, the method consists of heating a dry powdered specimen in a combustion tube, through which nitrogen or argon is passed to carry the gas into an adsorbent such as silica gel or a suitable molecular sieve. The weight of these adsorbents is accurately known beforehand, so the increase in their weight is taken as a direct measure of the water content of the sample.

RILEY (1958) carried out a comprehensive investigation of

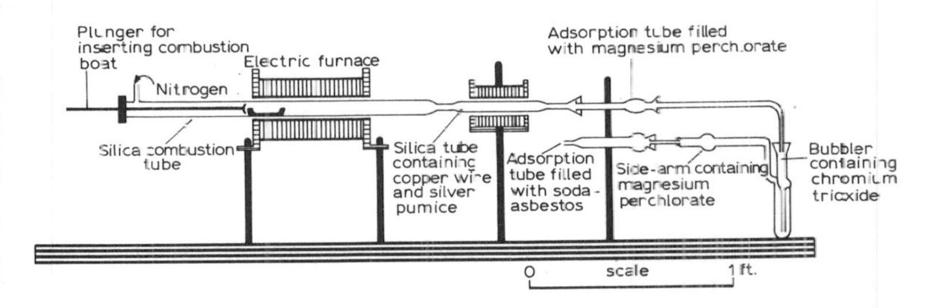

Fig.24. Apparatus for the simultaneous determination of carbon dioxide and water (by RILEY, 1958).

the various methods of water determination and eventually devised a dual set of furnaces fitted with adsorption tubes which selectively entrain water and CO_2 derived from heating rock specimens. The two furnaces and adsorption tubes are described with Fig.24. The water adsorption tube is filled with anhydrous magnesium perchlorate. The carbon dioxide adsorption tube is packed with a layer of soda asbestos and a layer of anhydrous magnesium perchlorate. For samples containing more than 0.5% of sulphur, a bubbler-tube filled with chromium trioxide is interposed between the water and CO_2 adsorption tubes. The side arm of the bubbler is packed with magnesium perchlorate. For most rock samples, the specimen should be heated to 1,100°C. Minerals such as staurolite, topaz, epidote and talc require temperatures of at least 1,200°C to release their water content. Prior to use the whole apparatus must be purged with nitrogen for about half an hour. At the outset of each determination, remove the two adsorption tubes, wipe them dry, and weigh them as accurately as possible. The procedure then consists of inserting an accurately weighed sample of rock powder (0.5–1.5 g), contained in a previously ignited 2 ml alumina boat, lined with a piece of nickel foil. If the specimen contains excessive quantities of sulphur or fluorides, cover it with a layer of freshly ignited magnesium oxide. The boat is then inserted into the end of the combustion tube. The seal is replaced, and when the tube has been cleared of air, by the nitrogen carrier gas, the boat is pushed forward with a stainless steel rod into the hot region of the furnace for about half an hour. The adsorption tubes are then removed, carefully wiped, and weighed. The increase in weight of each respective tube gives the water and CO_2 content of the rock sample. A serious error in the determination of carbon dioxide results from the presence of graphite or organic carbon in the samples as these often burn to carbon dioxide in the water laden nitrogen carrier gas. This is reduced to almost negligible quantities if the nitrogen carrier gas is replaced by argon. Long experience of the use of this apparatus suggests that it is superior to any other employed for the dual determination of CO_2 and H_2O in rocks.

THE ANALYSIS OF CARBONATE ROCKS

In many respects the methods employed for silicate rocks are applicable to calcareous sediments and carbonate-rich rocks. Consequently, only the modifications are indicated in the following procedures.

Silica

The determination of SiO_2 in carbonate rocks needs only be modified to take into account the larger sample needed for this purpose. A 200 mg sample is required for such rocks, although the use of a 50 mg sample of the standard is retained for calibration purposes.

Alumina

As in the case of silica, the determination of Al_2O_3 in carbonate rocks employs a 200 mg sample. In every other way, the analysis is carried out as for silicate rocks.

Titanium oxide

The determination of TiO_2 in carbonate rocks is identical with that described for silicate rocks.

Total iron

The total iron in carbonate rocks is the same as that used for silicate rocks.

Manganese oxide

In carbonate rocks the determination of MnO is identical with that described for silicate rocks.

Phosphorus pentoxide

If the rock contains less than 2.5% of P_2O_5 the procedure adopted for silicate rocks is quite adequate. Where the content exceeds this percentage the following procedure has been established by SHAPIRO and BRANNOCK (1962):

(*1*) Prepare a standard solution of P_2O_5 by using 0.4430 g of phosphate rock No.56b of the National Bureau of Standards. Place in a 400 ml beaker, and add 50 ml of distilled water and 10 ml conc. HNO_3. Cover the beaker, and boil for about 5 min. Allow to cool, and add 200 ml distilled water and 16 ml of 1 + 1 H_2SO_4 and filter. Dilute the filtrate to 500 ml in a graduated flask.

(*2*) Prepare solutions of samples in the same way or in the manner used for preparing solution B from silicate rocks.

(*3*) Transfer 15 ml of the standard solution to two 100 ml volumetric flasks and, likewise, 15 ml of each sample solution.

(*4*) To each flask add 5 ml of 1 + 7 H_2SO_4.

(*5*) Repeat this procedure to produce a similar volume of reagent blank.

(*6*) Add 25 ml of molybdivanadate solution to each flask. This solution is prepared by dissolving 5 g of sodium orthovanadate ($Na_3VO_4 . 16H_2O$) in 400 ml of 1 + 1 HNO_3 and adding this to a solution of 66 g of sodium molybdate ($Na_2MoO_4 . 2H_2O$) in 400 ml of water. Mix well and dilute to 2 l with water.

(*7*) Dilute each flask to 100 ml and allow to stand for at least 15 min, before determining the percentage transmission in the spectrophotometer, at 430 mμ, using the reagent blank as the reference solution.

(*8*) During this operation, prepare three volumes of the standard solution for the construction of an absorbance graph of percentage concentrations of P_2O_5. If 15, 30 and 45 ml of the standard solution are processed they will represent 14.0, 28.0 and 42.0% P_2O_5.

The percentage of P_2O_5 can then be obtained by reference to the graph after the absorbances have been derived from Fig.16.

Sodium oxide and potassium oxide

The same procedure is used as described for silicate rocks in determining the Na_2O and K_2O content of carbonate rocks.

Calcium oxide

In carbonate rocks the CaO content often represents the major part of the sample. From experience it has been found advisable to obtain the percentage of CaO by subtracting the quantity of MgO from the joint determination of these two oxides.

Calcium oxide and magnesium oxide

Owing to the higher concentrations of calcium and magnesium in carbonate rocks, than in silicate, they are jointly determined by visual titration, or by means of the new automatic titrator (Fig.21). The procedure described by SHAPIRO and BRANNOCK (1962) is as follows:

(*1*) Prepare a standard solution of Ca + Mg, from a solution of calcium carbonate low in alkali content. From this one might assume that pure $CaCO_3$ contains 56% CaO. The standard solution is produced by dissolving 0.500 g of $CaCO_3$ in 25 ml of distilled water and 5 ml of HNO_3 by boiling in a beaker. Allow to cool and add 8 ml of 1 + 1 H_2SO_4 and make up the solution to 250 ml in a volumetric flask. A magnesium sulphate solution is made from 6 g of $MgSO_4.7H_2O$ in 1 l of distilled water.

(*2*) Transfer two 25 ml aliquots of the standard solution together with aliquots of the sample solutions to a series of 400 ml beakers.

(*3*) To each beaker add 5 ml of a 10% solution of hydroxylamine hydrochloride and allow to stand for at least 5 min.

(*4*) Add 200 ml of distilled water to each beaker, and then introduce 5 ml of a complexing solution composed of 64 g of KCN dissolved in 600 ml of water to which has been added 400 ml of triethanolamine.

(*5*) To each beaker add 25 ml of a buffer solution composed of 66 g of NH_4Cl in 500 ml of water to which is added 500 ml of concentrated NH_4OH.

(*6*) Add to each beaker 2 ml of a magnesium sulphate solution composed of 6 g of $MgSO_4.7H_2O$ in 1 l of water. This provides a standard solution of CaO + MgO and adds a known quantity of MgO to each sample solution. This assists the determination of the colorimetric end-point.

(*7*) Using the automatic titrator, with 1 ml of 0.05% eriochrome black T solution and 1 ml of 0.02% methyl red solution as indicators, titrate the samples with EDTA.

Percent CaO + MgO =

$$\frac{56.0}{\text{av. vols. for 25 ml stds.}} \times \text{titration vol. of 25 ml. samples}$$

The percentage of CaO + MgO obtained in this way can be used to calculate the CaO percentage by subtracting the percentage of MgO (estimated separately) multiplied by 1.39.

Magnesium oxide

When the concentration of MgO exceeds about 2%, it is determined by titration with eriochrome black T solution after the precipitation of CaO as calcium oxalate. Where the percentage is less than this the procedure employed for silicate rocks proves more satisfactory.

THE ANALYSIS OF PHOSPHATE ROCKS

In many ways the procedure adopted for silicate rocks is acceptable. On the other hand, it must be remembered that phosphate rocks normally contain fluorine which tends to lower the values obtained for alumina. For rocks rich in phosphates the P_2O_5 content is determined in the manner described for carbonate rocks, but sample solutions must be diluted to appropriate levels for titration purposes. In general the methods covered by the analysis of silicate and carbonate rocks will provide the analytical data required by the geochemist.

CHAPTER 4

The Chemical Analysis of the Minor Elements

INTRODUCTION

In addition to the conventional requirements of chemical data concerning the principal oxides there arises from time to time the need for analytical information concerning the so-called minor elements. Consequently, the following methods of analysis have been selected to cover most of the elements and to provide qualitative as well as quantitative information concerning them. From a wide variety of analytical methods the selection in each case has been conditioned by a combination of requirements. Firstly, the method should provide a simple but reasonably accurate procedure which will yield reproducible data for geochemical purposes. Secondly, the method should not involve the use of equipment which would be outside the range of apparatus normally found in a laboratory equipped for silicate analysis. Finally, it should not involve the use of chemicals which are difficult or expensive to obtain. By applying these conditions the choice of analytical procedures was reduced considerably. On the other hand, care has been taken to provide references to alternative methods which will enable the analyst to test the selected form of analysis against others. Finally it was decided to list the elements alphabetically so as to enable the analyst to locate the relevant procedure as simply as possible and to avoid a considerable amount of descriptive duplication. Moreover, the qualitative and quantitative approach to analytical work has been maintained.

ANTIMONY (Sb)

At.wt. = 121.76; s. g. = 6.62; m.p. = 630°C;
b.p. = 1,440°C

Occurrence

Occurs free, but commonly as a grey or black iridiscent sulphide called stibnite (Sb_2S_3), which is the main commercial source of antimony. Other antimony minerals include senarantimonite (Sb_2O_3), valentinite (Sb_2O_3), cervantite ($Sb_2O_3.Sb_2O_5$), kermesite ($2Sb_2S_3.Sb_2O_3$), jamesonite ($2PbS.Sb_2S_3$), famatinite ($3Cu_2S.Sb_2S_5$) and bournonite ($3(PbCu_2)S.Sb_2S_3$).

Detection

H_2S precipitates orange-coloured sulphides of antimony from fairly concentrated solutions in HCl. Arsenic, which precipitates at the same time can be removed by boiling, since $AsCl_3$ is fairly volatile. When heated with fusion mixture (K_2CO_3 and Na_2CO_3), minerals containing antimony yield dense white fumes. Addition of antimony salts to aqueous solutions of tetraethyl-rhodamine (rhodamine B) causes the red dye to change to violet or blue, with the separation of a finely divided precipitate. Antimonous salts yield green precipitates with chromic acid, whilst antimonic salts liberate iodine from potassium iodide solutions.

Estimation

The rapidity and accuracy of volumetric methods in determining antimony has resulted in little use being made of gravimetric methods. In the analysis of rocks and minerals, some antimony is lost by volatilization, some remains in the silica as the oxychloride. It also contaminates the alumina precipitates, unless previously separated by H_2S in acid solution and then determined as follows:

(*1*) Precipitate arsenic by H_2S and filter. Filtrate contains the remainder of the hydrogen sulphide group of metals.

(*2*) Dilute the filtrate with double its volume of warm water and saturate with H_2S to precipitate the antimony and other metals. Filter and dry the precipitate.

(*3*) Dissolve antimony from the precipitate with 5–10 ml of an alkaline sodium sulphate solution (60 g Na_2S + 40 g NaOH in 1 l of distilled water).

(*4*) Filtrate, containing the antimony, is made strongly acid with concentrated HCl and estimated volumetrically, using 0.1 *N* potassium bromate ($KBrO_3$) solution with four drops of methyl orange as indicator. Titration is carried out at the boiling point of the solution until the colour of the indicator is destroyed. The end-point is slow to develop, so that considerable care has to be taken in the addition of potassium bromate. Traces of iron and copper affect the colour of the solutions. (1 ml 0.1 *N* potassium bromate = 0.006088 g Sb). The method using lead tetra-acetate developed by BERKA et al. (1960) is also recommended.

ARSENIC (As)

At.wt. = 74.91; s.g.cryst. = 5.73; s.g.amorph. = 4.72; m.p.cryst. = 850°C; sublimates at 554°C

Occurrence

Found native and also occurs in most pyritic minerals and in sulphur. The principal minerals consist of white arsenic (As_2O_3), orpiment (As_2S_3), realgar (AsS), mispickel (FeAsS), cobaltite (CoAsS), kupfernickel (NiAs) and smaltite ($CoAs_2$).

Detection

Arsenic compounds, when heated on charcoal give a white encrustation and the evolution of a garlic odour. Heated in a closed tube with sodium carbonate, a black mirror of arsenic is formed. Hydrogen sulphide precipitates As_2S_3 from solutions in strong hydrochloric acid. Traces of arsenic can be detected by means of filter paper, sensitized with mercuric chloride or bromide. As_2O_3 produces a pink stain.

Estimation

Acid solutions of arsenic are neutralised by sodium or potassium hydroxide, or carbonate using phenolphthalein as an indicator. The solution is then made slightly acid and to this, 2 or 3 g of sodium carbonate is added and titrated with an 0.1 *N* iodine solution. Using starch as an indicator the end-point is marked by a permanent blue colour (1 ml 0.1 *N* iodine ≡ 0.003746 g As or 0.004946 g As_2O_3)

BARIUM (Ba)

At.wt. = 137.36; s.g. = 3.78; m.p. = 850°C; volatile at 950°C

Occurrence

Commonly occurs as barytes ($BaSO_4$), witherite ($BaCO_3$) and barytes–calcite ($BaSO_4.CaCO_3$). It occurs also in feldspathic rocks associated with strontium, and is never found native.

Detection

Barium is precipitated as the carbonate, along with strontium and calcium, by the addition of NH_4OH and NH_4CO_3 to the filtrate of the ammonium sulphide group of metals. Separation from strontium and calcium is attained by precipitation, as yellow barium chromate ($BaCrO_4$), from a dilute acetic acid solution.

Estimation

It is now usual to estimate barium by means of the flame photometer. For volumetric analysis, the solution containing barium is saturated with ammonia and then heated to 70°C. To this, 0.1 *N* potassium dichromate solution is added, whilst stirring until all the barium chromate is precipitated. The supernatant solution should become faintly yellow, owing to the presence of a slight excess of the reagent. Filter and wash the precipitate, then remove it by solution using excess of 0.1 *N* ferrous ammonium sulphate solution containing free sulphuric acid. The excess of ferrous salt is titrated with standard 0.1 *N* potassium permanganate solution. The volume of 0.1 *N* ferrous ammonium sulphate solution, minus the volume of permanganate titration, multiplied by 0.004579 gives the weight of barium in solution.

The barium chromate can also be dissolved in 50–100 ml of dilute hydrochloric acid, to which about 2 g of solid potassium iodide is added. The iodine liberated in this way can be titrated with 0.1 *N* thiosulphate using starch solution as an indicator (1 ml 0.1 *N* thiosulphate = 0.004579 g).

BERYLLIUM (Be)

At.wt. = 9.02; s.g. = 1.85 at 20°C; m.p. = 1,350°C

Occurrence

Mainly found in beryl ($3BeO.Al_2O_3.6SiO_2$) and also in chrysoberyl ($BeO.Al_2O_3$), helvite—$(ZnFe)_2(Fe_2S)Be(SiO_4)_3$, leucophane—$NaCaBeF(SiO_3)_2$, hambergite ($Be_2(OH)BO_3$). It is also associated with yttrium, zirconium and cerium in gadolinite–$Be_2Fe(YO)_2(SiO_4)_2$.

Detection

Dilute alkaline solutions of beryllium are coloured cornflower-blue by the addition of a few drops of dilute quinalizarin in alcohol, or alkali. Interferences from phosphates, tartrates, iron and magnesium must first be eliminated by their removal from such solutions.

Estimation

MOTOGIMA (1956) has used 2-methyl-8-hydroxyquinoline in the extraction and determination of beryllium, but bismuth, cadmium, cobalt, copper, indium, iron, nickel, tin, titanium and zinc interfere. All of these can be removed on a mercury cathode with the exception of titanium. Cadmium, copper, iron, nickel and zinc can also be masked by cyanide, or most of them partially removed by extraction with chloroform at pH 4.5–5.0. In practice, to a solution containing 30 μg of beryllium contained in about 35 ml of a slightly acid solution, add 3 ml of 1% 2-methyl-8-hydroxyquinoline in 2% acetic acid and 5 ml of 10% NH_4Cl solution. Adjust the pH to 8.0 with $2M$ NH_4OH and allow to stand for 30 min before extracting with chloroform. Dry the extract with anhydrous Na_2SO_4 and determine the extinction against a blank at 380 mμ.

BISMUTH (Bi)

At.wt. = 209.0; s.g. = 9.7474; m.p. = 271°C;
b.p. = 1,450°C

Occurrence

Occurs free, but usually associated with silver, gold, copper, lead and other metals. The principal minerals are bismuthite or bismuth ochre (Bi_2O_3), and bismuthine (Bi_2S_3). The oxide occurs as a weathered product of other ores whilst the sulphide is mainly found associated with tin.

Detection

Bismuth is precipitated from acidic solutions by hydrogen sulphide as a brown sulphide Bi_2S_3, which is insoluble in ammonium sulphide. A filter paper impregnated with cinchonine, treated with a drop of a slightly acid solution of bismuth, yields an orange-red spot (FEIGL and NEUBER, 1923).

Estimation

A solution of bismuth, free from appreciable quantities of tin and antimony, is warmed gently and the free acid neutralised with ammonia so that all traces of precipitation are reduced to a faint opalescence. Acidify with 1–3 ml of dilute HCl and stir thoroughly. The mixture is then allowed to stand for an hour or so on a steam bath and the bismuth oxychloride, so formed, filtered through a weighed Gooch crucible. This is dried at 100°C and weighed as BiOCl (BiOCl $\times$ 0.8024 = Bi). CARTWRIGHT (1960) has shown that if bismuth formate is formed by slow urea hydrolysis, dense easily filtered precipitates are formed which are virtually free from any lead in the solution. To 200 ml of solution, add 2.5 ml of formic acid, 10 g of urea and a drop or two of bromocresol-green indicator. The latter changes blue-green at pH 5.0 and at this level precipitation is complete. Filter, wash and dry at 105–110°C then thoroughly ignite. Heat this residue to 500°C to achieve constant weight (mg Bi_2O_3 $\times$ 0.8970 = mg Bi).

Volumetrically the cinchonine potassium iodide method has proved satisfactory and depends upon the fact that bismuth nitrate yields crimson or orange colours with this reagent. The solution must be free from lead, arsenic, antimony and tin. These are removed with H_2SO_4 and H_2S. The residual sulphides are dissolved in hot dilute nitric acid, followed by neutralisation of the free acid with a continuous addition of dilute ammonia until a faint cloudiness appears in the solution. Then 10–15 ml of 10% ammonium carbonate is added and stirred thoroughly. The mixture is then digested on a steam bath for at least 3 h and the precipitate recovered by filtration is washed with hot water. The residue of bismuth basic carbonates is then dissolved in the smallest quantity of dilute nitric acid and made up to 50 ml or 200 ml according to the bulk of precipitate involved. The solution is then titrated against a standard solution of bismuth nitrate using 3 ml of cinchonine solution.

BORON (B)

At.wt. = 10.82; s.g. amorph. = 2.45; s.g. cryst. = 2.55; m.p. amorph. = 2,200°C; m.p. cryst. = 2,500°C; sublimates at 2,200°C (amorph.); b.p. cryst. = 3,500°C

Occurrence

As borax, which is sometimes called toncalenite or tincal ($Na_2O.2B_2O_3.5H_2O$), sassoline or native boric acid (H_3BO_3 or $3H_2O.B_2O_3$), ulexite ($NaCaB_5O_9.8H_2O$), colemanite ($Ca_2B_6O_{11}.5H_2O$), boracite ($Mg_7Cl_2B_{16}O_{30}$) and kramite ($Na_2O.2CaO.5B_2O_3.10H_2O$). It is also associated with calcium and aluminium in several silicates such as tourmaline (a boro-silicate of aluminium), datolite (a boro-silicate of calcium) and axinite (a calcium aluminium boro-silicate).

Detection

On the addition of borates to a solution of p-nitrobenzene-azochromotropic acid in concentrated sulphuric acid, the blue solution changes to greenish blue. This colour change is adversely affected by the presence of oxidizing agents and fluorides (KOMAROWSKY and POLNEKTOFF, 1933/1934). Another test consists of adding a few drops of acetic acid with two or three drops of an alcoholic turmeric solution to an alcoholic extract of the sample. The mixture is then diluted with water and evaporated to dryness in a porcelain dish on a water bath. If boron is present, down to 1/1,000 mg of boric acid, the residue will be coloured reddish-brown. A drop of sodium hydroxide added to the residue will turn it bluish-black.

Estimation

Boron, precipitated as barium borotartrate, was developed as a gravimetric method by GANTIER and PIGNARD (1951). Transfer a 10 ml aliquot of a barium-free solution to a 100 ml graduated flask. Make up to graduation mark with a mixture of a barium chloride reagent (13 g of barium chloride plus 240 g of ammonium chloride and 14 g of tartaric acid in 1 l of distilled water) containing one tenth of this volume of concentrated ammonia. Allow to stand for 2 h, and filter and weigh as barium borotartrate.

Volumetrically the barium can be determined by estimating the excess barium in the filtrate of the above gravimetric procedure. Take 10 ml of this filtrate and dilute to exactly 100 ml with distilled water. Transfer 10 ml of this solution to a conical flask and add 10 ml of concentrated ammonia, 15 ml of distilled water, 50 ml of ethanol and a few drops of phthalein purple indicator (0.1% phthalein in ethanol). Titrate with standard EDTA (1 ml of 0.01 M EDTA = 0.04328 mg of boron = 0.38143 mg of borax).

BROMINE (Br)

At.wt. = 79.916; s.g. = 3.1883; m.p. = −7.3°C;
b.p. = 58.7°C

Occurrence

Only occurs in combination and generally associated with alkaline earths and evaporates. It also occurs in coal. Mainly obtained as a by-product of the salt industry and the processing of sea water.

Detection

Silver nitrate gives a light-yellow precipitate of silver bromide. The magenta test uses a stock solution, consisting of 10 ml of a 0.1% solution of magenta diluted with 100 ml of a 5% solution of sulphurous acid. This solution is allowed to cool until it becomes colourless. Of this stock solution 25 ml is mixed with 25 ml of glacial acetic acid and 1 ml of sulphuric acid. On test, this reagent is used in aliquots of 5 ml per 1 ml of the solution to be tested. Bromine produces a reddish violet colour, whilst chlorine yields a yellow colour.

Estimation

Standard chlorine water is commonly used for volumetric determinations of soluble bromides. The estimation is based on the liberation of bromine. A more satisfactory method is the same as that used for chlorine (see p.157). The bromine solution is treated with excess 0.1 *N* silver nitrate solution and the amount of this reagent in excess of the formation of silver bromide is determined by titration with ammonium thiocyanate, using either ferric alum as an indicator or the eosin adsorption indicator (1 ml of 0.1 *N* silver nitrate = 0.007992 g of bromine).

CADMIUM (Cd)

At.wt. = 112.41; s.g. = 8.65; m.p. = 320°C; b.p. = 767°C

Occurrence

Greenockite (CdS) is a rare mineral of cadmium which is usually associated with zinc ores.

Detection

It is precipitated from solution by hydrogen sulphide. The yellow precipitate is distinguished from arsenic, antimony and stannic tin by its solubility in ammonium hydroxide or colourless ammonium sulphide. In an alcoholic solution of diphenylcarbazide, cadmium salts yield a red-violet precipitate or colouration, but this is not a strictly specific reaction. Likewise dinitro-diphenylcarbazide yields a brown coloured compound with cadmium which changes in time to a greenish-blue colour. This colour change is accelerated for some unknown reason by the addition of formaldehyde. In order to depress the action of copper salts with dinitro-diphenylcarbazide, it is usual to add an excess quantity of potassium cyanide.

Estimation

Gravimetrically cadmium is usually estimated as cadmium sulphide by precipitation with hydrogen sulphide and removal of arsenic antimony and tin by means of ammonium hydroxide. As it usually occurs in zinc ores, these are usually dissolved in aqua regia, and the insolubles removed by filtration and the filtrate diluted to at least ten times its original volume. The zinc sulphide is removed by hydrochloric acid. In some cases it is convenient to determine the quantity of cadmium electrolytically with a potassium cyanide electrolyte.

MOORE and ROBINSON (1960) have shown that the reaction of cadmium with 1-phenyltetrazolene-5-thione, yields a precipitate which is readily filterable, and can be dried at 100°C without decomposition. Although the reagent is not entirely selective, the high sensitivity of the cadmium reaction offers a promising method for gravimetric determination (mg of ppt. × 0.2408 = mg of Cd).

MAHR and OHLE (1937) developed a method of separating cadmium from zinc by the precipitation of the former with a 5% solution of thiourea and Reinecke's salt —$NH_4[Cr(HN_3)_2(SCN)_4]H_2O$—in distilled water. The acidity of the cadmium solution is adjusted to 0.1–1.0 *N* with sulphuric, or hydrochloric acid and the solutions of thiourea and Reinecke's salt are added until a strong pink colour develops in the supernatent solution. The solution is cooled in an ice bath,with frequent stirring to complete precipitation. The precipitate is filtered and dried at 110°C for 1 hour (mg of ppt. × 0.1247 = mg of Cd). This method has been further investigated by DEVOE and MEINKE (1959) and they have shown that it offers a simple and reliable means of estimating cadmium in zinc ores.

CARBON (C)

At. wt. = 12.01; s.g. amorph. = 1.75–2.10;
s.g. graphite = 2.25; s.g. diamond = 3.47–3.5585;
sublimates at 3,500°C

Occurrence

As diamond, graphite and amorphous carbon it occurs free. Occurs in combination as carbonate and with hydrogen and nitrogen as hydrocarbon in bitumen, coal and petroleum.

Detection

It is usually recognized by its appearance and inertness towards reagents. By combustion with oxygen, or by oxidation, using chromic and sulphuric acids to form carbon dioxide, which forms a white precipitate of calcium carbonate when bubbled into lime water.

Estimation

It is usually determined as carbon dioxide (see p.133).

CERIUM and RARE-EARTH ELEMENTS

This is a group of elements lying between lanthanum (at.no.57) and lutetium (at.no.71). They are often divided into the following three groups.

Cerium Group
Lanthanum (La), at.no.57, at.wt. = 138.92
Cerium (Ce), at.no.58, at.wt. = 140.13
Praseodymium (Pr), at.no.59, at.wt. = 140.92
Neodymium (Nd), at.no.60, at.wt. = 144.27
Illinium (Il), at.no.61, at.wt. not known
Samarium (Sm), at.no.62, at.wt. = 150.43

Terbium Group
Europium (En), at.no.63, at.wt. = 152.0
Gadolinium (Gd), at.no.64, at.wt. = 156.9
Terbium (Tb), at.no.65, at.wt. = 159.2

Yttrium Group
Dysprosium (Dy), at.no.66, at.wt. = 162.46
Holmium (Ho), at.no.67, at.wt. = 163.5

Yttrium (Y), at.no.39, at.wt. = 88.92
Erbium (Er), at.no.68, at.wt. = 167.64
Thulium (Tm), at.no.69, at.wt. = 169.4
Ytterbium (Yb), at.no.70, at.wt. = 173.04
Lutetium (Ln), at.no. 71, at.wt. = 175.0

This classification is based upon the solubilities of the double salts of potassium or sodium rare-earth sulphates in concentrated solutions of potassium or sodium sulphate. The cerium group is only slightly soluble, whilst the yttrium group is readily soluble, and the members of the terbium group occupy an intermediate position. Even so, the complete separation of the individuals of each group on a solubility basis has proved to be almost impossible to achieve.

Occurrence

The rare-earth elements are fairly widespread, but the minerals which contain significant quantities are principally the silicates, phosphates and tantalates. Among the silicates, gadolinite contains the yttrium group, with small amounts of the cerium and terbium groups, whilst cerite consists mainly of silicates of the cerium group.

The several varieties of albanite are silicates of calcium, iron, aluminium and cerium salts. Among the phosphates, monazite is the most important source of cerium, whilst xenotime is an orthophosphate of ytterbium. Fergusonite is a tantalate and columbate of the yttrium earths, euxenite is a columbo-titanate of the yttrium group, whilst samarskite is a columbo-tantalate of yttrium, iron and calcium as well as being one of the best sources of the terbium group.

Detection

Cerium salts are precipitated by ammonia and hydrogen peroxide as yellow or reddish brown cerium perhydroxides. The cerium group can be distinguished from the terbium and yttrium groups by the colour of their platinocyanides—$R_2Pt(CN)_{12}$. The members of the cerium group form yellow monoclinic crystals with $18H_2O$, whilst those of the terbium and yttrium groups yield red rhombic crystals with $21H_2O$. Solutions of the rare earths can also be judged by colour. Those of lanthanum, cerous cerium, gadolinium, terbium, yttrium, ytterbium and lutetium are usually colourless, whilst solutions of europium, erbium and neodymium yield pale pink to reddish violet colours in this order. Ceric cerium solutions are deep reddish orange in colour, whilst samarium and holmium yield yellow solutions. Solutions of praseodymium, dysprosium and thallium are green.

Estimation

Spectrographic methods provide the most reliable means of estimating cerium and its associated rare-earth elements (see Chapter 5).

CHLORINE (Cl)

At.wt. = 35.457; D(air) = 2.491; m.p. = −101.6°C;
b.p. = −34.6°C

Occurrence

In combination with sodium, potassium and magnesium as chlorides such as halite, or rock salt (NaCl), sylvine (KCl), and carnallite ($KCl.MgCl_2.6H_2O$). It is also combined with copper, lead, silver, etc., in rocks usually high in sodium and low in silica, although quartz-bearing ore-deposits also contain metallic chlorides.

Detection

The gas is easily characterized by its odour and yellow colour and it liberates iodine from iodides, bleaches litmus, indigo and many other organic colouring agents. Chlorides, in the absence of bromides and iodides, form white, opalescent or curdy precipitates in the presence of silver nitrate, manganese dioxide, potassium permanganate and other similar oxidizing agents, which liberate chlorine from free hydrochloric acid. Chlorides, in the presence of concentrated sulphuric acid and ammonium hydroxide, produce white fumes.

Estimation

Finely ground rocks or ores are fused with about five times their weight of potassium carbonate. The melt is extracted with hot distilled water, and the solution cooled and acidified with nitric acid, using methyl orange as an indicator. The mixture is allowed to stand, preferably overnight. Any precipitate of silicic acid is treated with ammonia and the mixture boiled and filtered. The filter is washed with hot water. The cooled filtrate is acidified with nitric acid and to this is added a solution of silver nitrate (4.8 g of $AgNO_3$ per 100 ml of distilled water). Filter, either through paper or a Gooch crucible, and weigh the filter as silver chloride ($AgCl \times 0.2474 = Cl$). Avoid exposing the filter to strong light, as this will result in the exposed surfaces of the precipitate decomposing to the subchloride, Ag_2Cl, with the liberation of chlorine.

CHROMIUM (Cr)

At.wt. = 52.01; s.g. = 7.1; m.p. = 1,615°C;
b.p. = 2,200°C

Occurrence

Usually in rocks of high magnesia and low silica content. The principal minerals are chromite, or chrome iron ore ($FeO.Cr_2O_3$), and crocoisite ($PbCrO_4$).

Detection

With ammonium hydroxide, chromium is precipitated as a bluish-green precipitate of $Cr(OH)_3$, along with hydroxides of iron and aluminium. Alizarin acid RC produces orange coloured laths with chromic salts. Chromates can be recognized by the formation of red silver salts with silver acetate, or yellow lead salts with lead acetate. Chromic salts can be converted into chromates by means of bromine in alkaline solution.

Estimation

Gravimetrically, chromium, as a chromic salt in a solution free from iron and aluminium, can be precipitated with NH_4OH. The hydroxide—$Cr(OH)_3$—is ignited to constant weight and weighed as Cr_2O_3 (mol.wt. = 152). Hydrochloric acid and sulphuric acid do not interfere with the precipitation (wt. of $Cr_2O_3 \times 0.6843$ = wt. Cr). Volumetrically, chromium is usually determined by the reduction of the chromate, in acid solution, by the addition of potassium iodide and the liberated iodine titrated with standard sodium thiosulphate. The presence of large quantities of Mg, Ca, Ba, Sr, Zn, Cd, Al, Ni, Co, does not interfere even in the presence of HCl and H_2SO_4. During titration, when the green colour of the reduced chromate begins to predominate over the brownish-red colour of the free iodine, a little starch solution is added and the thiosulphate titration continued until the blue colour of the starch is just destroyed (1 ml of 0.1 *N* $Na_2S_2O_3$ = 0.001734 g of chromium).

A solvent extraction method, by MCKAVENY and FREISER (1958), uses acetylacetone, adjusted to pH 2 and shaken out with chloroform. With the usual practice of reflushing and washing, the combined organic extracts are made up to 50 ml over 2 g of anhydrous Na_2SO_4 and measured absorptiometrically at 560μ .

COBALT (Co)

At.wt. = 58.94; s.g. = 8.7918; m.p. = 1,480°C; b.p. = 2,900°C

Occurrence

The principal minerals are smaltite ($CoAs_2$), cobaltite (CoAsS), erythrite—$Co_3(AsO_4)_2.8H_2O$ and asbolan, or "earthy cobalt", which is a mechanical mixture of manganese oxide containing a variable percentage of cobalt oxide.

Detection

Filter paper shows a brown stain when spotted with a neutral or weakly acid solution and a drop of nitrosonapthol. Iron, copper and uranyl and palladium salts also give similar coloured compounds. In the presence of these compounds the test becomes more specific if all the compounds are transferred as phosphates (by the addition of sodium phosphate) prior to the addition of nitrosonapthol.

Estimation

Gravimetrically cobalt may be precipitated from strongly acid solutions, by adding a hot solution of potassium nitrite.

Filtered through a Gooch crucible, the precipitate is washed at least six times with either 10% potassium acetate or 5% potassium nitrite. The precipitation takes at least 6 h to reach completion and filtration is sometimes troublesome. The precipitate is weighed as potassium cobalti-nitrite ($2K_2Co(NO_2)_6 \cdot 3H_2O$). Cobalt is also determined by means of electrolysis. Volumetric methods have not proved to be as efficient as gravimetry owing to problems of interference. KOPANICA and DOLEZAL (1956) developed a titrimetric method employing potassium ferricyanide. Barium carbonate is added to an aliquot of the cobalt solution in dilute hydrochloric acid in order to remove iron if present. The precipitate, if formed, is centrifuged down and about 150 ml of solution is treated with 0.5 g of glycerine and adjusted to pH 10 with sodium hydroxide. This is titrated to a potentiometric end-point with 0.05 *M* potassium ferricyanide (1 ml of 0.05 *M* $K_3Fe(CN)_6$ = 1.1788 mg of cobalt). The method is applicable to solutions containing large amounts of As, Ba, Ca, Cr, Cu, Pb, Mo, Ni, Wo, U and Zn and it is not subject to oxidation problems connected with other ferricyanide procedures for cobalt.

COLUMBIUM (Cb) and TANTALUM (Ta)

Cb: at.wt. = 93.1; s.g. = 8.4; m.p. = 1,950°C
Ta: at.wt. = 181.5; s.g. = 16.6; m.p. = 2,900°C

Occurrence

The important ores consist of an isomorphous series of oxides which range from tantalite—$(Ta,Cb)_2O_5.(FeMn)O$—into columbite or niobite—$(Fe, Mn)Nb_2O_6$.

Detection

It is usual to detect columbium and tantalum jointly as so-called "earth acids". Finely powdered minerals are digested in concentrated hydrochloric acid, followed by an addition of nitric acid, and evaporated to dryness. The residue is then moistened with hydrochloric acid and then taken into solution by dilution with distilled water. The solution is then boiled and filtered, and the residue washed and digested with warm dilute ammonia and filtered to remove tungsten. The residue is ignited in a silica crucible and fused with potassium bisulphate. It is then re-fused with a few drops of strong sulphuric acid at a low temperature. The fusion mixture dissolves in 20% tartaric acid and, when freed from insoluble siliceous matter, the addition of strong hydrochloric acid yields a white flocculent precipitate which specifically indicates the presence of even small quantities of "earth acid". To detect small amounts of columbium, ignite the precipitate and fuse with potassium carbonate. When the clear melt has cooled, dissolve in a little water. Treat the solution with excess phosphoric acid and heat until the solution is clear. Add zinc dust to the solution. If columbium is present a brown to inky-black colouration is formed. Tantalum produces no such colouration on reduction with zinc dust. To detect small amounts of tantalum in columbium use oxalic acid instead of phosphoric acid and add tannic acid. A sulphur-yellow precipitate or colouration indicates the presence of tantalum.

Estimation

Several gravimetric methods have proved to be satisfactory, but volumetric methods all suffer from incomplete reduction of the compounds. Gravimetric determinations have been satisfactorily achieved using the procedures described by POWELL and SCHOELLER (1925) and by SCHOELLER (1932). These can be used for all proportional mixtures of columbium and tantalum and also for micro work. Mixed oxides of "earth acid" are fused with bisulphate and the melt dissolved in hot, saturated ammonium oxalate. The solution is filtered to remove siliceous matter. The filtrate is boiled and treated with a 2% solution of tannic acid and titrated with 0.5N ammonia until a strong turbid solution persists. Add more tannic acid according to the following requirements:

10 ml for less than 0.03 g Ta_2O_5
15–20 ml for 0.03–0.06 g Ta_2O_5
25–30 ml for 0.06–0.12 g Ta_2O_5
35–40 ml for 0.12–0.18 g Ta_2O_5
45–50 ml for 0.18–0.25 g Ta_2O_5

If the precipitate first becomes yellow the addition of tannin should be interrupted to allow a permanent colouration to persist. When complete addition of tannin is achieved add a strong solution of 5 g of ammonium chloride, boil and allow the precipitate to settle before filtration. Wash the orange or red precipitate with 2% ammonium chloride and filter. The precipitate contains the columbium which is strongly ignited and weighed as Cb_2O_5. The filtrate contains the tantalum which is precipitated by the drop-wise addition of 0.5N ammonia and allowed to stand overnight. The precipitate is strongly ignited and weighed as Ta_2O_5.

COPPER

At.wt. = 63.57; s.g. = 8.92; m.p. = 1,083°C;
b.p. = 2,310°C

Occurrence

It occurs native and in minerals as copper pyrites

($Cu_2S.FeS_3$), copper glance (Cu_2S), erubescite (Cu_3FeS_3), tetrahedrite ($4Cu_2S.Sb_2S_3$), tennanite ($4Cu_2S.As_2S_3$), famatinite ($3Cu_2S.As_2S_5$), bournonite—$3(PbCu_2)S.Sb_2S_3$, chalcanthite ($CuSO_4.5H_2O$), malachite—$CuCO_3.Cu(OH)_2$, azurite—$2CuCO_3.Cu(OH)_2$, cuprite (Cu_2O), tenorite (CuO), chrysocolla ($CuO.SiO_2.2H_2O$), dioptase ($CuO.SiO_2.H_2O$), atacamite—$CuCl_2.3Cu(OH)_2$.

Detection

The sulphide precipitated by H_2S gas is separable from arsenic, antimony and tin by being insoluble in sodium sulphide. When taken up in nitric acid along with lead, bismuth and cadmium, the lead is precipitated by sulphuric acid, and the bismuth by the addition of ammonium hydroxide—the copper remains in solution, colouring it blue. A delicate test for copper is achieved by using p-dimethylaminobenzylidene-rhodamine as an indicator. To about 10 ml of the solution of copper add a few drops of 2 % hydrazine sulphate followed by a slight excess of 6 *N* ammonia. Add 0.2 ml of rhodanine, acidify with 30% acetic acid and allow to stand. The cuprous copper reacts to form a reddish-violet complex with the rhodanine. Traces of copper in water can be detected in the absence of chlorides by rendering it slightly acid with dilute sulphuric acid. To this 1 g of ammonium persulphate is added and dissolved and then treated with 1 ml of a saturated solution of dimethylglyoxime in alcohol containing silver nitrate and pyridine. The reaction is specific for copper and yields a reddish-violet colouration in solution down to 0.1 mg/l.

Estimation

The number of methods for the determination of copper probably exceeds that for almost any other single element. Copper minerals and ores are best decomposed with acids.

solution containing copper add 5 ml of nitric acid and boil Treatment with hydrochloric acid followed by nitric acid is an effective approach. Gravimetrically copper is precipitated from acid solutions by H_2S gas. It is often estimated, on the other hand, as cuprous thiocyanate. The method proceeds from a cold, concentrated, slightly acid copper solution, to which sulphur (as a gas or a saturated solution) is added in excess of the quantity estimated to reduce all the copper and ferric iron present. The liquid is kept cool and stirred constantly until precipitation ceases. Filtration is carried out through doubled filter papers and the precipitate washed with cold water until the filtrate fails to show traces of thiocyanate when tested with a ferric salt. The filter paper and precipitate are then dried and incinerated. The weighed residue is treated as cuprous thiocyanate (CuCNS). Electrodeposition is a convenient method for the determination of copper concentrations varying between 0.5 mg and 5 g. To the solution add 3 ml HNO_3 with 5 ml of 1:200 H_2SO_4 and a drop of 1:99 HCl. If lead is present add 5 ml of 10% sulphuric acid. Electrolyse at 0.5 A/dm^2 and 2 V for 12 h. Wash the electrodes with distilled water and ethyl alcohol, and dry them at 100°C. The increased weight of the electrode is a measure of the quantity of copper in solution.

Volumetrically, copper is often determined by the potassium iodide method, which depends upon the ability of cupric salts to liberate iodine when treated with potassium iodide. The cuprous iodide so formed is insoluble in dilute acetic acid, but soluble in excess potassium iodide. The liberated iodine is titrated with standard thiosulphate to complete the following reaction: $2Na_2S_2O_3 + I_2 = Na_4S_2O_6 + 2NaI$. Only iron, antimony, selenium, trivalent arsenic and hexavalent molybdenum interfere with this reaction.

The thiosulphate method also provides a reliable volumetric determination when standardized against pure copper. A standard 0.1*N* solution of sodium thiosulphate ($Na_2S_2O_3.5H_2O$,24,82 g/l) is used for this procedure. To the until all brown fumes are expelled. Dilute and add ammonia drop by drop until the precipitate which forms redissolves. The deep blue coloured solution is again boiled until only a trace of ammonia odour persists, then neutralize with glacial acetic acid added drop by drop until the precipitate which forms dissolves. Again boil and then allow to cool. Add solid potassium iodide until the precipitate of copper iodide redissolves. Titrate this solution with the 0.1 *N* standard thiosulphate using starch as an indicator. The solution should be stirred vigorously especially as the colourless end-point is approached. Of 0.1 *N* thiosulphate solution 1 ml is equivalent to 6.357 mg of copper.

AGTERDENBOS and TELLINGEN (1961) developed a titrimetric method for solutions containing similar amounts of copper and iron. They found that aluminium ions proved to be more satisfactory than boric acid, and that thiocyanate can be added without producing any deleterious effects. The solution (25 ml) is transferred to a flask and ammonia added until the blue colour of the cuprammonium salts is pronounced. To this 0.8–2.0 g of solid potassium hydrogen fluoride is added and the initial precipitate is allowed to redissolve before adding 2–3 g of potassium iodide. Titrate with 0.1*N* standard thiosulphate solution using starch as an indicator. When the starch end-point is reached add 10 ml of a 100% potassium thiocyanate solution and complete the titration to obtain the value for copper (1 ml standard thiosulphate = 6.357 mg of Cu). To obtain the value for iron add 25 ml of a solution of aluminium chloride (8 g of $AlCl_3.6H_2O$ in 100 ml of distilled water) together with solid sodium bicarbonate to suppress oxidation. Titrate with standard thiosulphate to obtain the value for iron (1 ml of standard thiosulphate = 5.585 mg of Fe).

FLUORINE (F)

At.wt. = 19.00; D(air) = 1.31 at 15°C; s.g. = 1.11 at −187°C; m.p. = −223°C; b.p. = −187°C

Occurrence

Not found free but in combination as fluorspar (CaF_2), cryolite (Na_3AlF), apatite—$Ca_5(ClF)(PO_4)_3$. It also occurs with phosphorus in all animal and vegetable matter.

Detection

Fluorides boiled in concentrated sulphuric acid liberate fluorine which etches a cover glass held lightly over a platinum crucible. With care this test will detect reasonably small quantities of liberated fluorine.

Estimation

Gravimetrically fluorides can be determined by the precipitation of lithium fluoride (CAYLEY and KAHLE, 1959). A volume of solution containing 20–200 mg of fluoride is made up to 15–40 ml, depending upon the estimated quantity present in solution. Heat the solution in a borosilicate vessel to 70°C and slowly add, by stirring, an equal volume of lithium chloride and an equal volume of ethanol to complete precipitation of lithium fluoride. Allow to stand until the supernatant liquid is clear, filter, and wash the precipitate with ethanol. Dry the precipitate for at least 1 h at 110°C. Weigh as lithium fluoride (mg LiF × 0.7325 = mg of fluorine).

GALLIUM (Ga)

At.wt. = 69.72; s.g. = 5.95; m.p. = 30°C

Occurrence

Mainly found as traces in zinc blendes, iron ores and aluminium minerals such as bauxite and kaolin. The mineral germanite found in South Africa contains 8.7% germanium and 0.8% gallium.

Detection

Usually detected by spectrographic methods.

Estimation

Gravimetrically gallium is estimated as the oxide derived by ignition of the hydroxide, the tannin complex, or a cupferron precipitate. The oxide is hygroscopic and should be weighed immediately. The precipitate should be ignited in a porcelain crucible as gallium attacks platinum. Using brilliant green, benzene extracts from solutions of gallium in 6*M* hydrochloric acid form a specific chlorogallate ion with which there is no interference from a wide range of elements. Interferences from antimony, gold, iron and thallium are prevented by reduction with titanous chloride.

GERMANIUM (Ge)

At.wt. = 72.6; s.g. = 5.47; m.p. = 958°C

Occurrence

It occurs in topaz and certain zinc ores and blendes and is sometimes associated with silver and tin sulphides and in tantalum and niobium minerals. Argyrodite ($GeS_2.4Ag_2S$) was the original source of the metal.

Detection

A delicate test for germanium is provided by using what is called "triselenomethylene", which is made by heating H_2Se and HCHO. This gives a yellow complex with germanium down to 0.5 p.p.m.

Estimation

It is inadvisable to use solutions in HCl or aqua regia owing to the volatility of the tetrachlorides. The mineral is fused with sodium carbonate and sulphur (1:1) and the melt extracted with water. It is advisable to re-fuse the residue and re-extract with water. The filtrate is neutralized with dilute sulphuric acid and then acidified until in contains about 25% of sulphuric acid. The germanium is then precipitated with hydrogen sulphide and weighed as GeS_2.

GOLD (Au)

At.wt. = 197.2; s.g. = 19.33; m.p. = 1,063°C;
b.p. = 2,600°C

Occurrence

Found native in a wide variety of rocks and also alloyed with silver or copper and occasionally with bismuth, mercury and other metals. Gold amalgam (with mercury and silver), sylvanite and calcaverite (tellurides of gold and silver) and nagyagite (telluride and sulphide of lead and gold) are the principal known mineralogical forms of the metal.

Detection

With p-dimethylamino-benzylidene rhodanine gold salts give

a red violet precipitate in neutral or weakly acid solutions. Silver, mercury and palladium salts give similar colourations. Benzidine produces a blue dye with gold salts in the presence of acetic acid.

Estimation

Gold is readily displaced from acid solution by the addition of base metals, aluminium, etc. The tedious nature of the complete separation of gold from associated metals, and the difficulty of obtaining it in a form suitable for accurate weighing, inhibits the use of direct precipitation methods. It is often so finely divided that it is not completely retained by the filter. HARVEY and YOE (1953) have shown that gold can be precipitated quantitatively as a complex by the sodium salt of N-(N–bromo-C-tetradecyl-betainyl)-C-tetradecyl betaine as a reagent. (Manufactured by Jackson Laboratories of du Pont de Nemours Co.) This reagent has the empirical formula $C_{38}H_{78}N_2O_4NaBr$ and the complex can be weighed as such. The procedure consists of adding to the test solution sufficient hydrobomic acid to give a concentration of 2*N* at 20–25 ml. Dilute the whole to 20–25 ml and add the reagent dropwise from a burette whilst stirring. At the point where the red flocculent precipitate becomes finely divided, 5–10% more reagent is added rapidly. Allow to settle for a few minutes and filter through a fine silica or glass filter crucible. Wash with distilled water and dry to a constant weight at 85°C. It usually takes 1–2 h to achieve constant dryness (mg of ppt. $\times$ 0.1801 = mg of Au). The gold can also be weighed as metal by igniting the crucible to the full heat of a Meker burner for at least 15 min.

Several volumetric and colorimetric methods have been used for the estimation of gold (see FURMAN, 1942, pp.436–441). Ascorbic acid as a reductimetric titrant has been investigated for noble metals (ERDEY and BUZAS, 1954) and a remarkably accurate method has been developed for gold by ERDEY and

RADY, 1958. An aliquot of the gold solution ($c = 0.01N$) is titrated with enough hydrochloric acid to give a concentration of not more than $0.1N$ at a pH of 1.3–3.0. The solution is then diluted to 100 ml, warmed to 50°C, and titrated potentiometrically with $0.01N$ ascorbic acid using a platinum–silver–silver chloride electrode system. A $0.1N$ solution of ascorbic acid is prepared by dissolving 8.806 g of ascorbic acid in distilled water containing 0.1 g of disodium ethylenediaminetetra-acetate dihydrate with 4 ml of 60% formic acid and the whole diluted to 1 l. A glass electrode may be used instead of the chloride electrode which has to be renewed after every three or four titrations. If iron is present then 1 ml of H_3PO_4 is added to the test solution.

HYDROGEN (H)

At.wt. = 1.008; s.g. = 0.07; b.p. = −253°C;
m.p. = −259°C

Occurrence

Occurs free in small quantities in volcanic gases and certain petroleum and gas wells. All other hydrogen is found combined with oxygen as water.

Detection

Its lightness and its combustibility with oxygen to form water as well as its combination with chlorine to form hydrogen chloride are the usual methods used for detection.

Estimation

A reasonably quantitative method uses the combustion of hydrogen in oxygen to form water as a basis of estimation.

The advent of chromatography into gas analysis is rapidly replacing most other methods of hydrogen determination (see Chapter 9).

INDIUM (In)

At.wt. = 114.76; s.g. = 7.12; m.p. = 155°C

Occurrence

Found in minute quantities in zinc blende, some tungsten and most tin ores. It also occurs in pyrite, siderite and galena.

Detection

Ignition of the hydroxide precipitates from indium solutions by ammonia, produces a characteristic pale yellow deposit of indium sesquioxide.

Estimation

Indium as chloride is precipitated quantitatively by ammonia, filtered, washed and weighed after ignition as indium sesquioxide (ppt. × 0.8270 = In).

Precipitated from solutions as the sulphide the precipitate is washed with a weak solution of ammonium acetate. The precipitate is dried in a stream of hydrogen sulphide at 350°C, cooled and weighed as indium sulphide (In_2S_3 × 0.7047 = In).

IODINE (I)

At.wt. = 126.92; s.g. = 4.948 at 17°C; m.p. = 113°C; b.p. = 184.4°C

Occurrence

Found as iodides of copper, lead and silver and as the iodate of calcium, called lanterite, in Chili saltpetre.

Detection

It vaporizes at ordinary temperatures yielding a characteristic odour. The fumes colour the skin brown and turn a cold starch solution blue.

Estimation

A solution of an iodide readily forms a precipitate of silver iodide by the addition of silver nitrate. The precipitate is filtered through a Gooch crucible, washed and dried. It is then gently ignited and the residue weighed as silver iodide ($AgI \times 0.5405 = I$). A potentiometric method (MINCZEWSKI and GLABISZ, 1960), consists of taking an aliquot of solution (ca. $0.02M$ with respect to iodide) to which is added an equal volume of $1M$ sulphuric acid. This is titrated with $0.1M$ sodium chlorite to a potentiometric end-point using a platinum–calomel electrode. The stabilization of the potential occurs immediately between 800 and 1,100 mV.

IRIDIUM (Ir)

At.wt. = 19.31; s.g. = 22.4; m.p. = 2,350°C

Occurrence

Found native associated with platinum and it is usually alloyed with osmium as osmiridium.

Detection

Boiling solutions produce dark blue precipitates with caustic alkalis, which are insoluble in all acids except HCl. Potassium chloride forms a black double salt (K_2IrCl_6) which is only slightly soluble in water. Lead acetate yields grey-brown precipitates.

Estimation

Solutions of iridium usually contain platinum. Both metals are precipitated by the addition of pure granulated zinc in hydrochloric acid. The iridium and platinum is precipitated as fine black metal which is washed free of impurities. The platinum can be removed by solution in aqua regia and the insoluble residue dried, ignited and reduced with hydrogen. The residue is weighed as metallic iridium. Hydroquinone has been used for the selective titration of iridium in the presence of platinum, palladium and rhodium (BERKA and ZYKA, 1956). It has proved reliable in the presence of copper, cobalt, iron, manganese, nickel, titanium and tungsten.

LEAD (Pb)

At.wt. = 207.21; s.g. = 11.34; m.p. = 327.5°C; b.p. = 1,620°C

Occurrence

Rarely occurs native, but usually combined and associated with other minerals. Common minerals consist of galena (PbS), cerussite ($PbCO_3$), anglesite ($PbSO_4$), pyromorphite—$Pb_4(PBCl)(PO_4)_3$, minium (Pb_3O_4), wulfenite ($PbMoO_4$), crocoite ($PbCrO_4$), mimetite ($3Pb_3As_2O_8.PbCl_2$), vanadinite ($3Pb_3V_2O_8.PbCl_2$), and the rare mineral cromfordite or phosgenite ($PbCl_2.PbCO_3$).

Detection

The chloride is distinguished from silver and mercury by its solubility in hot water. A filter paper moistened with ammoniacal solution of 3% hydrogen peroxide will convert the lead in the test solution to lead peroxide. When a drop of benzidine in acetic acid solution is added the lead peroxide oxidizes the reagent to blue meriquinoid oxidation products. In the absence of bismuth this test is specific for lead.

Estimation

Lead sulphate may be separated from a number of elements whose sulphates are soluble in dilute acid solutions. A small quantity of alcohol added to the acids further inhibits the solution of lead sulphate. Lead may be separated from silica, barium, tin and antimony by extraction of these latter elements with excess quantities of ammonium acetate. Gravimetrically, lead is usually determined as a sulphate. The ore or mineral when brought into solution is treated with excess quantities of H_2SO_4 to complete the precipitation of $PbSO_4$. The impure precipitate is filtered and washed, first with 10% H_2SO_4, and then with a 50% alcohol solution to remove the free acid. The lead sulphate is further stripped of impurities by repeated extraction with a strong solution of ammonium acetate. The acetate solution is evaporated to dryness and the residue taken up in water. A large excess of H_2SO_4 is added and the precipitated $PbSO_4$ filtered and thoroughly washed, and finally dried at 110°C, or by gentle ignition of the filter paper ($PbSO_4 \times 0.6833 = Pb$). Gravimetric determinations of lead are mainly as accurate as volumetric procedures.

The most favoured titrimetric method uses potassium ferrocyanide and uranium acetate as an external indicator. The excess ferrocyanide yields a brown colour with the drop of uranium acetate which is placed as an outside indicator on a white tile. Antimony, bismuth, barium, strontium and calcium interfere to a slight extent in the determination. An alternative method employs the separation of lead as a sulphate and its conversion into lead carbonate. The carbonate in acetic acid solution is then precipitated as the oxalate, which is in turn decomposed by sulphuric acid to release the oxalic acid. The oxalic acid is titrated in the usual way with standard permanganate. Similar procedures using ammonium molybdate, potassium dichromate or the reduction by iodine or ferrous sulphate solutions have been suggested for the determination of lead in minerals and alloys.

LITHIUM (Li)

At.wt = 6.94; s.g. = 0.53; m.p. = 186°C; b.p. = 1,400°C

Occurrence

Rare, but widely disseminated in all igneous rocks and also in many mineral springs. Important constituent of lepidolite, spodumene, petalite, amblygonite, triphylite, lithiphylite and certain tourmalines. It frequently occurs in feldspars, muscovite, and beryl.

Detection

Lithium salts yield a carmine-red flame, but this is obscured by sodium and large amounts of potassium.

Estimation

The Gooch method is still widely used. It is based on the

ready solubility of lithium chloride in amyl alcohol and the insolubility of sodium and potassium chlorides. To a solution cleared of all constituents other than the alkali metals, add a small amount of amyl alcohol and heat cautiously in a 50 ml Erlenmeyer flask over an asbestos plate. When all the water has been expelled at the boiling point of amyl alcohol (132°C) the sodium and potassium chlorides with some lithium hydroxide will separate from solution. Decant through a filter and wash the residue several times with hot amyl alcohol. Then moisten the residue with dilute hydrochloric acid, and dissolve in water and repeat the extraction with amyl alcohol. Evaporate the combined filtrates and washings to dryness. Dissolve in a little dilute sulphuric acid, filter off the carbonaceous matter and evaporate the solution to dryness. Ignite the residue and cool in a desiccator and weigh as lithium sulphate ($Li_2SO_4 \times 0.1262 = Li$).

The use of standardized solutions of lithium chloride in 2-ethyl-1-hexanol against standard solutions of silver nitrate has proved to be an effective titrimetric method for the determination of lithium (WHITE and GOLDBERG, 1955). Evaporate gently to dryness a 20 ml aliquot of lithium chloride solution. Dissolve the residue in 3–5 ml of distilled water. Add 10–15 ml of 2-ethyl-1-hexanol. Insert one or two glass beads and slowly heat to a controlled temperature of 135°C until the aqueous phase has volatilized. Cool and filter into a 250 ml titration flask. Add 50 ml of ethanol and cautiously follow this with 2 ml of nitric acid and afterwards with 2 ml of a ferric alum solution, consisting of 20 g of ferric alum in 100 ml of distilled water. To the flask add 1 ml of 0.1*N* potassium thiocyanate and titrate with silver nitrate. When the colour of the ferric-thiocyanate complex begins to fade, stopper the flask, shake vigorously, allow the precipitate to settle and back titrate with thiocyanate until a permanent pink colour develops (1 ml of 0.1*N* $AgNO_3$ = 0.694 mg of Li).

MANGANESE (Mn)

At.wt. = 54.93; s.g. = 7.2; m.p. = 1,260°C;
b.p. = 1,900°C

Occurrence

The important minerals are pyrolusite (MnO_2), manganite ($Mn_2O_3.H_2O$), psilomelane (hydrous manganese manganate), rhodochrosite ($MnCO_3$), rhodonite ($MnSiO_3$), and spessartite—$Mn_3Al_2(SiO_4)_3$.

Detection

Minerals are usually incinerated with nitric acid and then brought into solution by adding perchloric acid. The solution is then evaporated until strong fumes of $HClO_4$ are given off. At this point nitric acid (s.g. = 1.135) is added and the solution boiled to liberate chlorine. Ammonium persulphate and silver nitrate is then added. On boiling, a pink colouration indicates the presence of manganese. Manganese compounds heated with borax in an oxidizing flame produce an amethyst-red bead. This colour is destroyed by a reducing flame. When fused with sodium carbonate and sodium nitrate on platinum foil, manganese compounds produce a green coloured fusion mixture.

Estimation

For the evaluation of manganese ores, the gravimetric potassium bromate method of KOLTHOFF and SANDELL (1929) has become a standard procedure. The ore is brought into solution by means of HCl and $KClO_3$. About 50 ml of the test solution is adjusted to 0.8–1.0*N* with respect to free sulphuric, or free nitric acid. Potassium bromate (1–2 g) is then added and the solution boiled for about 20 min. The dioxide precipitate is filtered off and washed with hot water. The ashless filter paper is ignited and the residue weighed as Mn_3O_4.

The volumetric "bismuthate method" of analysis depends upon the ability of manganese to form permanganic acid which can then be titrated with reducing agents such as sodium thiosulphate, arsenious acid, or ferrous sulphate (CUNNINGHAM and COTTMAN, 1924). Using diphenylamine as an indicator, in the presence of potassium ferricyanide (1% solution freshly prepared), manganese can be determined titrimetrically (CHENG, 1955). To an aliquot of the manganese solution containing 0.5–2.0 ml of manganese, add 20 ml of a 5% solution of EDTA, plus 70 ml of distilled water, 20 ml of glacial acetic acid, four drops of diphenylamine (1% solution in acetic acid, with a four drops of 1% freshly prepared potassium ferricyanide). After the addition of an excess of standard potassium ferrocyanide solution, it is back-titrated with 0.1*M* manganese chloride until the colourless, or slightly yellow solution turns purple. Direct titration may be employed but the end-point is not as good as that obtained by back-titration.

MERCURY (Hg)

At.wt. = 200.61; s.g. = 13.595; m.p. = −38.9°C; b.p. = 357.33°C

Occurrence

The chief source is cinnabar (HgS). Other minerals include calomel (Hg_2Cl_2), coloradoite (HgTe), amalgam (Ag.Hg), livingstonite ($HgSb_4O_7$) and tiemannite (HgSe). It also occurs as a chloride in horn silver.

Detection

It is the only metal which is a liquid at ordinary temperatures. In its mercurous form it is precipitated as white mercurous chloride by HCl, which turns black on treatment with ammonium hydroxide. Mercuric chloride is not precipitated by HCl, but forms a black precipitate with H_2S gas. It is distinguished from the other elements of the group by the insolubility of the sulphide in yellow ammonium sulphide, or in dilute nitric acid. Mercury salts in acid solutions give a violet to blue precipitate with diphenyl-carbazide. This test is specific for mercury in a 0.2*N* nitric acid solution of diphenylcarbazide.

Estimation

In all preparations the volatility of mercury and its compounds must be recognized. Fusion of compounds with sodium carbonate will completely volatilize mercury. Gravimetrically mercury may be determined by the precipitation of mercuric sulphide with ammonium sulphide. The acid solution of the mercuric salt is neutralized by sodium carbonate and then heated with excess ammonium sulphide. Sodium hydroxide is added until the dark solution is clear. Any undissolved lead is removed by filtration. After adding ammonium nitrate, the clear solution is boiled until most of the ammonia has been expelled. The precipitate of mercuric sulphide is dried at 110°C and weighed as HgS (HgS $\times$ 0.8622 $=$ Hg). WALTER and FREISER (1953) indicate that 2-(0-hydroxyphenyl)-benzimidazole is more selective than the azo derivatives of 8-hydroxy-quinoline, thionalide, tetraphenylarsonium chloride, or mercapto-benzimidazole, as a selective reagent for mercury. It can be used for gravimetric and titrimetric determinations. Gravimetric determinations are carried out in solutions free from traces of iron, to which 2 g of sodium citrate has been added. The solution is warmed to

60°C and adjusted with acetic acid or sodium hydroxide to pH 5. Add excess of the reagent (1 g of 2-(0-hydroxyphenyl)-benzimidazole in 100 ml of 95% ethanol). The pH is then raised to pH 5 with 0.3*N* sodium hydroxide and the suspended precipitate maintained as such for about 15 min at 60°C before allowing it to cool to room temperature. The precipitate, on filtration, is washed several times with ethanol and dried to a constant weight at 130–140°C. In composition the precipitate corresponds to $(C_{13}H_9N_2O)_2H$ and is weighed as such (mg of ppt. × 0.3240 = mg of Hg). For titrimetric determination, dissolve the precipitate in 50 ml of hot glacial acetic acid and transfer the solution to an iodine flask and dilute with distilled water. Add 60 ml of 0.1*N* solution of potassium bromate and potassium bromide and stopper the flask immediately. Add a few ml of potassium iodide solution to the reservoir of the iodine flask to prevent the loss of bromine. After maintaining the iodine flask at 35°C for over an hour, add 1.5 g of potassium iodide and titrate the liberated iodine with a 0.1*N* solution of sodium thiosulphate (1 ml of 0.1*N* potassium bromate = 1.672 mg of Hg).

MOLYBDENUM (Mo)

At.wt. = 96.0; s.g. = 10.2; m.p. = 2,620°C

Occurrence

The principal minerals are molybdenite (MoS_2), wulfenite (MoS_2) and molybdine (MoO_3).

Detection

Solutions of molybdates are coloured yellow by potassium thiocyanate and hydrochloric acid. The addition of a small piece of zinc, acting as a reducing agent, produces a deep carmine-red solution. Molybdate solutions in mineral acids yield deep red-blue colourations with potassium ethyl xanthogenate. Sodium thiosulphate, added to a slightly acid ammoniacal solution of a molybdate, yields a blue precipitate or a blue solution. Addition of acid turns the precipitate brown. Sulphur dioxide produces a bluish-green solution, or a precipitate if large quantities of molybdenum are present.

Estimation

To 150 ml of an alkaline, or ammoniacal, solution of molybdate add a drop of methyl orange and sufficient sulphuric acid to change the colour to red. Dissolve 1 g of $NaC_2H_3O_2.3H_2O$ in the solution and heat to boiling before adding $AgNO_3$ solution in order to precipitate silver molybdate. Filter the precipitate and wash with a solution of 5 g $AgNO_3$ in 1 l of distilled water. The excess $AgNO_3$ is removed by washing with ethyl alcohol. Dry the precipitate to constant weight at 250°C and weigh as Ag_2MoO_4.

The determination of molybdenum as oxide, when in small quantities in ores, is often effected by separation as mercurous molybdate. One gram of ore, fused with 4 g of fusion mixture, is extracted with hot water. A coloured solution often indicates the presence of manganese which is precipitated by reduction with alcohol and removed by filtration. Neutralize the alkaline molybdate solution with nitric acid. To the nearly neutralized solution add a faintly acid solution. The precipitate is boiled and allowed to stand for 15 min. Remove the black precipitates by filtration and wash with a dilute solution of mercurous nitrate.Dissolve the precipitate with hot nitric acid, then evaporate the solution to dryness in a porcelain crucible. Heat the residue cautiously over a low flame until the mercury has completely volatilized and weigh the oxide ($MoO_3 \times 0.6667$ = wt. of Mo).

Molybdenum can be determined volumetrically by means

of iodometric reduction, making use of the principle that molybdic acid and potassium iodide in the presence of hydrochloric acid liberates iodine. The test solution is placed in a 150 ml Erlenmeyer flask, with about 25 ml of hydrochloric acid and 0.2–0.6 g of potassium iodide. Evaporate the solution to exactly 25 ml and then immediately dilute to a volume of 125 ml . Allow to cool and add a solution of tartaric acid (equivalent to 1 g of the solid) and then neutralize the free acid with sodium hydroxide using litmus or methyl orange as an indicator. Finally, add an excess of sodium carbonate. Introduce a measured volume of 0.1 *N* iodine and allow the solution to stand in the dark for 2 h to complete oxidation. The excess iodine is then titrated with 0.1*N* sodium arsenite (1 ml 0.1*N* iodine = 0.0144 g MoO_3 = 0.0096 g Mo).

NICKEL (Ni)

At.wt. = 58.69; s.g. = 8.6–8.9; m.p. = 1,452°C

Occurrence

Occurs native in meteorites, whilst the principal minerals include kupfernickel (NiAs), chloanthite ($NiAs_2$), millerite (NiS), emerald nickel—$NiCO_3.2Ni(OH)_2.4H_2O$, nickel vitriol ($NiSO_4.7H_2O$), annabergite or nickel bloom ($Ni_3As_2O_8.8H_2O$) and garnierite (a hydrated nickeliferous magnesium silicate).

Detection

In neutral acetic acid, or ammoniacal solutions, dimethylglyoxime and a number of other dioximes produce bright-red insoluble salts.

Estimation

To an ammoniacal solution of nickel, add a slight excess of a benzildioxime reagent, which consists of 0.2 g of the salt in 1 l of ammonium hydroxide in alcohol (s.g. = 0.96). Heat the mixture for a few minutes on a water bath to produce rapid precipitation. Filter through a Gooch crucible. Wash the precipitate with 50% alcohol followed by hot water. Dry the precipitate at 110°C (wt. of $C_{28}H_{22}N_4O_4Ni \times 0.1093 = Ni$)

Solutions free from copper and cobalt can be determined volumetrically using the potassium cyanide method, which is rapid and accurate. No interferences are developed by iron, manganese, chromium, vanadium, molybdenum or tungsten. The method depends upon the selective action of potassium cyanide on nickel in the presence of silver iodide, which is used as an indicator. To an acid solution of nickel, add 100 ml of a citrate solution consisting of 200 g of ammonium sulphate, 150 ml concentrated ammonium hydroxide and 120 g of citric acid in 1 l of distilled water. Dilute the solution to about 250 ml. Add 5 ml of 0.1*N* silver nitrate, followed by a few drops of ammonium hydroxide to remove cloudiness. Add 2 ml of a potassium iodide solution (13.5 g of pure KCN in distilled water plus 5 g of KOH, made up to 1 l). The solution is then titrated with the standardized solution of potassium cyanide, with constant stirring, until the turbidity disappears. Deduct the volume of potassium cyanide, which is equivalent to the total amount of silver nitrate used, from the total volume of potassium cyanide used for titration. The remainder represents the volume of potassium cyanide required to form the double cyanide of nickel. This volume of KCN multiplied by the factor for nickel derived from a standardized solution gives the weight of nickel in the sample.

$$\text{nickel factor} = \frac{\text{wt. of Ni in standardized solution}}{(\text{ml of KCN}) - (\text{ml of KCN eq. to 5 ml of AgNO}}$$

NITROGEN (N)

At.wt. = 14.008; D(air) = 0.9674; m.p. = –210°C; b.p. = –195.8°C

Occurrence

Occurs free in air (78.1% by volume and 75.47% by weight). In addition to its combined ocurrence in coal, nitrogen is mainly concentrated in minerals such as nitratine or Chili saltpetre ($NaNO_2$), and nitre or saltpetre (KNO_3).

Detection

Nitrogen, in gas, is usually detected by spectroscopic or chromatographic techniques. Moist red litmus paper is turned blue by ammonia. The red colour is restored on heating the paper. Nessler's reagent (20 g KI in 50 ml distilled water, plus 32 g HgI in 150 ml water, plus 134 g KOH in 260 ml distilled water), produces a brown precipitate of $NHg_2I.H_2O$ in a solution containing combined or free ammonia. Diphenylamine dissolved in sulphuric acid produces a blue colour on heating it in a nitrate solution on a watch glass. Copper placed in a solution containing nitric acid produces brown fumes. Acetic acid added to a nitrite in an inclined test tube produces a brown ring which is not developed by nitrates.

Estimation

In general nitrogen is more accurately determined as ammonia, so that reduction methods are usually employed for this purpose. The ferrous volumetric method for nitric acid and nitrates devised by KOLTHOFF et al. (1933) is carried out in a 250 ml Erlenmeyer flask. The sample of nitrate (0.1–0.2 g) is placed in the flask and to this is added 25 or 50 ml of 0.18*N* ferrous iron solution plus 70 ml of 12*N* hydrochloric acid. Solid sodium bicarbonate (3–5 g) is then carefully added in small portions. This displaces the air from the flask, into which a suspension of 50 g of sodium bicarbonate in 100 ml of water is lead through a stopper. The stopper also carries a dropper containing 3 ml of 1% ammonium molybdate solution, which is added after the solution has boiled for a few minutes. Boiling is continued for about 10 min and then the flask is immersed in cold water. When cold the flask is unstoppered and to it are added 35 ml of 6*N* ammonium acetate per 50 ml of solution with 3–5 ml of 85% phosphoric acid. This reduces the concentration of acid to between 1 and 2*N* in a volume of about 100–150 ml of solution. This is slowly titrated with a 0.1*N* dichromate solution. Six to eight drops of diphenylamine sulphonate are used as an indicator (1 ml of 0.10*N* iron = 3.370 g of potassium nitrate or 2.067 mg of NO_3).

OSMIUM (Os)

At.wt. = 191.5; s.g. = 22.4; m.p. = 2,700°C

Occurrence

In platinum it is found native or in osmiridium as an alloy with iridium.

Detection

In the presence of strong acids, hydrogen sulphide precipitates dark brown osmium sulphide which is insoluble in ammonium sulphide. Potassium hydroxide precipitates reddish-brown osmium hydroxide. A deep violet colouration and a dark blue precipitate are developed by sodium sulphite. B-naphthalamine hydrochloride produces a blue colour in solutions of sodium or potassium osmates. Thiourea added to solutions acidified with hydrochloric acid and heated for a few minutes develops a red or rose colouration.

Estimation

Gravimetrically, it has been determined by the use of benzo-1,2,3-triazole which appears to yield a stoichiometric compound which provides a direct means of weighing the osmium complex (BEAMISH, 1958). To a flask containing 1 ml of ethanol transfer 10 ml of the osmium solution, and add 25 ml of benzo-1,2,3-triazole as a 2% aqueous solution. Heat on a steam bath for 15 min and then adjust to pH 3 with acetic acid. Boil for a further 15 min to coagulate the precipitate. Filter the precipitate through a sintered glass crucible and wash several times with hot water. Dry to a constant weight at 110°C (mg of ppt. × 0.3178 = mg of Os).

OXYGEN (O_2)

At.wt. = 16.00; b.p. =−182°C

Detection

Free oxygen is usually recognized by its activity with heated substances. Carbon combines with oxygen at dull red heat to form CO_2 which is detected by bubbling through limewater.

Estimation

The determination of combined oxygen is extremely difficult and seldom attempted. It is frequently estimated by difference. Gas chromatography has greatly simplified the estimation of oxygen and by pyrochromatography, oxides can be determined by direct reduction in an inert gas stream (see Chapter 9).

PALLADIUM (Pd)

At.wt. = 106.7; s.g. = 11.9; m.p. = 1.549°C

Occurrence

Usually found associated with platinum and iridium and also with ruthenium, rhodium and osmium. It also occurs in small quantities in cupriferous pyrites and in particular those containing nickel and pyrrhotine.

Detection

Concentrated solutions yield a flesh-coloured precipitate with ammonia. Mercuric cyanide precipitates a yellowish white gelatinous precipitate which is insoluble in dilute acids. Black palladium iodide is precipitated by potassium iodide.

Estimation

A selective reagent for palladium is 2-thiophene-trans-aldoxime (TANDON and BHATTACHARYA, 1960) in hydrochloric acid. This solution yields a light yellow insoluble heat-stable complex of definite composition. Dilute the test solution to 100-150 ml in a borosilicate glass beaker and adjust the pH to 0.2–0.8 by means of about 5 ml of concentrated hydrochloric acid. From a burette, add excess of 2-thiophene-trans-aldoxime as a 2% solution in 95% ethanol. Digest for 15 min on a steam bath, whilst stirring occasionally. Allow to cool for 2–3 h and then filter through a fine sintered glass crucible. Wash with about 25 ml of cold 1% hydrochloric acid followed by about 100 ml of cold water. The precipitate is dried at 110°C for an hour (mg of ppt. × 0.2470 = mg of Pd).

PLATINUM (Pt)

At.wt. = 195.23; s.g. = 21.48; m.p. = 1,755°C

Occurrence

The principal form is native platinum, usually alloyed with base metals such as copper and iron and sometimes with native gold. Rare minerals containing platinum include sperrylite ($PtAs_2$), which is mainly found in the nickel ores of Sudbury, Ontario, Canada.

Detection

Potassium iodide yields a precipitate of platinum iodide which re-dissolves readily into a pink or dark-red solution. Hydrogen sulphide precipitates black disulphide in hot solutions. Metallic zinc, iron, magnesium, aluminium and copper precipitate metallic platinum from chloride solutions.

Estimation

If a solution of platinum in aqua regia contains very few impurities it is precipitated as $(NH_4)_2PtCl_6$. The solution is first gently heated and additional aqua regia is added to complete solution. The volume is then reduced by evaporation and additional volumes of hydrochloric acid are added to expel the nitric acid. Filter and then evaporate the filtrate in order to concentrate the solution from which the remaining chloride is precipitated with ammonium chloride. Stir and allow to stand overnight. Filter and wash with alcohol. Ignite the precipitate and weigh as metallic platinum. If the sample contains large quantities of iron, magnesium, nickel etc., direct precipitation with metals is recommended. The solution is made acid with hydrochloric acid, then add Zn, Fe or Mg in small quantities until the solution becomes colourless. Filter through ashless filter paper and wash out excess metal with hydrochloric acid. Weigh as metallic platinum. Platinum solutions acidulated with sulphuric acid yield platinum as a bright deposit upon the electrode of an electro-analysis bath using a feeble current.

RHENIUM (Re)

At,wt. = 186.31; D = 21.04; m.p. = 3,167 ± 40°C

Occurrence

Exceedingly rare, it occurs in a number of minerals but principally in molybdenite.

Detection

A particle of caesium chloride added to a concentrated solution of rhenium yields highly refracting bi-pyramid crystals. A trace of potassium permanganate stains the crystals.

Estimation

The quantitative determination of rhenium is usually accomplished by isolating the element from interfering materials by distilling it from hot concentrated sulphuric acid in a stream of moist hydrochloric acid. Rhenium hepta-sulphide is precipitated in the distillate. This is converted into a weighable precipitate of per-rhenate with 5% NaOH and 30% hydrogen peroxide. Colorimetrically, small quantities of rhenium are determined by the formation of a yellow complex by the addition of $SnCl_2$ and KSCN. The yellow complex is extracted with ether and the rhenium estimated by means of coloured standards made up in ether.

Rhenium is selectively extracted down to 0.4 μg per ml at pH 3.5–5.0 by methyl violet, when the solution contains forty-fold quantities of molybdenum. The absorption of the extract is measured at 330 mμ or at 600 mμ (PILIPENKO and OBOLONCHIK, 1959)

Using 8-mercaptoquinoline (BARKOVSKII and LOBANOVA, 1959) a chelate of rhenium, which is soluble in polar organic solvents, can be used for absorptiometric determinations. The concentrated test solution of per-rhenate is taken up in a solution of 37 ml of concentrated hydrochloric acid and 5 ml of 6% 8-mercaptoquinoline in concentrated hydrochloric acid. It is then boiled on a water bath for about 3 min and then cooled rapidly. The cold solution is then extracted with 5–10 ml of chloroform. The extract is shaken with 22 ml of 9 *M* HCl for 2 min. A further 2–5 ml of 8-mercaptoquinoline solution is then added and the extract shaken again. The chloroform layer is then separated and dried over calcium chloride. Absorption is measured at 438 mμ .

RHODIUM (Rh)

At.wt. = 102.91; s.g. = 12.5; m.p. = 1,955°C

Occurrence

So far, it has only been found in platinum ores.

Detection

To detect small quantities of rhodium in the presence of other metals add sodium hypochlorite and evaporate the solution nearly to dryness. A yellow precipitate is formed which is soluble in excess acetic acid and yields a solution which changes to an orange-yellow colour on long agitation. A grey precipitate eventually settles out and the solution turns sky-blue.

Estimation

A solution containing rhodium is treated with zinc and the residue washed with hot water acidulated with hydrochloric acid. The residue is then finally cleaned with dilute aqua regia and the black metallic rhodium filtered, dried and ignited in hydrogen. It is then cooled and weighed as metallic rhodium. Rhodium is also conveniently precipitated by magnesium from solutions of its salts. The precipitate is then washed with dilute sulphuric acid to remove excess magnesium and dried and ignited in a current of hydrogen. It is cooled in a current of carbon dioxide and weighed as metallic rhodium.

Colorimetrically, 2-mercapto-4,5-dimethylthiazole has been found to be an excellent reagent for rhodium (RYAN, 1950; WESTLAND and BEAMISH, 1956). To a 40 ml solution containing 50–80 μg of rhodium add 10 ml of hydrochloric acid. Add the reagent as a 0.5% solution in 50% ethyl alcohol and boil for 1 h with an occasional topping up with water. Cool and dilute with water to 100 ml and compare the colour with standardized solutions (1 ml of reagent represents 0.1 mg of rhodium).

RUBIDIUM (Rb) and CAESIUM (Cs)

Rb: at.wt. = 85.44; s.g. = 1.53; m.p. = 38.5°C; b.p. = 700°C

Cs: at.wt. = 132.91; s.g. = 1.90; m.p. = 26°C; b.p. = 670°C

Occurrence

Rubidium is mainly found in lepidolite and in mineral springs. It also occurs as a trace element in leucite, spodumene, triphylite, petallite, carnallite, mica and orthoclase.

Caesium, which is a rare metal, is found in pollucite ($H_2Cs_4Al_4(SiO_3)_9$) and is occasionally found in lepidolite and beryl.

Detection

Rubidium and caesium are usually detected spectroscopically.

Estimation

Nearly all the methods used for potassium can be used for rubidium and caesium. Only a few specific methods exist for the estimation of caesium, whilst quite a number are available for rubidium. Gravimetrically, or colorimetrically, $K_2Bi_2I_9$ can be used for caesium. Treat the solid chloride with a small amount of acetic acid, or water, and add the reagent consisting of 5 g BiO_3 plus 17 g KI boiled in 50 ml acetic acid. Filter and weigh as $Cs_3Bi_2I_9$. No interference occurs with Pb, Na, K, Mg, Li, Ca, Fe, Al, NH, SO_2 or SO_4 (PLYUSHCHEV and KORSHUNOV, 1955). A more accurate determination of the precipitate can be obtained colorimetrically by the use of dithizone (ISHIBASHI and HARA, 1953).

RUTHENIUM (Ru)

At. wt. = 101.7; s.g. = 12; m.p. = 2,450°C

Occurrence

Found as laurite (Ru_2S_3) and as an alloy in platinum ores.

Detection

Potassium hydroxide precipitates a black hydroxide which is

readily soluble in hydrochloric acid. Hydrogen sulphide slowly yields a brown precipitate, and ammonium sulphide a brownish-black precipitate. Zinc precipitates metallic ruthenium, causing the solution to rapidly turn blue. Silver nitrate yields a rose-red precipitate, whilst mercurous nitrate produces a bright-blue precipitate. Zinc chloride yields a bright-yellow precipitate which darkens on standing.

Estimation

The quantitative determination of ruthenium is usually accomplished by distilling it from hot sulphuric acid solutions in a stream of moist hydrochloric acid. The distillate is then evaporated to a most condition and to this 10 ml of hydrochloric acid is then added. The covered beaker is then warmed on a water-bath for at least 1 hour. A further 50 ml of distilled water is then added and the solution boiled and filtered. The precipitate is washed with 1:99 hydrochloric acid. The filtrate is diluted to 200 ml and boiled. When boiling, add 10% sodium carbonate until a precipitate begins to appear. At this point, add bromocresol purple and adjust to pH6 (blue) with the reagent. Boil for about 6 minutes and filter. Wash the precipitate with hot 1% ammonium sulphate until no chlorine remains and then with cold 2.5% ammonium sulphate three or four times. Transfer the filter paper to a porcelain crucible and dry below 100°C. Heat the contents slowly in air and then in hydrogen. Allow to cool in hydrogen. Extract the precipitate on the paper with water and ignite and weigh as ruthenium metal (AYRES and YOUNG, 1950).

Colorimetrically, ruthenium at dilutions of 25 μg in 10 ml of 4*N* hydrochloric acid, responds to yield a reddish colour with 5ml of 1% phenylthiosemicarbazide in ethyl alcohol. Measure the colour with a blue-green filter (KOLTHOFF and SANDELL, 1952, p.481).

SCANDIUM (Sc)

At.wt. = 45.10

Occurrence

In minute quantities it is widely distributed but mainly concentrated in monazite, micas, wolframite, auxenite, keilbanite and wiikite. Thortveitite—$(Sc,Y)_2Si_2O_7$—is possibly the only mineral which contains scandium as an essential element.

Detection

Scandium is usually detected by spectrographic methods.

Estimation

ALIMARIN and YUNG-SCHAING, (1959, 1961) developed the following method for scandium. Add 30 ml of a 20% ammonium acetate solution to 100 ml of the test solution, together with a few ml of 1:1 acetic acid, to bring the pH to about 5.4 after precipitation. Add a small excess of 0.4–0.5 g of n-benzoyl-n-phenylhydroxylamine in 100 ml of boiling water. Allow to stand, then filter and wash the precipitate with 50 ml of hot water. Ignite the precipitate at 600°C and weigh as Sc_2O_3 (mg of ppt. × 0.6527 = mg of Sc). To obtain a measure of the rare earths, add ammonia to the filtrate and precipitate the hydroxides on a fine filter. Dry and ignite to a constant weight at 800–900°C (mg of residue × 0.86 = mg of rare earth elements).

A titrimetric determination in EDTA provides another reliable method (CHENG and WILLIAMS, 1955). Add sodium hydroxide to the test solution containing 3–10 mg of scandium until it becomes turbid. Adjust to pH 2.5 with acetic acid and add an excess of 0.01*M* EDTA solution with six drops of 0.01% solution of 1-(2-pyridylazo)-2-naphtol in ethyl alcohol as an indicator. Titrate with 0.01*M* copper sulphate from yellow to red.

SELENIUM (Se) and TELLURIUM (Te)

Se: at.wt. = 78.96; s.g. = 4.26–4.82; b.p. = 690°C
Te: at.wt. = 127.61; s.g. = 6.24; m.p. = 45.2;
b.p. = 1,390°C

Occurrence

Selenium and tellurium occur in native sulphur and in all pyrite ores. Selenium also occurs native in such places as cavities in the lavas of Vesuvius. The principal minerals are clausthalite (PbSe), berzeliamite (Cu_2Se), lehrbachite (PbSe.HgSe), onofrith (HgSe.4HgS), encairite ($CuSe.Ag_2Se$), crookesite—(CuTlAg)Se, tetradymite (Bi_2Te_2S), hessite (Ag_2Te), calaverite ($AuTe_2$), sylvanite ($AuAgTe_4$), nagyagite—$(Pb, Au)_{16}Sb_3(S_1Te)_{24}$, petzite ($Ag_3AuTe_2$), coloradolite (HgTe), and telluride or tellurium ochre (Bi_2Te_3).

Detection

Solid, or dissolved, thiourea precipitates a characteristic red powder but nitrates and large amounts of copper interfere. Tellurium and bismuth yield yellow precipitates. Selenius acid or alkali selonite with 1-8-naphthylene-diamine in acetic acid produces a brown precipitate. This reaction is specific for selenites.

Alkali stannites reduce tellurite and tellurates to the black metal and they are without effect on analogous selenium compounds.

Estimation

Potassium iodide provides an excellent titrimetric method for selenites (WILLIS, 1942). To 20 ml of an approximately 0.05*N* selenite solution add 150 ml of water with 15 ml of 1% starch solution, 10 ml of 1*M* potassium iodide and 10 ml of 5*M* hydrochloric acid. The reaction mixture is allowed to stand for 15 min to complete the reduction. Neutralize the mixture to pH 7–7.4 with sodium bicarbonate and titrate with standardized 0.1*N* hydrazine hydrochloride to a sodium starch glycolate end-point, which is marked by the disappearance of the blue colour. No interference takes place with Te, Pb, Cu, and most other metals.

The gravimetric method for tellurium using N_2H_4 is the oldest but still the most widely used. Boil 50 ml of a 3*N* hydrochloric acid solution of tellurium and then add 15 ml of a standard SO_2 solution with 10 ml of 15% $N_2H_4.HCl$. Add a further 25 ml of saturated SO_2 solution and boil for 15 min. Filter the precipitate on a Gooch crucible and wash with hot water followed by ethyl alcohol. Dry at 120–130°C and weigh as tellurium metal. Some tellurium is usually oxidized but this can be avoided by heating at 105°C for 45 min in carbon dioxide (SEATH and BEAMISH, 1937). For small amounts of tellurium, use potassium iodide as a colorimetric agent (JOHNSON and KWAN, 1951). Add 5 ml of 2*N* hydrochloric acid to 30 ml of the tellurium iodide and dilute the mixture to 50 ml and immediately measure the red-yellow colour at 335 mμ.

SILVER (Ag)

At.wt. = 107.88; s.g. = 10.50–10.57; m.p. = 960.5°C

Occurrence

As metallic silver, but more commonly in combination as argenite (Ag_2S), hessite (Ag_2Te), proustite (Ag_3AsS_3) and pyrargyrite (Ag_3SbS_3). Horn silver or cerargyrite (AgCl) resembles the bromide and iodide of silver in possessing a characteristic horny appearance. Silver also occurs as a trace element in sea water.

Detection

One tenth of a milligram of silver chloride in 200 ml acid solution produces a perceptible opalescence. Bright-red silver chromate is produced by an ammoniacal silver solution with potassium chromate and acetic acid when spotted on to filter paper. Silver is superficially attacked by hydrochloric acid, and the action of alkali hydrates and carbonates is negligible. In the absence of oxidizing agents dilute H_2SO_4 has practically no action and only begins to attack silver at a concentration of 75%. Hot dilute and concentrated HNO_3 attacks silver readily, but the presence of a soluble chloride, iodide or bromide in the solvent retards solution. As a nitrate, silver is precipitated quantitatively as a chloride, but the presence of lead, uni-valent mercury, copper and thallium causes interference. Preliminary oxidation prevents the co-precipitation of Hg, Cu and Tl. $PbCl_2$ is soluble in hot water whilst AgCl is only slightly soluble.

Estimation

Silver sulphide is quantitatively precipitated by H_2S in acid or alkaline solutions. The presence of cyanides and thiosulphates in the solution interferes by retaining any silver chloride in solution. Gravimetrically, silver can also be determined as an iodide or bromide but being more light sensitive and less stable than the chloride, the latter is more commonly employed. Heat the solution to boiling and from a burette add 5 ml of dilute HCl (1 vol. of conc. HCl in 5 vols. of H_2O contains 0.074 g of HCl which is equivalent to 0.219 g of Ag). This will precipitate over 1 g of silver, but the excess of acid employed ensures complete precipitation of AgCl. (AgCl ×

0.7527 gives the weight of Ag in the salt). REMINGTON and MOYER (1937) recommend the use of benzo-1,2,3-triazole as a precipitant for silver. In an ammoniacal solution containing EDTA, this reagent precipitates silver specifically and when dried out at 110°C it corresponds to the formula $AgC_6H_4N_3$ (wt. of ppt. × 0.4774 = wt. of Ag). For the titrimetric determination of this precipitate, dissolve the precipitate in 10 ml of 1:1 nitric acid in a 250 ml beaker. Dilute the solution to about 50 ml with water and neutralize the acid with ammonia. Dissolve the so-called precipitate by neutralization with excess 0.05 *M* cyanide. Back-titrate with silver nitrate using 1 ml of potassium iodide as indicator (BELCHER and WILSON, 1964, p.114).

STRONTIUM (Sr)

At.wt. = 87.63; s.g. = 2.6; m.p. = 800°C

Occurrence

The chief minerals are coelestine ($SrSO_4$) and strontianite ($SrCO_3$), but it is also found in calcite, aragonite and baryte. In brewsterite it occurs with barium as an aluminium silicate–$Al_2O_3.H_4(BaSr)O_3.(SiO_2)_6.3H_2O$.

Detection

Strontium and barium salts react with a neutral solution of sodium rhodizonate to form brown precipitates (FEIGL, 1924). Strontium chloride produces a crimson flame test.

Estimation

Gravimetrically, strontium is determined by adding a ten-fold excess of dilute sulphuric acid to a neutral solution (BALL-

CZO and DOPPLER, 1956). Add an equal volume of ethyl alcohol and then allow to stand overnight. Filter the precipitate on a Gooch crucible and wash with 50% ethyl alcohol, containing some sulphuric acid. Finally, wash with ethyl alcohol, dry and ignite the residue at about 100–300°C and weigh as strontium sulphate ($SrSO_4 \times 0.4770 = Sr$).

Titrimetric determinations can be carried out after precipitation with oxalate, or after fusion with boric acid. Colorimetrically, chloranilic acid is used for the determination of 0.2 mg of Sr per ml. The solution is adjusted to pH 5.7 and 5 ml transferred to a centrifuge tube with 5 ml of 0.05% chloranilic acid. Centrifuge and store the solution overnight and measure at 530mμ (LUCCHESI, 1954).

SULPHUR (S)

At.wt. = 32.06; s.g. = 2.07; m.p. = 112.8°C;
b.p. = 446.6°C

Occurrence

Occurs free as native sulphur, but mainly as mineral sulphides and sulphates.

Detection

Sodium nitroprusside—$Na_2(Fe(CN)_5NO)$—in alkaline solutions gives a violet-red colouration with sulphides.

Estimation

Gravimetrically, the usual method is to precipitate as barium sulphate (HILLEBRAND and LUNDELL, 1929, p.570). To a neutral solution of 25 mg of sulphur per 100 ml, add 1 ml of hydrochloric acid and boil. Pour into a boiling solution of

10% barium chloride and place on a steam bath for at least 6 h. Filter through paper and wash with hot water until the filtrate is chlorine-free. Ignite below 900°C and weigh as barium sulphate.

THALLIUM (Tl)

At.wt. = 204.39; s.g. = 11.85; m.p. = 303°C

Occurrence

Occurs mainly in small quantities in association with alkali metals, zinc, iron and lead, and also in crookesite—$(Cu, Tl, Ag)_2Se$—and lorandite ($TlAsS_2$).

Detection

Bright-yellow iodide is precipitated by potassium iodide but this is affected by the presence of mercury, silver and lead. Excess potassium iodide re-dissolves any mercury precipitates. With silver and lead digest with sodium thiosulphate after precipitation with potassium iodide; the thallous iodide remains unchanged. Thallic compounds, converted to $Tl(OH)_3$, are detected by the oxidation of benzidine to benzidene blue.

Estimation

Potassium chromate, applied gravimetrically, provides one of the most accurate procedures known (CHRETIEN and LONGI, 1944), for 0.1 g Tl in 100 ml of solution. To 100 ml of the solution, add 3 ml of 2:1 ammonium hydroxide and heat to 70–80°C. Add 10% K_2CrO_4 in excess, cool and allow to stand overnight. Filter on a Gooch crucible. Wash the precipitate with 1% reagent solution and then with 50% ethyl alcohol. Dry at 120–130°C and weigh as Tl_2CrO_4. Interferences from Ag, Hg and Cu are removed by the addition of KCN. A 50% solution of sulpho salicylic acid suppresses Ga, In, Al, Fe and Cu. Tartaric acid prevents interference from Zn, Cd, Ni, Co, and Mo, or N, if more NH_4OH is added.

Rhodamine-B is a well established colorimetric reagent for 1–10 μg solutions of thallium in sulphuric acid. To 0.5 ml of the test solution, add 5 ml of 2*N* hydrochloric acid with 1 ml of bromine water and heat to remove bromine. When cool, dilute with 2*N* hydrochloric acid to 10 ml and add 1 ml of 2 % rhodamine-B together with 10 ml of C_6H_6. Shake for 1 min and centrifuge. Measure at 560 mμ (ONISHI, 1957).

THORIUM (Th)

At.wt. = 232.12; s.g. = 11.0–12.2; m.p. = 1,450°C

Occurrence

Mainly in monazite, which is essentially an orthophosphate of cerium-earths containing up to 10% of thorium oxide. The silicate is called thorite and the uranate is known as thorianite.

Detection

A sensitive test for thorium is the precipitation of thorium iodate in a nitric acid solution by means of potassium iodate. Spectrum analysis and radioactive methods are also useful for detecting thorium.

Estimation

Thorium is precipitated quantitatively between pH 4.5 and 5.5 by n-benzoyl-n-phenylhydroxylamine. Cerium is retained in solution at pH 5 (SINHA and SHOME, 1959). To a boiling test solution, add 10 ml of 5% hydroxylamine hydrochloride, then cool and dilute to 250 ml. Add a further 10 ml of hydroxylamine hydrochloride, followed by 10% ammonium acetate solution, to adjust the pH to 5.0. Add 0.3–0.4 g of n-benzoyl-n-phenylhydroxylamine in 10–15 ml of ethanol and stir occasionally for 15 min. Filter and wash the precipitate with a freshly prepared 0.1% solution of hydroxylamine hydrochloride followed with distilled water. Dry, ignite and weigh the precipitate as thorium oxide (mg of $ThO_2 \times 0.8788$ = mg of Th).

Thoron is an excellent colorimetric reagent for solutions of 5–80 μg of thorium in hydrochloric acid. Adjust the pH to 0.5–0.6 and add 1 ml of 0.1% thoron. Dilute with water to 10 ml and measure the red complex at 545 mμ (BANKS and EDWARDS, 1955).

TIN (Sn)

At.wt. = 118.7; s.g. = 6.56; m.p. = 232°C; b.p. = 2,275°C

Occurrence

Occurs native to a limited extent, but usually as oxides and sulphides. Cassiterite (SnO_2) is the principal source of the metal, but is also sometimes derived from tin pyrites or stannine ($Cu_2S.FeS.SnS_2$).

Detection

From dilute acid solutions, hydrogen sulphide precipitates brown or yellow SnS_2, which is soluble in yellow ammonium sulphide. This distinguishes it from the sulphides of Hg, Pb, Bi, Cu, and Cd. Mercuric chloride is reduced by stannous chloride to white Hg_2Cl_2 or grey Hg and $HgCl_2$. This reaction is not particularly sensitive but if carried out in alkaline solution the whole of the mercuric chloride is reduced to black metallic mercury. Aniline is recommended, as its alkaline reaction is so small that the presence of antimony salts does not interfere with the reaction. Acid solutions of tin produce a red circle when a drop is placed on a slightly moist filter paper which has been saturated with cacotheline.

Estimation

Tin is separated from aluminium, iron and chromium by its insolubility in dilute hydrochloric acid, and it is isolated from lead, mercury, copper, cadmium and bismuth by the solubility of its sulphide in yellow ammonium sulphide. A rapid and accurate volumetric method of analysis of tin is based on the action of iodine in the presence of stannous chloride in excess hydrochloric acid. A solution of tin, produced by dissolving a pure sample of tin in boiling hydrochloric acid in an Erlenmeyer flask can be standardized by titration against a standard solution of iodine using starch (5 g in 1 l of water) as an indicator.

In a solution containing 1–8% of concentrated hydrochloric acid, tin is precipitated quantitatively by n-benzoyl-n-phenylhydroxylamine (RYAN and LUTWICK, 1953). To the test solution of stannic chloride add 10 ml of concentrated hydrochloric acid and dilute to 200 ml. Dropwise, whilst stirring, add 5 ml of a 1% ethanol solution of n-benzoyl-n-phenylhydroxylamine for every 10 mg of tin present plus 8 ml in excess. Stand in an ice bath for at least 4 h, then filter and wash with a little ice-water. Dry at 110°C (mg of ppt. × 0.1927 = mg of tin).

Dithiol is an excellent colorimetric reagent for tin. It forms red precipitates with tin, bismuth and molybdenum (MATSUURA, 1953). The acidity ranges of the solution are 0.4–0.8N H_2SO_4 or 0.4–0.6N HCl. Add one drop of thioglycollic acid plus 0.5 ml of Santomerse S (1:19) and 0.5 ml of 0.1 g dithiol in 50 ml of 1% NaOH containing 0.3–0.5 ml of thioglycollic acid. For 5 min, heat to 50°C, then add 0.5 ml of 1:3 Santomerse S, cool and dilute to 10 ml with water. At a wavelength of 350 mμ tin is detectable from 0.28 to 6 p.p.m.

TITANIUM (Ti)

At.wt. = 47.90; s.g. = 4.5; m.p. = 1,795 ± 15°C

Occurrence

The principal minerals are ilmenite ($FeTiO_3$), titanite ($CaTiSiO_5$), perovskite ($CaTiO_3$), rutile (TiO_2), anatase (TiO_2) and brookite (TiO_2).

Detection

A finely powdered, pin-head sized sample, mixed with three to four times its amount of potassium bisulphate, is melted on a platinum lid and when cool transferred to a watch glass. Mix with a little water and deposit the suspension on a filter paper saturated with chromotropic acid. Titanium yields a red to brown diffusion ring.

Estimation

The following method has been devised by KAIMAL and SHOME (1962). Dilute the test solution containing about 0.1 g of titanium to 400 ml and neutralize with ammonia. Add 5 ml of concentrated hydrochloric acid. Precipitate the titanium as a complex by the slow addition of a 10% ethanolic solution of n-benzoyl-n-hydroxylamine. Stir continuously. Allow the precipitate to stand for 1 h. With occasional stirring the ethanol content is reduced to something less than 5%. Filter the complex on paper and wash with a 0.1% aqueous solution of the complexing reagent. Ignite carefully in a platinum crucible and weigh as titanium oxide (mg of ppt. × 0.5995 = mg of Ti). Cupferron has also been used as a suitable reagent (BANDISCH, 1909). More recently, the use of 3-hydroxyl-1-p-chlorophenyl-3-triazine has been recommended by SOGANI and BHATTACHARYA (1956).

Until recently, no simple method had been developed to determine titanium, iron and aluminium in a mixture of all three. When it was found that salicylic acid, or p-amino-salicylic acid, masks aluminium in an acid medium the following procedure was developed (PRIBIL and VESELY, 1963). The acid test solution is diluted to 100 ml and 10 ml of 10% sodium salicylate solution is added along with 20 ml of 20% triethanolamine solution. Two drops of phenolphthalein are added as an indicator and the whole treated with 2*M* sodium hydroxide until the solution turns intense violet, but not to red. Boil for 1 min and filter off the precipitate. Wash the precipitate five times with hot 1% triethanolamine solution and then twice with hot water. Dissolve the precipitate in 25–30 ml of 1:3 nitric acid and bring the pH to between 1 and 2 by means of pH indicator paper. Stir the solution with 10 ml of hydrogen peroxide and cool to below 20°C. Add a measured amount of 0.05*M* EDTA solution in excess of the estimated quantity of titanium, together with a few drops of xylenol orange indicator. Back-titrate with a 0.05 *M* solution of bismuth nitrate solution to the intense red end-point.

TUNGSTEN (W)

At.wt. = 184.0; s.g. = 19.3; m.p. = 3,370°C

Occurrence

The principal minerals are ferberite ($FeWO_4$), hubnerite ($MnWO_4$), wolframite–(Fe,Mn)WO_4, scheelite ($CaWO_4$), cupro-scheelite–$(CaCu)WO_4$, tungstenite (WS_2) and stolzite ($PbWO_4$).

Detection

Stannous chloride reduces tungstenates to blue tungsten oxide.

Estimation

On a quantitative basis, easily filtered precipitates of tungsten complexes are formed by the use of tri-n-butylamine. The reagent must be purified by double distillation. To 50 ml of a test solution, containing 10–200 mg of tungsten as sodium tungstate, add about 85 mg of disodium hydrogen phosphate. When this has dissolved slowly, add hydrochloric acid and stir until the pH is about 2, then add 4 ml in excess. With water, dilute to 100 ml and bring the solution to the boil. Stir in 10 ml of tri-n-butylamine, still holding the temperature below boiling point, until the precipitate has coagulated. Cool in ice water and filter after 1 h through a sintered glass crucible. Wash the precipitate with a small amount of 0.1% (w/v) solution of tri-n-butylamine in 0.1*N* hydrochloric acid. Finally wash with a little cold water. Heat the crucible for 2 h at 210°C. Cool in a desiccator and weigh as the tungsto-phosphate–$\{(C_4H_9)_3NH_3\}PW_{12}O_{40}$ (mg of ppt. × 0.6424 = mg of tungsten).

URANIUM (U)

At.wt. = 238.07; s.g. = 18.7; m.p. = 1,850°C

Occurrence

The chief minerals are uranic ochre (U_2O_2), pitchblende

(U_3O_4), torbenite ($CuO.2UO_3.P_2O_5.8H_2O$), autunite ($CaO.2UO_3.P_2O_5.8H_2O$) and carnotite, which is a vanadate of uranium and potassium.

Detection

Neutral, or acetic acid solutions of uranyl salts give a red-brown precipitate with potassium ferrocyanide. The test is specific in the absence of ferric and copper salts.

Estimation

DAS and SHONE (1962) have developed the following method for uranium. Dilute the test solution to 200 ml and heat to 40–50°C. Slowly add 10 ml of an ethanol solution, consisting of 0.3 g of n-benzoyl-n-phenylhydroxylamine in 95% ethanol. Dropwise, add 2*N* ammonia solution until the pH is adjusted to 5.4. Allow to stand on a water bath at 40–45°C for nearly 2 h, stirring from time to time. Maintain the pH at 5.4, by adding more ammonia if necessary. Filter and wash the precipitate. Dry and ignite the precipitate to oxide. Dissolve the oxide in concentrated nitric acid. Evaporate to dryness and re-ignite to oxide and weigh as U_3O_8 (mg of ppt. × 0.8480 = mg of U).

VANADIUM (V)

At.wt. = 50.95; s.g. = 6.025; m.p. = 1,720°C

Occurrence

The principal minerals are vanadinite—$(PbCl)Pb_4(VO_4)_3$, and patronite, which is a sulphide of vanadium associated with pyrites and carbonaceous matter. Roscoelite is a vanadium mica. Vanadium occurs to the extent of 19–20% V_2O_5 in carnotite.

Detection

Vanadium solutions, containing sulphuric acid, turn red-brown to blood-red on the addition of hydrogen peroxide.

Estimation

Tannin has been used for gravimetric determinations by adjusting the solution to a slightly acidic level with ammonium hydroxide and acetic acid. Heat and add 5–10 g of ammonium chloride, 5–10 g of ammonium acetate, then add a tenfold quantity of tannin. Filter and wash with 2% ammonium nitrate, containing a little tannin. Ignite the precipitate and weigh as V_2O_5. Silver nitrate has also been used gravimetrically (KROUPA, 1944) for the estimation of vanadium.

"Redoxal" (3,3-dimethoxybenzidine-N,N-di-o-benzoic acid) is a highly sensitive indicator in alkaline solutions (FURMAN and MUSTAFIN, 1960) of vanadium. Fuse a powdered sample of rock or mineral (1 g) in 2 g of potassium carbonate and 4 g of sodium carbonate. Dissolve the fused mass in hot water. Filter and wash, then dilute the filtrate to 100 ml. Acidify a 10–20 ml aliquot with sulphuric acid, add 10 ml of a solution of sodium potassium tartrate and boil for a few minutes. With three drops of "redoxal", add 10 ml of 40% sodium hydroxide solution and one drop of a 1% copper sulphate solution. With a 0.01*N* potassium ferricyanide solution titrate to a red coloured end-point.

ZINC (Zn)

At.wt. = 65.38; s.g. = 7.133; m.p. = 419°C; b.p. = 905°C

Occurrence

The principal minerals are zincite (ZnO), franklinite—(Fe,Zn, Mn)O.$(Fe,Mn)_2O_3$, blende (ZnS), calamine ($ZnCO_3$), willemite ($2ZnO.SiO_2$), hemimorphite ($2ZnO.SiO_2.H_2O$), and goslarite ($ZnSO_4.7H_2O$).

Detection

Fifteen drops of potassium ferricyanide mixed with ten drops of diethylaniline in sulphuric or phosphoric acid, produces a red or dark-brown cloudiness or precipitate with a zinc solution. A purple to red colour is produced by zinc salts with dithizone. Rinmann's test (BENEDETTI-PICHLER, 1932) is obtained by heating oxides, or salts which are easily converted to oxides, with crystals of cobalt oxide. A distinctive green ash, or powder is produced on ignition.

Estimation

Gravimetric separations with hydrogen sulphide are well established for acidic solutions (HILLEBRAND and LUNDELL, 1929, p.328). To the solution, placed in an Erlenmeyer flask, add ammonium hydroxide until precipitation takes place. At this stage, add 25 ml of 20% citric acid and neutralize with ammonium hydroxide, using methyl orange as an indicator. Add 25 ml of a mixture of 200 ml of 23.6*N* formic acid, 250 g ammonium sulphate and 30 ml of ammonium nitrite. Dilute to 200 ml with water, to adjust the pH between 2 and 3. Warm the solution and saturate with hydrogen sulphide. Filter and wash the precipitate with 1*N* formic acid and ash below 500°C. Finally heat to 950–1,000°C and weigh as ZnO.

The acetylhydrazone of salicylaldehyde has been developed as a convenient fluorescent indicator for the EDTA titration of zinc (HOLZBECHER, 1958). To 10–20 ml of the test solution containing 4–170 mg of zinc, add 10 ml of a sodium acetate buffer (pH4.5) and 2 ml of the indicator as a 0.1% solution in ethanol. Titrate, with constant shaking, under ultraviolet light until the intense blue fluorescence disappears.

Alizarin fluorine blue has been established as a complexi-metric indicator for zinc and lead, when screened with xylene cyanol FF (LEONARD and WEST, 1960). The indicator consists of 500 mg of solid alizarin fluorine blue suspended in water to which two drops of ammonia and seven to eight drops of 20% ammonium acetate are added and the whole diluted to 100 ml. This complexing reagent is stable for considerable periods of time. The xylene cyanol FF screen is made up as a 0.1% aqueous solution. When titrating for zinc or lead, add five drops of the indicator and two to three drops of the screen, to the test solution. To this, add 3 ml of a sodium acetate buffer (pH 4.3) and titrate with standard EDTA to a red-to-green end-point.

ZIRCONIUM (Zr) and HAFNIUM (Hf)

Zr: at.wt. = 91.22; s.g. = 6.4; m.p. = 1,700°C
Hf: at.wt. = 178.6; s.g. between 12.1 and 13.3; m.p. = 2,200°C

Occurrence

The main sources of zirconium are baddeleyite (mainly zirconium oxide) and zircon (a zirconium orthosilicate). Almost all zirconium compounds contain hafnium.

Detection

Turmeric paper, moistened with HCl or H_2SO_4 and dried, produces a reddish brown spot-test for zirconium solutions. Borates and titanium give similar colourations, but if the test solution is reduced by the addition of zinc foil the test is specific for zirconium.

Estimation

Mandelic acid has proved to be a good gravimetric reagent for zirconium (HAHN and BAGINSKI, 1956). To the test solution, add sufficient concentrated hydrochloric acid to yield a final acidity of 5–6*N* when diluted to 50 ml. Heat to 85–90°C and dropwise add 25 ml of 15% , 1*M* mandelic acid solution, with constant stirring for about 45 min. Allow to cool and filter through a sintered-glass crucible. Wash the precipitate with a saturated solution of zirconium mandelate, followed by 10 ml portions of 95% ethanol and finally with two or three 10 ml volumes of ethyl ether. Dry at 110°C to a constant weight and weight as zirconium tetra mandelate (wt. of ppt. × 0.1311 = mg of Zr).

ALIMARIN and YUNG-SCHAING (1959) have shown that n-benzoyl-n-phenylhydroxylamine is a sensitive gravimetric reagent for zirconium. Hafnium in the presence of zirconium proved difficult, until Pulnektov and his co-workers (KONONENKO et al., 1960) used arsenazo as a reagent. Hafnium and zirconium form coloured complexes with arsenazo, which respond differentially to varying concentrations of hydrochloric acid. Dissolve 15 mg of mixed oxides of hafnium and zirconium in 2–3 ml of hydrofluoric acid. Add 2 ml of concentrated perchloric acid and 10 ml of 4% boric acid and evaporate to dryness. Dissolve the residue in 250 ml of 1*M* hydrochloric acid. Dilute two 5 ml aliquots with 30 ml of water. To one add 1.88 ml of 4*M* hydrochloric acid and to the other, 11.25 ml of the acid. To each add 2.5 ml of 1% gelatin and 5 ml of a 1% aqueous solution of arsenazo. Dilute both to 50 ml and measure their optical density at 570 mμ in a 1 cm cuvette against a reagent blank. Calculate the ratio of the extraction and read off the hafnium content from a calibration curve.

CHAPTER 5

Emission Spectrography

INTRODUCTION

Spectrography is an analytical technique whereby the concentration of a number of elements in a sample may be determined simultaneously. The method is highly sensitive, of reasonable accuracy and provides positive qualitative identification of components in addition to quantitative estimation. It can be applied to about 70 elements without recourse to special procedures and, above all, complete analyses can be performed rapidly.

Because of these points the technique has long appealed to those concerned with the analysis of geochemical and allied samples and some of the most outstanding contributions to the subject as a whole have been made by these workers. It must be emphasized now, however, that the optical spectrograph on its own is not by any means the complete answer to all analytical problems. Whilst it is a very powerful tool it is most efficiently used in conjunction with other methods.

THE ORIGIN OF SPECTRAL LINES

When an atom receives energy some of its outer electrons may move into an orbit further away from the nucleus. The return of an electron to its normal orbit is accompanied by the emission of energy in the form of radiation, the wavelength of which is determined by the energy difference between the two levels. When all electrons are in their normal orbits the atom is said to be in the "ground state". When at a given instant one or more electrons exist at a higher energy level the atom is in an "excited state". Thus if E_0 and E_1 represent the energy levels of an electron when the atom is in the ground state and an excited state respectively, the wavelength of the light emitted as the electron falls back from the E_1 level to the E_0 level is defined by the expression:

$$E_1 - E_0 = h\nu$$

where h = Planck's constant and ν = frequency of emitted radiation.

Frequency and wavelength are related as follows: $c = \nu\lambda$, where c = velocity of light.

The return of an electron from a higher to the normal energy level is not always accomplished in a single step, a number of intermediate orbits may be involved and each jump is accompanied by the emission of light of the appropiate wavelength. In addition, a number of electrons may be involved so that the result of imparting energy to an atom is the production of radiation at a large number of discrete wavelengths, the greater the number of electrons and energy levels the more complex the spectrum. Emitted radiation at one wavelength is known as a spectrum line and the amount of energy needed to excite any particular line is the "excitation potential".

Many methods of exciting spectra supply enough energy to ionize some atoms, that occurs when an electron is removed completely from the influence of the nucleus. The energy required is the "ionization potential". The remaining electrons, excited as in a neutral atom then give rise to a complete new series of lines, the distribution of which resembles the spectrum of the preceding element in the periodic table. These are known as ion lines.

Fig.25. Arc and spark spectra of iron. An example of the increase in the number of lines produced by the increased energy of the source (d.c.).

It can be seen, therefore, that the total spectrum of an element may be of considerable complexity, the degree depending on the element itself and on the energy available in the source. Fig.25 is a portion of the iron spectrum excited by two methods and illustrates the number of lines produced as the supply of energy is increased.

SPECTROGRAPHIC EQUIPMENT

The equipment required for spectrographic analysis comprises a source unit, a spectrograph and the means of recording and evaluating the relative intensities of spectrum lines. A wide variety of instrumentation is available and the choice depends on the type of material being analysed, the number of samples to be examined and the accuracy required.

The source unit

In the source unit the sample is vaporized and atoms of the elements in the sample are excited. For geochemical specimens flames and direct current arcs have been most commonly employed though other techniques are also of considerable interest and are being increasingly used.

Flame sources

Flames of various kinds have been used to excite spectra which, because of the low available energy, are relatively simple. Because of this the rest of the equipment used to examine flame spectra can be less complex than when large numbers of lines are produced. In the hotter flames over 50 elements are excited, fewer in the cooler flames, and extensive use has been made of the method for the determination of certain elements. Flame photometers are normally used nowadays for the exploitation of this method of excitation and a fuller description of the technique will be given in another section.

Direct current arc sources

In this method of excitation an electric arc is maintained between two carbon or graphite electrodes usually placed between 4 and 10 mm apart. The sample in powder form is packed into the lower electrode which may be made either the anode or cathode. Fig.26 shows a typical arrangement and Fig.27 shows a number of different shapes which have been used for the lower electrode.

The electrodes become incandescent when the arc is running and the sample is vapourized. Various kinds of particles exist in the arc column. They may be molecules, atoms, ions and electrons, all of them moving rapidly as a result of the thermal energy imparted to them. In addition, the charged particles move under the influence of the electric field. Since the electric gradient in the main body of the arc is low, the voltage drop between the electrodes being normally between 30 and 80 V, most of the movement of the particles is thermal.

Impact of these moving particles with atoms of an element gives rise to excitation of that element and in the centre part of the arc the excitation can be considered as being entirely of thermal origin.

A special state can exist in the vicinity of the cathode. According to MEGGERS (1941) when a carbon arc is struck in air most of the voltage drop occurs very close to the cathode. Consequently, more energy is imparted to electrons in

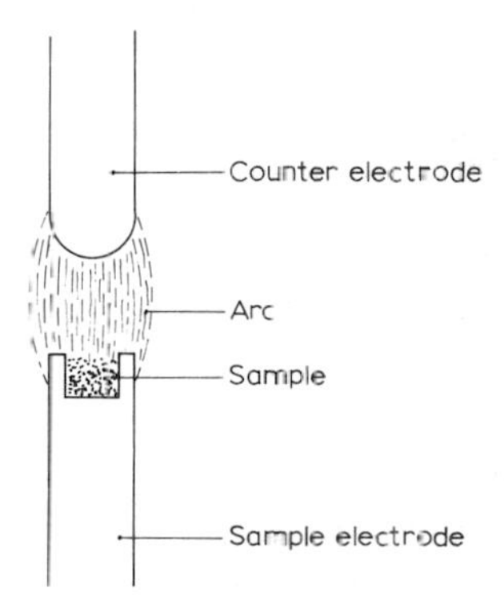

Fig. 26. Electrode arrangement for arc excitation.

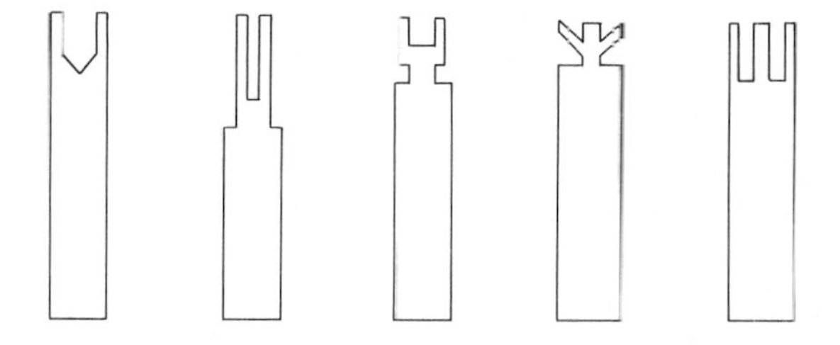

Fig.27. Electrode shapes for arc excitation.

that region than in any other part of the arc. If very small amounts of a sample are introduced, a concentration of ions exists around the cathode and because of the short life of ions this gives rise to a concentration of atoms. As a result, the emission intensity of elements from the sample is enhanced considerably in this region. It is essential that small amounts of sample are present in the arc at any time because the temperature must be kept high to ionize the elements and the potential gradient must be high to accelerate the ions to the cathode. Larger concentrations of sample reduce both temperature and voltage, particularly when elements having low ionization potentials such as the alkalis or alkaline earths are present.

This phenomenon was exploited by MANNKOPF and

PETERS (1931) in their "cathode layer" technique. The required conditions are obtained by packing the sample into a long narrow carbon electrode such as that shown in Fig.26, and thus is made the cathode. When an arc is struck between two carbon electrodes the cathode is always cooler than the anode so the small sample (about 10 mg) is volatilized relatively slowly. The gap between the electrodes is kept fairly large (about 10 mm) both to increase the volume of the arc and to maintain as high a potential difference as possible. In this way the concentration of the sample in the arc is kept low and the condition best suited to cathode layer enrichment maintained. In addition, the light entering the spectrograph is limited to that emitted from the region just around the cathode and the lower third of the arc column.

This technique is eminently suited to qualitative analysis as the sensitivity obtainable is extremely high. It is also suitable for quantitative analysis and has been used extensively for the analysis of geochemical specimens by GOLDSCHMIDT (1954) and by MITCHELL (1948).

When the sample is packed into the anode the higher temperature causes it to be volatilized much more rapidly and, as a result, it is distributed fairly evenly in the arc. In the technique of "anode excitation" a shorter arc of about 5 mm length is used and light taken from the centre of the column. Care is exercised to exclude from the spectrograph any emission from the electrodes themselves. The quantity of sample volatilized is normally much greater than when the "cathode layer" technique is used. Usually, upwards of 50 mg is packed into the electrode. This method is nowadays used extensively for the analysis of geochemical specimens, largely as a result of the work of AHRENS (1950).

Spark sources

A high voltage spark gives much higher levels of excitation energy than arcs, because the charged particles are accelerat-

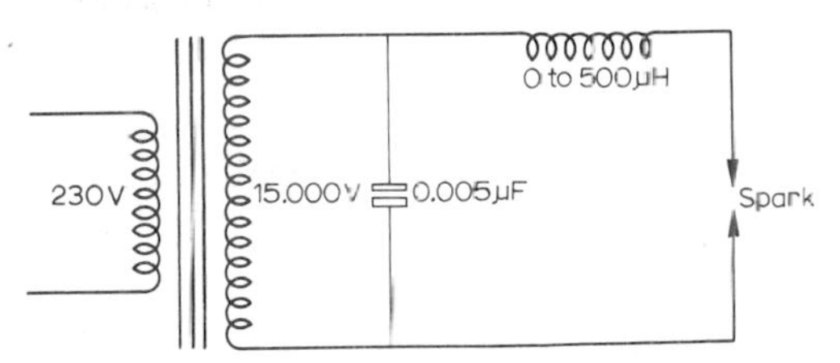

Fig.28. Circuit for high voltage spark.

ed to high velocities under the influence of large potential gradients. The high voltage spark is produced when electrodes are connected to the secondary terminals of a high voltage transformer with a condenser in parallel. A self inductance coil may be included in the lead to one of the electrodes. The secondary voltage of the transformer is usually between 8,000 V and 15,000 V. A typical circuit is shown in Fig.28.

The spectra produced by this type of high voltage spark unit are very different from those arising from an arc. They are much more complex and include both atomic and ionic lines.

Many methods have been used to introduce the sample into the spark and an extensive review of the available techniques has been published by YOUNG (1962). The most important for the examination of geochemical specimens are those utilizing solutions of the samples. Of these, three methods merit special attention.

Porous cup technique

This technique was devised by FELDMAN (1949) and has been used extensively for metal analysis, although it has also been applied to the analysis of soils by SCOTT and URE (1958). The porous cup is made by drilling a hole down the centre of a carbon or graphite rod as shown in Fig.29. The exact dimension of the hole depends on the sample size required and some workers prefer to finish the bottom with a

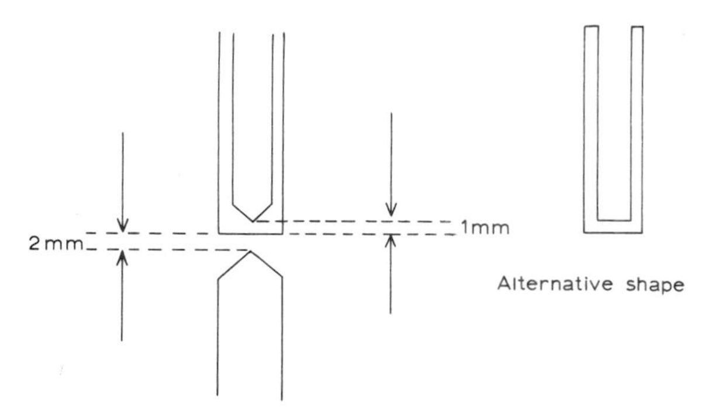

Fig.29. Porous cup electrodes.

square ended drill (see Fig.29) but the first type shown in Fig.29 has been found satisfactory for most purposes and is easier to produce. The lower electrode comprises a pointed graphite rod and a 2 mm gap between the two is normally used. The base of the cup electrode is made porous by heating to incandescence in a flame furnace or arc. The solution sample is introduced with a long narrow pipette inserted to the bottom of the bore. It is important to ensure that no air bubbles are trapped in the cup. The solution seeps slowly through the porous end and presents a continuously replenished solution surface to the spark generated between the two electrodes.

Rotating disc technique

This equipment which was described by HARVEY (1950) comprises a disc of graphite $^1/_2$ inch diameter and $^1/_8$ inch thick which is rotated slowly in such a way that it passes through the sample solution contained in a shallow boat. The upper electrode is positioned over the disc so that a spark passes from it to the top of the disc (see Fig.30).

For some types of sample this apparatus is superior to the porous cup, and it is especially suitable for emulsions, oils and other viscous solutions. On the other hand, the quantity of so-

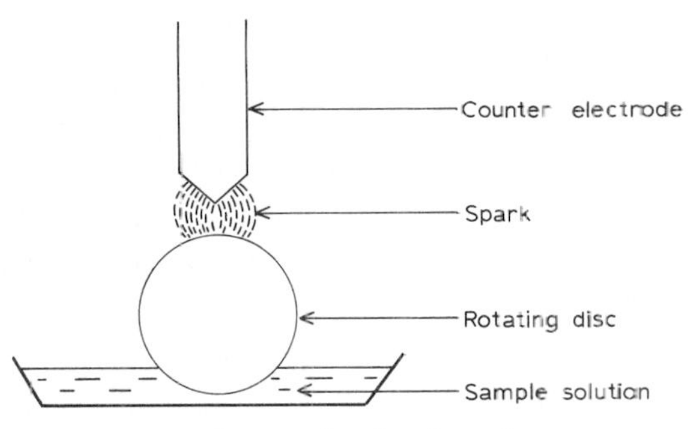

Fig.30. Rotating disc electrode.

lution actually consumed in the analysis is not easily reproducible. Geochemically, this type of equipment has been used for the determination of iron, aluminium and calcium in magnesite by MOSHER et al. (1951) and in rocks by GRABOWSKI and UNICE (1958). Likewise, calcium and magnesium have been determined in the lake waters by MELOCHE and SHAPIRO (1954) who used a silver disc for introducing the sample into the spark.

Vacuum cup technique

This method is illustrated in Fig.31. The lower graphite electrode has a small diameter hole drilled axially which is entered radially by another hole as shown. A plastic cup fits closely over a tapered portion of the electrode so that liquid can be held in the cup at such a level as to cover the radial

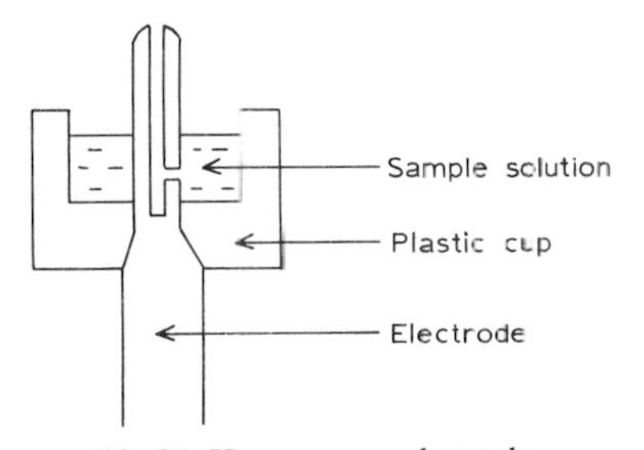

Fig.31. Vacuum cup electrode.

hole. When a discharge is passed from the top of the electrode to a counter electrode held above it, solution is drawn up into the spark. One advantage of this equipment is that any fine precipitate which would be filtered from solution in the porous cup is taken into the spark. Although not yet widely used in the field of geochemistry it does possess certain potentialities.

THE SPECTROGRAPH

Spectrographs are in principle simple pieces of equipment and many commercial models are available. Both gratings and prisms are used for dispersion and the choice between the two is largely a matter of preference of the user, but it is safe to say that prism instruments have predominated in the past in this country whilst gratings have been used extensively in the U.S.A. The important consideration in choosing an instrument for the examination of geochemical specimens is to ensure that adequate dispersal is available. Since the spectra are often highly complex it is advisable to use large spectrographs having a linear dispersion of 2.5–7.0 Å/mm.

Photographic methods are most commonly used for the recording and evaluation of spectrographic analyses. Direct reading instruments are available in which photomultipliers are placed to receive the light from selected isolated lines and to give direct meter readings or recordings of the intensities of these lines. Since these instruments are expensive, their use can only be justified when very large numbers of similar samples are to be examined.

PREPARATION OF SAMPLES AND STANDARDS

The sample which is presented for analysis may be in one of

several forms. On the one hand it may comprise large pieces of rock or at the other extreme may be a very small inclusion. In the latter case the portion actually used for the analysis may be all, or a large proportion, of the total sample and it is necessary only to powder it in an agate mortar.

When a large sample is received the main problem is to ensure that the small portion used is representative of the whole.The initial stages of preparation are little different from the standard techniques which precede any analysis —crushing and powdering in a percussion mortar being utilized with "quartering" or "splitting" at appropriate times. The final grinding must be done carefully to reduce the sample to a very fine powder. Agate pestles and mortars are normally used at this stage, either motorized or manually operated. The final powder should all pass a 150 mesh sieve and should not be separated by sieving.

Throughout the procedure care must be taken to avoid contamination from the materials of the equipment used or from previous samples which have passed through the equipment.

Aids to the initial powdering of various types of sample have been proposed. For example, silicate rocks and minerals can be heated and chilled in water to provide a suitable breakdown. On the other hand, micas are extremely difficult to grind by normal methods and the cutter designed by NEUMANN (1956) is probably the most efficient method yet available.

Soils, coals and other materials which contain organic matter should always be ashed before grinding and in order to avoid losses of the more volatile inorganic components the temperature of the furnace should be kept below 420°C. This temperature is high enough to ensure the destruction of all organic matter so long as air is allowed into the muffle furnace during the early stages. Oxidation is usually complete in about 12 h. The samples are best contained in shallow platinum basins during this operation.

The necessity to obtain uniform mixing of the sample cannot be overemphasized. The material used for analysis may not weigh more than 5 mg and this often has to be representative of many pounds of the original sample.

Frequently powder samples are mixed with carbon powder or other materials before use. This must be done with the same care as is used in the preparation of the sample. Weighings must be accurate and contamination avoided at all stages.

When solution methods are used it is not normally necessary to reduce the sample to such a fine powder when plenty of material is available. The reason for this is that about 1 g of sample can be taken into solution and it is much less of a problem to ensure that this amount is representative of the original material than when much smaller amounts are used.

Any of the standard methods can be used to produce solutions. The most useful technique is to evaporate off silica with sulphuric and hydrofluoric acids. Fusion techniques with various mixtures can also be used but in this case special attention must be paid to the purity of the fusion mixtures.

PREPARATION AND FILLING OF ELECTRODES

The electrodes commonly used are made from carbon or graphite. Suitable grades of either material can be obtained in rods 10 or 12 inches long and about $\frac{1}{4}$ inch diameter. It is important to ensure that the material used is sufficiently pure for the purpose particularly when trace levels of some elements are being determined. The most common impurities are boron, calcium, titanium and silica.

Several methods have been described for purifying the rods and these are reviewed by MITCHELL (1948) but during recent years better grades of carbon and graphite having very low impurity levels have become available at moderate prices.

The shape required depends on the technique to be used. Electrode cutters are available commercially and are eminently suitable for preparing the type with shallow holes of a fairly large diameter such as those illustrated in Fig.27. They are also suitable for preparing a pointed end on the counter electrode. When shapes such as that used for the cathode layer method are required in which a deep hole of small diameter is drilled then a lathe must be employed. MITCHELL (1948) has described a suitable cutter with which the shoulder and hole are cut in one operation. Alternatively a small capstan lathe can be used and this method can be recommended when a number of different shapes are required. After setting up such a lathe one person can cut upwards of 200 electrodes in about 5 h. The tools used for cutting and drilling must be of the highest quality since carbon in particular is highly abrasive. Tungsten carbide tipped cutters are most useful and they must be prepared with a very sharp edge. Even so, such a tool is dulled after cutting about 200 carbon electrodes to the shape shown in Fig.27.

It is important to ensure that the electrodes are not contaminated with lubricating oil or other substances during the preparation. Gloves should be worn by the operator or all electrodes handled with forceps.

The packing of the sample into the electrode must be done with care. This is easily accomplished when the wide bore variety is used. The electrode is pressed repeatedly into the powder or the latter is poured into the crater and compressed with a suitable tool.

It is rather more difficult when the narrow bore electrodes are used. The powder must fill the hole and be compressed uniformly. During this investigation it was found that the best results were achieved by wrapping a cylinder of glazed paper around the electrode which was held in a hole drilled in a block of perspex (see Fig.32).

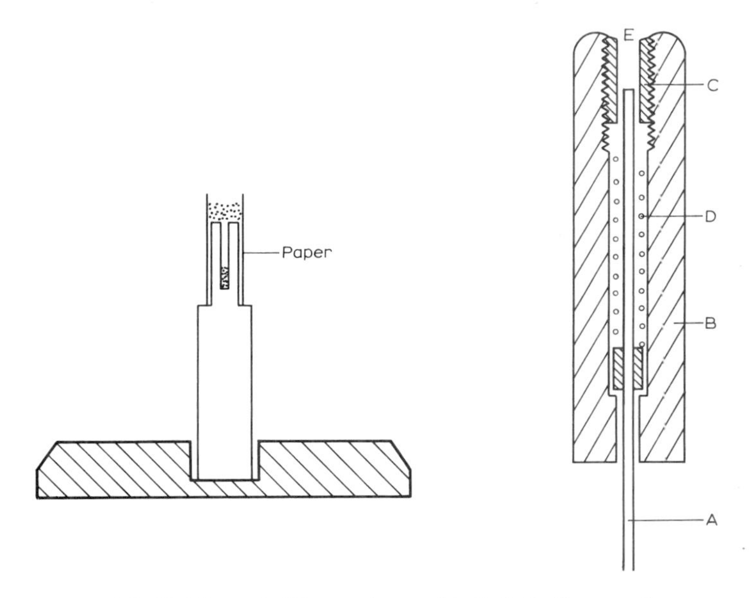

Fig.32. Electrode filling equipment.(*A, B, C, D, E*: see text.)

A small amount of powder was poured into the paper cylinder and introduced into the bore by tapping the perspex block gently on the bench. It was then compressed by pushing the well fitting needle *A* (Fig.32) into the bore. Reproducible pressure was obtained by placing the forefinger over the end *E* of the tool and depressing the body *B* until the top of the needle could just be felt projecting at *E*. The pressure transmitted by the needle to the powder was adjusted by the positioning of the hollow screw *C* which controlled the tension on the spring *D*.

QUALITATIVE AND SEMI-QUANTITATIVE ANALYSIS

The qualitative analysis of a sample may be required for many different purposes. It may be that one or two metals are of interest in a complex sample or a fairly complete evaluation of 30 or 40 elements at concentrations ranging from a few parts per million to tens of per cent may be required. The spectra of samples of geological origin are highly complex and may comprise of several thousand lines. For example, the arc spectrum of iron alone may have two or three thousand lines. Consequently, it is clearly impossible to identify all the lines in a spectrum, but fortunately this is not necessary so the normal procedure is to look for certain lines most suitable for the purpose. If a given element is present as a trace the most sensitive lines of that element must be looked for. Sometimes the identification of a single line may be enough proof of the presence of an element in a sample but it is more usual to look for two or three lines.

A worker will soon become familiar with groups and combinations of lines in the spectra with which he is dealing and will "read" plates as another person will "read" a map. Many aids to the identification of lines are available. Detailed spectrograms of many elements have been published by GATTENER and JUNKES (1947) and complete tables of the wavelengths of lines of all elements were prepared by HARRISON (1939). AHRENS (1950) also listed the most sensitive lines of the elements together with those of other elements which occur in close proximity and which might interfere with one another. On the other hand, MITCHELL (1948) lists only those lines considered most useful for certain elements. Undoubtedly, the job of identifying lines is a laborious task during the early period of spectrochemical work in a laboratory but gradually plates are accumulated with elements identified and marked for future purposes. It is, therefore, a simple and speedy task to compare a sample spectrum with that of a standard plate making use of a projection comparator or a Judd Lewis comparator (see Fig.33, 34). These are instruments which bring spectra on two plates into juxtaposition for easy comparison.

Semi-quantitative analysis of samples can be accom-

Fig.33. The comparator projector with inbuilt microphotometer (by Hilger and Watts).

plished by visually comparing the densities of spectral lines on sample and standard plates. Whilst qualitative analysis is important it is of very great value to be able to give quickly some idea of the concentration of a number of elements in a sample. Many techniques have been reported, but one based on that described by Mitchell has been found most useful for the examination of a wide variety of samples.

The method depends on the direct comparison of the spec-

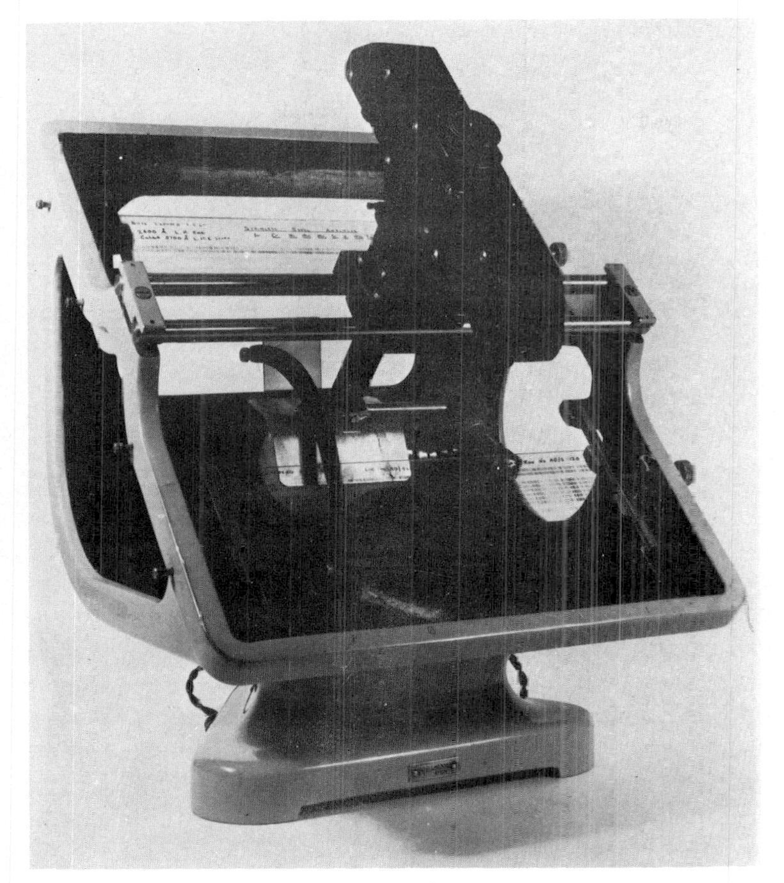

Fig.34. The Judd Lewis comparator.

trum of the sample with a standard plate and to assess the concentration of elements in the sample from the relative densities of suitable spectrum lines. The accuracy of the method is of the order 30–50% and suitable for the estimation of trace elements in the range 1 p.p.m. to 1%.

It is necessary first of all to prepare a series of standards comprising a matrix containing varying concentrations of trace elements. The matrix must be approximately of the

TABLE XI

TRACE ELEMENT MIXTURE

A		*B*		*C*		*D*	
LiF	0.3738 g	Co_3O_4	0.1362 g	SnO_2	0.1270 g	As_2O_3	0.1321 g
RbCl	0.1414 g	NiO	0.1272 g	PbO	0.1077 g	Sb_2O_3	0.1197 g
CaCl	0.1266 g	CuO	0.1252 g	Tl_2O_2	0.1116 g	SiO_2	0.1403 g
BaO	0.1117 g	MoO_3	0.1500 g	ThO_3	0.1138 g	TeO_2	0.1251 g
SrO	0.1182 g	$K_2Cr_2O_7$	0.2824 g	ZnO	0.1245 g	BeO	0.2773 g
CeO_2	0.1228 g	Ag_2O	0.1074 g	CdO	0.1142 g	HgO	0.1080 g
H_2WO_4	0.1350 g	V_2O_5	0.1784 g	Bi_2O_3	0.1114 g	GeO_2	0.1441 g
TiO_2	0.1666 g	ZrO_2	0.1351 g	Ga_2O_3	0.1344 g		
Mn_2O_3	0.1436 g	Y_2O_3	0.1271 g				
		La_2O_3	0.1173 g				
Total	1.4406 g		1.4863 g		0.9446 g		1.0466 g

same composition as the samples to be analysed since the nature and concentration of major constituents in a sample can control the intensity of lines arising from other elements in the sample. It is important that the materials used to make up the matrix are pure and uncontaminated with other elements which are being determined. Purification of these materials may be necessary. When the matrix mixture has been made it is wise to produce a spectrum and to check its purity.

Mixtures containing all the trace elements of interest are then made up. Mitchell has given the formula of such mixtures and these are reproduced in Table XI.

The total weight of each mixture contains 0.100 g of each of the elements. One tenth of the total weights, i.e., 0.1441 g of A, 0.1486 g B, 0.0945 g C and 0.1047 g D are then mixed with 0.5082 g of the matrix mixture thus producing a standard containing 1% of each of the 34 elements in the base. From this, subsequent mixtures are made in dilutions of 1 : 0.316, that is, if the 1% mixture is called M, then a full range of standards is produced as shown in Table XII.

TABLE XII

STANDARD DILUTIONS OF STANDARD MIXTURES

Mixture	*Trace element concentration (p.p.m.)*	*Mixture*
M	10,000	as above
N	3,160	0.316 g M + 0.648 g matrix mixture
O	1,000	0.316 g N
P	316	0.316 g O
Q	100	0.316 g P
R	32	0.316 g Q
S	10	0.316 g R
T	3	0.316 g S
U	1	0.316 g T

Each mixture is made by grinding the components carefully and thoroughly in an agate mortar. Great care must be taken to avoid contamination. Full details of the method of producing these standards is given because the principle can apply to any series for spectrochemical work.

The standards are mixed with carbon powder in a 1 : 1 ratio and then packed into carbon electrodes of the form and dimensions given in Fig.35.

The sample bearing electrodes are burned as cathode in a d.c. arc with a square ended 5 mm carbon rod as made. The separation is 10 mm and the arc current 9 A. The electrodes are maintained constantly in a suitable position by adjustment, using a projected image of the arc onto a screen as a means of positioning. Mitchell focussed the lower third of the arc around the cathode onto the spectrograph slit and doubtless this method gives the maximum sensitivity of detection, but an alternative procedure has been found to be more generally useful.

A condensing lens is placed near the slit so that when the slit is wide open an image of the arc is seen focussed on the spectrograph collimator. The arc is so positioned as to bring

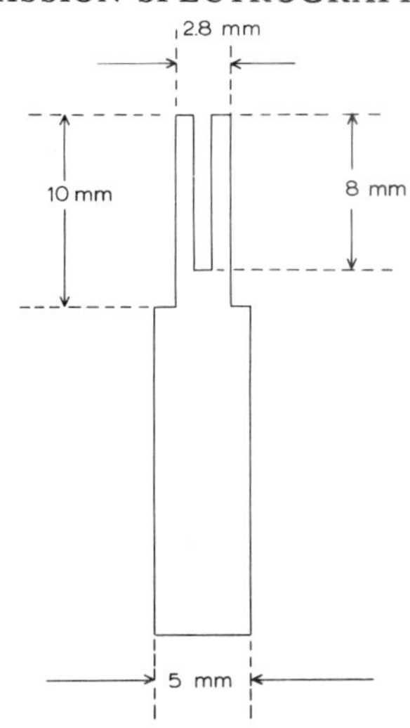

Fig.35. Details of electrode for cathode layer excitation. (Bore of cavity: 0.8 mm or 1/32 inch.)

the cathode just into the collimator and the focal length of the condensing lens is such that the collimator is filled with about one third of the arc column. Between the condensing lens and the slit the light beam can be considered parallel and in this space a step sector of the form shown in Fig.36 is rotated during each exposure.

The sample is completely consumed in about 3 min using this technique and the resulting spectrogram has the form shown in Fig.37. Effectively seven exposures are produced each adjacent exposure being a 2 : 1 ratio of its neighbour.

The standards and samples are thus burned by a strictly standardized method and estimations of the concentrations of elements in the sample are made by visual comparision with the standards. The comparisons of line density are facilitated by the form of the spectra, the step of exposure on which a line just disappears being an excellent guide.

It is necessary to pay strict attention to detail in the method. All variables should be eliminated from the electrode packing to the processing of the photographic plates.

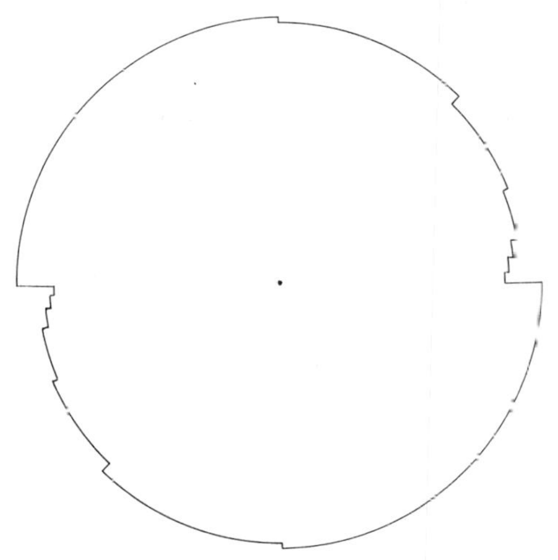

Fig.36. Form of rotating step sector.

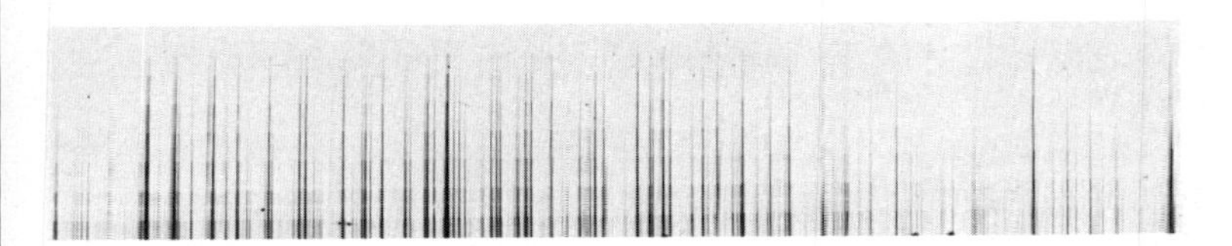

Fig.37. Spectrum of ironstone taken with step sector.

QUANTITATIVE ANALYSIS

Quantitative procedures are all based on the concept that for a given element the integrated intensity of the radiation at one wavelength is proportional to the quantity of an element excited in the source. Several features of spectrographic techniques make it very difficult to determine directly a value for the integrated intensity (hereafter called "intensity"). This is not strictly an accurate description but is a term commonly used in practice and in literature.

(*1*) It is not convenient to weigh out an accurate quantity of sample and pack it into an electrode.

(*2*) It is not possible to reproduce strictly the source conditions from sample to sample. Even if the electrical conditions supplied to the arc are controlled, there can be little control over the phenomena within the arc itself.

(*3*) The time taken to consume the sample cannot be controlled.

(*4*) The processing of the photographic plate cannot be accomplished in a way which will guarantee the exact reproducibility of a given density of image for a given amount of light falling on the plate.

(*5*) The conversion of image density to light intensity is inexact.

For these reasons the concept of an internal standard is always used in quantitative work. This was introduced by GERLACH (1925) and is based upon the use of the ratio of the image density of the line of the sample element to that of another element present at a known concentration in the sample as a working parameter.

The success of a quantitative technique depends upon the suitable choice of the internal standard. The element used may be present in the sample at a known concentration or more usually can be added to samples and standards alike at a fixed concentration. The following are the ideals at which to aim when making the choice:

(*1*) The internal standard should volatilize at about the same temperature as the element being determined.

(*2*) The spectrum lines of internal standard and element should have similar excitation and ionization potentials.

(*3*) They should be located fairly close together in the spectrum.

(*4*) The two elements should be similar chemically.

(*5*) The sample should not contain the element chosen as internal standard.

It can be seen immediately that no one element is suitable

as internal standard for all elements being determined and indeed it will be impossible often to find an ideal material, but certain elements have been used with success for a wide variety of work. For example, iron is suitable for a large number of elements. Of course, it is always present in the sample but a technique known as the "variable internal standard" method has been used successfully by MITCHELL (1948). Cadmium can be used in the determination of zinc, silver for copper, beryllium for a wide range of elements. Most books on the subject deal with the problem at length.

When the internal standard has been chosen and the excitation technique established, the common problem outstanding is to select some parameter which is proportional to the relative intensities of sample and internal standard lines from the photographed spectrum. The first stage in any method is to measure the line densities on the plate. This is accomplished with the aid of a microphotometer (or densitometer). A narrow beam of light is transmitted first through a clean portion of the plate, then through the line and is made to fall on a photocell. The responses from the photocell, as measured on a galvanometer, can be used to calculate the line density (D) using the expression:

$$D = \log_{10} (I_0/I)$$

where I_0 = response from clear plate, and I = response from line.

Sometimes the Seidel function (S) is preferred to density for certain purposes:

$$S = \log_{10} (I_0 /I - 1)$$

The response of a photographic emulsion to light takes the form shown in Fig.38 and the procedure for converting the measured density of a line to a relative intensity is known as "plate calibration". This can be accomplished in many ways

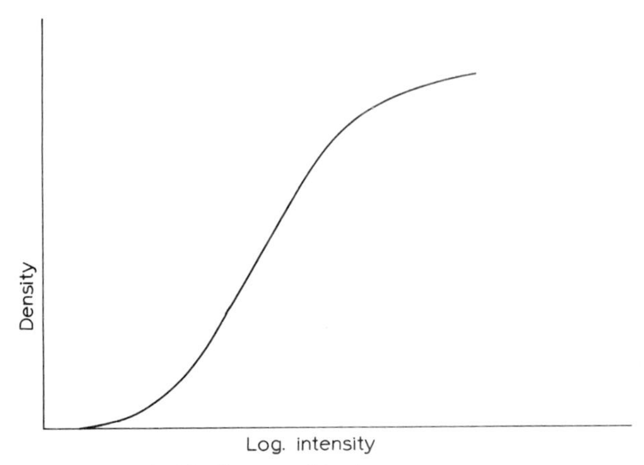

Fig.38. Photographic plate response.

and two methods illustrating the technique will be described briefly. For a full treatment of the subject reference should be made to HARRISON et al. (1948), NACHTRIEB (1950) or AHRENS (1950).

The first method known as "self calibration" is of particular value. The spectra are produced using a step sector in the manner described under "Semi-Quantitative Analysis" and the densities are measured on three or four suitable steps for the analysis and internal standard lines. The values obtained are plotted as shown in Fig.39. Log relative exposures are as shown by the vertical lines when the sector steps are cut in successive ratios of 2 : 1. The value *A–B* measured at a fixed density is a measure of the log relative exposures required to record the two lines at that density and this value is directly proportional to the log relative intensities of the lines. (The value may be positive or negative depending on whether the sample line is more dense than the internal standard line or vice versa.)

This procedure is followed for the standards and a concentration calibration curve produced for each element. This will normally take the form shown in Fig.40.

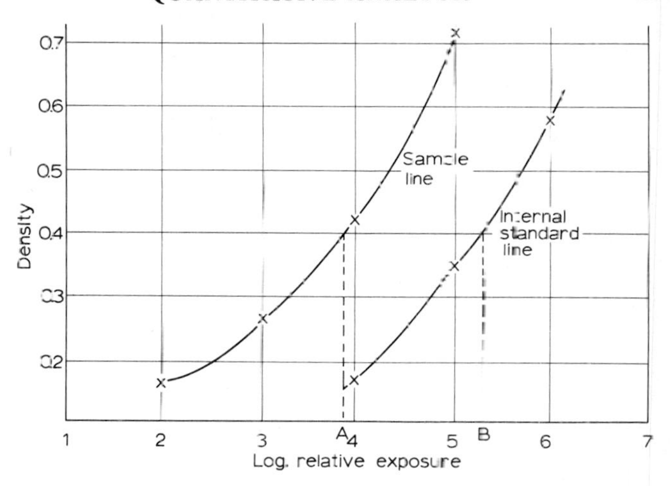

Fig.39. Internal calibration. (*A*, *B*: see text.)

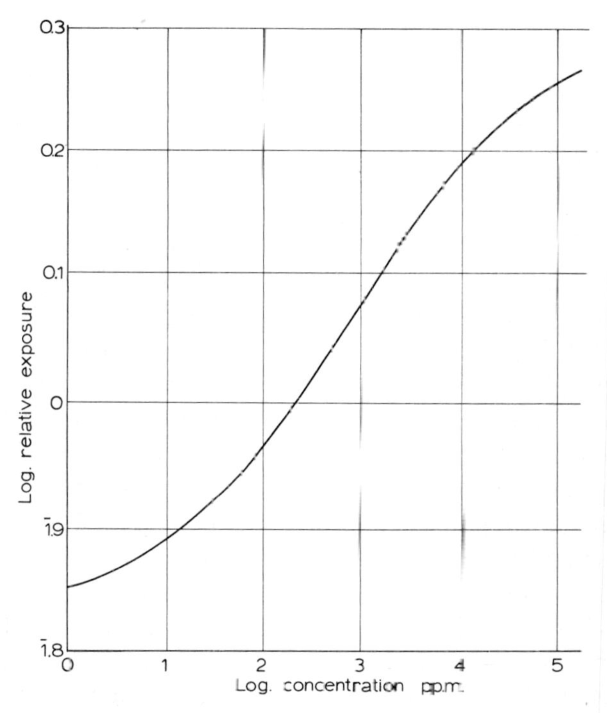

Fig.40. Typical calibration curve.

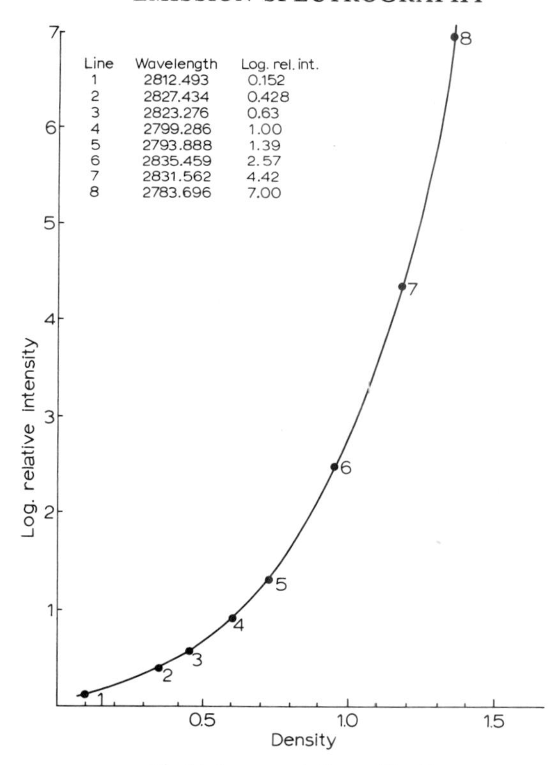

Fig.41. Iron spark line calibration.

If a large number of similar samples are recorded on one plate it may be more economical in time to use an alternative method of plate calibration. The spectra are recorded with a shorter slit and the step sector is not used. Apart from the spectra of the samples an iron arc or spark spectrum is recorded on the plate—usually in triplicate—and the density of about seven prechosen iron lines measured. It is necessary previously to have obtained the relative intensities always given by these lines under the conditions used and this can have been accomplished by the previous method. The lines are chosen to cover a good range of intensity and when the densities have been obtained and plotted against the relative intensities a curve as shown in Fig.41 is obtained.

This curve provides plate calibration *for that plate* and when the densities of all sample and internal standard lines have been obtained the values can be converted to log relative intensities, thus eliminating the plate characteristics completely.

This procedure for obtaining concentration standard curves is exactly the same as with the previous method.

THE ANALYSIS OF GEOCHEMICAL SAMPLES

Spectrographic methods of analysis hold many attractions to the geochemist for obvious reasons. One of the most important is that an overall picture of the composition of complex samples can be obtained with reasonable accuracy. Semi-quantitative methods give a very rapid method of estimating 30 or 40 elements within the concentration range 1 p.p.m. to 1% with about 30% accuracy. At best, excitation methods are capable of giving between 5% and 10% accuracy over the same concentration range by quantitative procedures. For higher concentrations, solution-spark methods can give 2–3 % accuracy.

The preliminary work required for any of these techniques is considerable, standards have to be made up and standard curves produced. Ideally, this procedure must be repeated for every type of sample for which an analytical procedure is required, but once done it is possible to perform routine analyses quickly and economically.

It would be tedious to quote all the various examples to which spectrography has been applied during this work. To illustrate its range the following have been selected to bring out its versatility and limitations.

The inorganic content of organic extracts from coal

Samples of vitrain were extracted with purified chloroform and the solvent evaporated in vacuo leaving a hard brown tarry residue. This was separated into fractions soluble and insoluble in purified ethyl alcohol. These two fractions were concentrated and analysed as a solution spectrographically. The results are given in Table XIII. Apart from the various

TABLE XIII

SPECTROGRAPHICAL ANALYSIS OF VITRAIN[1]
% alcohol insoluble = 77 % ash on this 85–91
% alcohol soluble = 23 % ash on this 50–60

Alcohol insoluble (p.p.m. or %)					
Ag	15	Fe	M.C.	Pb	45
Al	500–1,500	K	150–500	Rb	N.D. (150)
Ba	20–50	Li	5–15	Sb	N.D. (50)
Bi	N.D. (15)	Mg	1–2%	Si	5–15
Ca	500–1,500	Mn	500–1,500	Sn	N.D. (15)
Co	150–500	Mo	N.D. (5)	Sr	5–15
Cr	5–15	Na	M.C. 1%	Ti	30–100
Cu	150	Ni	500–1,500	V	20–50
Alcohol soluble (p.p.m. or %)					
Ag	10	Fe	M.C.	Pb	N.D. (100)
Al	1,000–4,000	K	100–400	Rb	N.D. (300)
Ba	10	Li	30–100	Sb	N.D. (100)
Bi	N.D. (100)	Mg	1%	Si	100
Ca	100	Mn	1,000	Sn	N.D. (30)
Co	400	Mo	N.D. (10)	Sr	30–100
Cr	10–40	Na	3,000	Ti	3000–4,000
Cu	300	Ni	300–1,000	V	10

[1]M.C. = major constituent; N.D. = not detectable.

differences in the quantities of calcium, lithium, strontium and titanium, the most significant result which emerged was the quantity of silicon in each entity of the gross extract. In the alcohol insoluble residue this ranged from 5 to 15 p.p.m. whereas in the soluble fraction it was about 100 p.p.m. This has led to a further investigation of the bonding of silicon in these complicated organic complexes and it may lead to the establishment of silicones in humic residues. If this proves to be the case it would be the first demonstration of the occurrence of natural silicones in geological deposits.

Spectrographic analysis of boxstones from the Northampton Ironstone

The reported need for developing techniques to explore the trace element content of sedimentary ironstones, prompted the use of spectrography and X-ray methods for this purpose at the University of Nottingham. Two samples of typical boxstone, from the Northampton Sand Ironstone (Jurassic) were analysed chemically and then explored by spectrographical methods. The location of the samples taken from the boxstones are indicated in Fig.42, which indicates that one sample was of a simple ring structure whilst the other consisted of a multiple series of diffusion rings. From Table XIV it will be seen that there is a very definite relationship between the migration of certain elements outwards from *A* to *D* in sample *1* (Fig.42), but in similar positions in sample *2* the picture has been confused by secondary diffusion products. It is premature to exploit these analyses for advancing theoretical considerations concerning boxstone formation, but it clearly points the way to research into this fascinating and widespread problem of secondary ironstone formation in sediments.

Of more significant interest is the range of trace elements in these ironstones. An example is the occurrence of detec-

TABLE XIV

SEMI-QUANTITATIVE ANALYSIS OF SAMPLES FROM BOXSTONE IRONSTONES

(Results in p.p.m. except where stated)

Sample	*Al*	*B*	*Ba*	*Be*	*Ca*	*Co*	*Cr*	*Cu*	*Fe*	*K*	*Li*	*Mg*	*Mn*	*Mo*	*Na*	*Ni*	*Pb*	*Rb*	*Si*	*Sn*	*Sr*	*Ti*	*V*	*Zn*
1*A*	M.C.	3,000	300	N.D. < 10	1%	300	300	N.D. < 10	M.C.	2,000	1%	1,000	100	N.D. < 10	300	N.D. < 30	N.D. < 10	N.D. 300	1%	N.D. < 100	3,000	1,000	30	N.D. < 100
1*B*	M.C.	3,000	100	N.D. < 10	1%	300	300	N.D. < 10	M.C.	2,000	1%	3,000	300	10	500	N.D. < 30	N.D. < 10	N.D. 300	M.C.	N.D. < 100	1,000	2,000	100	N.D. < 100
1*C*	1%	3,000	N.D. 30	N.D. < 10	1%	300	100	N.D. < 10	M.C.	1,000	1%	1,000	100	N.D. < 10	300	N.D. < 30	N.D. < 10	N.D. 300	1%	NRD. < 100	3,000	500	100	N.D. < 100
1*D*	1%	3,000	N.D. 30	N.D. < 10	1%	100	100	N.D. < 10	M.C.	10	100	3,000	300	10	100	N.D. < 30	N.D. < 10	N.D. 300	M.C.	N.D. < 100	100	1,000	100	N.D. < 100
2*A*	M.C.	1%	300	N.D. < 10	M.C.	1,000	1,000	N.D. < 10	M.C.	5,000	1%	1,000	300	N.D. < 10	1,000	1,000	N.D. < 10	N.D. 300	1%	N.D. < 100	1,000	1,000	30	N.D. < 100
2*B*	M.C.	1%	300	N.D. < 10	M.C.	1,000	1,000	N.D. < 10	M.C.	3,000	1%	1,000	300	10	3,000	1,000	N.D. < 10	300	1%	N.D. < 100	1,000	1,000	1,000	N.D. < 100
2*C*	M.C.	1%	300	N.D. < 10	M.C.	1,000	1,000	10	M.C.	5,000	1%	3,000	3,000	10	3,000	300	N.D. < 10	300	M.C.	N.D. < 100	1,000	3,000	1,000	N.D. < 100
2*D*	M.C.	1%	300	N.D. < 10	M.C.	1,000	1,000	10	M.C.	5,000	1%	3,000	3,000	10	3,000	300	N.D. < 10	300	M.C.	N.D. < 100	1,000	3,000	1,000	N.D. < 100
2*E*	M.C.	1%	300	N.D. < 10	M.C.	1,000	1,000	10	M.C.	5,000	1%	3,000	3,000	N.D. < 10	3,000	300	N.D. < 10	300	M.C.	N.D. < 100	1,000	3,000	1,000	N.D. < 100
2*F*	M.C.	1%	1,000	N.D. < 10	M.C.	1,000	1,000	10	M.C.	5,000	1%	1,000	1,000	N.D. < 10	3,000	300	N.D. < 10	300	M.C.	N.D. < 100	300	1,000	300	N.D. < 100

[1]M.C. = major constituent; N.D. = not detected (with lowest limit of detection in p.p.m.).

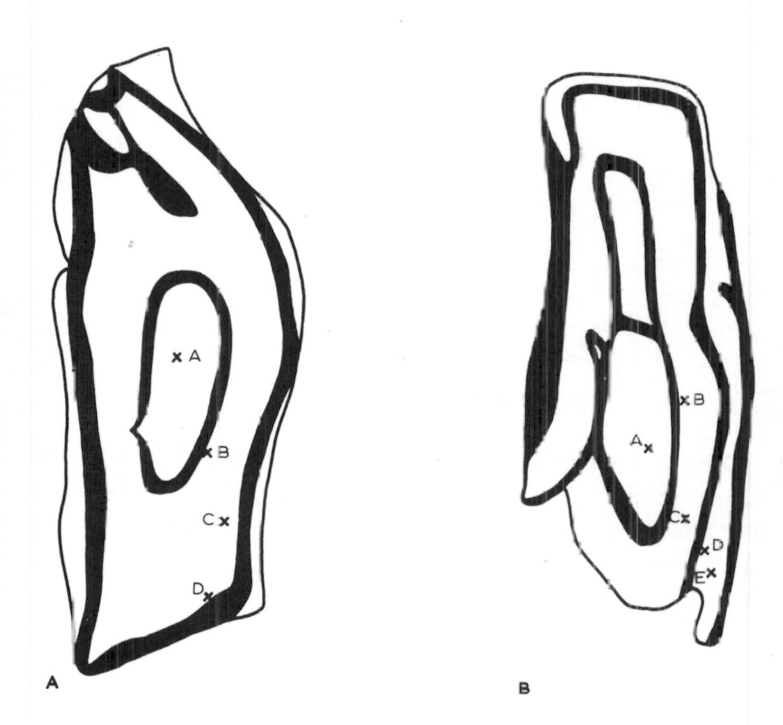

Fig.42. Tracings of the structures shown in the frontispiece, indicating the precise position of samples extracted for spectrographic analysis (see Table XIV). A. Sample *1*. B. Sample *2*. *X*=positions of samples (see text).

table traces of copper in sample *2* and its absence in sample *1*. The range of strontium, vanadium and titanium is also of considerable interest, as also is the significant quantities of boron which these ironstones appear to contain. Clearly, this is a field of fundamental research which merits attention as the use of such methods will shed a great deal of light on the geochemical controversy which still surrounds the origin of sedimentary iron ores.

CHAPTER 6

Flame Photometry

INTRODUCTION

There are many aspects of flame photometry which have yet to be appreciated by the geochemist. As will be shown later, it is a technique which can be used with considerable success in the analysis of underground waters, especially in tracing the infiltration of sea waters into maritime aquifers. Ultimately, it should prove possible to determine the sodium–potassium–calcium ratios in mineral grains and thereby dispense with the tedious and difficult optical methods for the determination of detrital felspars and chloritoid minerals.

It is common knowledge that when sodium and certain other elements enter a flame, light of a characteristic colour is emitted, the intensity of which varies with the amount of material added. Flame photometry is based upon the optical analysis of this emission. When a flame is used as a spectrographic source it is a simple matter to obtain a steady output of energy. A solution sample can be introduced continuously and at a constant rate and the flame itself can be stabilized by controlling carefully the gases which are combusted. Because of this feature an instantaneous measurement of light intensity can be used for analytical purposes. With most other sources the light output must be integrated over a period of time in order to obtain measurements suitable for analytical purposes.

The use of the flame dates from the work of TALBOT (1826) who, in the early 19th century, used an alcohol flame to excite spectra of lithium and strontium. KIRCHHOFF and BUNSEN (1860) reported the discovery of rubidium and caesium, as a result of their investigations into the flame spectra of the alkali metals, and in fact the bunsen burner was developed for this work. When the potentialities of the source were realized, workers invented various methods of introducing samples into their flames as a means of improving sensitivity. Bunsen and Kirchhoff, for example, at first used a platinum wire, moistened in a solution of the sample, but later formed a spray by adding zinc to an acid solution of the sample and then introduced the spray into a flame. MITSCHERLICH (1862) fed solutions by capillary action through a bundle of fine platinum wires contained in a glass tube. EDER and VALENTA (1893) used a platinum gauze disc which was turned by a clockwork motor, first through a solution of the sample then through the flame. GOUY (1879) sprayed solutions with a concentric atomizer using compressed air and led the spray into the flame.

From about 1900 to 1930 attempts were made with varying success to obtain quantitative results, first by visual comparison of spectra, then later by photographic methods, and it was the work of LUNDEGARDH (1929, 1934) which laid the foundation for the subject as it is known today. Lundegardh used an air–acetylene flame, the air for the flame being used to atomize the sample solution. The light from the flame was dispersed in a spectrograph and the spectrum was photographed. Later a particular line was isolated and its intensity measured by means of a photocell and amplifier. From that time, instruments in various forms have been developed, but it is during the last 15 years that most progress has been made.

The factors which have contributed to the development of the subject are:

(*1*) The comparative simplicity of flame-spectra, the

amount of energy available is low and only the lowest energy lines are excited.

(*2*) Many of the lines and molecular bands emitted are in the visible region and, therefore, glass optics and optical filters can be used to isolate the required lines.

(*3*) The elements which are easiest to determine by flame photometry are some of the most difficult to determine chemically, e.g., sodium and potassium.

(*4*) When the preliminary work is completed samples can be analysed in extremely short times.

EQUIPMENT

A large number of instruments has been built or is available commercially, and they vary considerably in design. Nevertheless, any equipment comprises three basic units: the source unit, the selection system, and the detector and readout system. These are treated separately.

The source unit

This consists of a flame into which a solution sample is introduced. The flame must be of constant thermal output and the sample must be introduced in such a way that a constant amount is added per unit time so that the emitted light is of constant intensity. In order to achieve a flame with a constant characteristic, the burner itself must be designed carefully and supplied with a gas mixture at a constant flow rate.

Two gases are required, one for combustion and the other to support combustion. The mixtures most commonly used are air or oxygen with coal-gas, propane, acetylene or hydrogen. Others have, however, been used for special cases. VALLEE and BARTHOLOMAY (1956) used an oxygen–cyanogen flame in order to excite a large range of elements. The range of flame temperatures available with different gas mixtures is given in Table XV.

TABLE XV

FLAME TEMPERATURES OF GAS MIXTURES
(DEAN, 1960)

	Temperature (°C)	
Fuel	*in air*	*in oxygen*
Coal gas	1,700	2,700
Propane	1,925	2,800
Butane	1,900	2,900
Hydrogen	2,100	2,780
Acetylene	2,200	3,050
Cyanogen		4,550

Two main types of source units are in common use:

(*1*) The air or oxygen, usually air, is used to atomize the sample into a vessel in which the gas velocity decreases and the larger drops fall out of the stream so that only the finest mist emerges. This mist is then carried into the flame. The atomizer is of paramount importance, it needs to be rigid so as to give a steady spray with as large a proportion of small particles as possible. It should not easily become blocked and, therefore, the tubing through which the solution passes should not be of too small a bore. It must be constructed so that the solutions, which are often highly acidic, will not come into contact with materials which will be corroded. Materials which have been used for the purpose are glass, stainless steel, noble metals, perspex, etc. A number of different designs have been used but nowadays the most common is the concentric type. An example which has given excellent service is shown in Fig.43.

The burner is usually of the Mekker type and the gas mixtures most commonly used with the unit are air and hydrogen, propane or acetylene. A suitable unit is illustrated in Fig.44.

A large proportion of the sample sprayed by the atomizer is

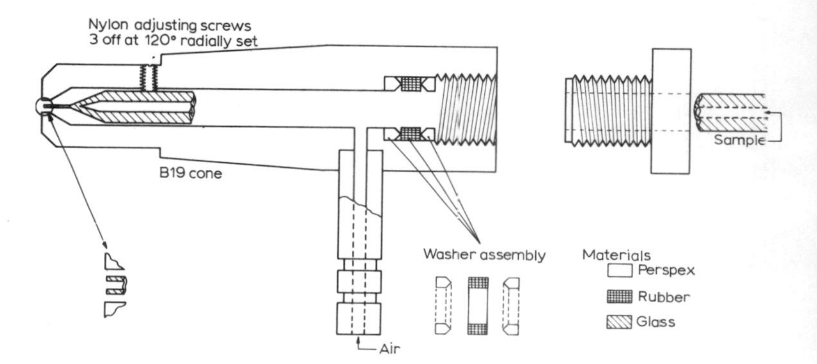

Fig.43. Flame photometer atomizer. (Twice full size.)

rejected in the expansion vessel and only a few percents reach the burner as a very fine mist. Many designs have been used but one based on a suggestion by Still (private communication) has been found most suitable (see Fig.45). Because of the small drop-size all the water is easily evaporated and a very stable and steady flame is produced. The combination of the units shown in Fig. 43–45 results in a source unit which was developed between 1950 and 1954 and has given excellent service since then. It is now used with only minor modifications in a commercial instrument.

(*2*) In the second type of source unit the atomizer and burner are combined so that the sample is sprayed directly into the flame. A typical example is shown in Fig.46. These burners usually utilize oxygen for the spraying of the sample and, therefore, the flames produced are very much hotter than in the previous type. The gas mixtures most commonly employed are oxygen and hydrogen or acetylene. It would appear at first sight that a much larger proportion of the sample which is sprayed is available for excitation and indeed this type of unit is often called the "total combustion burner". This, however, is a misnomer because the spray particles are ejected into the flame at a high velocity and are of varying sizes. The larger ones are projected right through the flame,

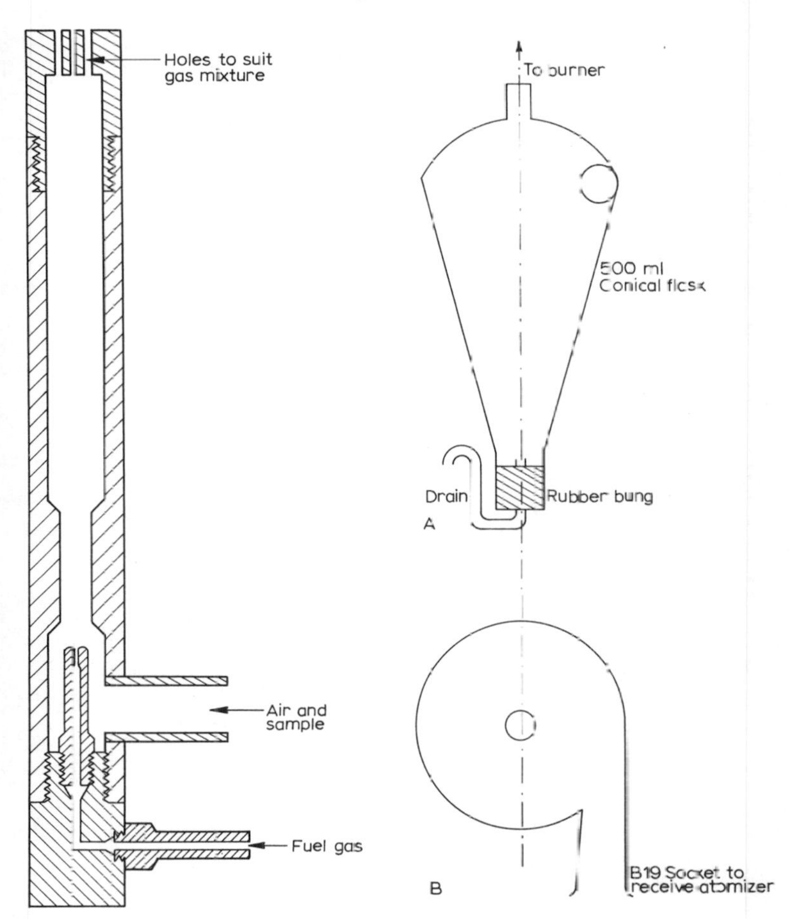

Fig.44. Mekker type burner. (Constructed in stainless steel.)

Fig.45. Cyclone expansion chamber. A. Elevation. B. Plan.

so that the water from only the smallest ones is evaporated completely and that proportion of the sample excited.

With any type of burner assembly it is essential to ensure that the gases are supplied at constant pressure so that the flame remains steady. This is particularly important in the

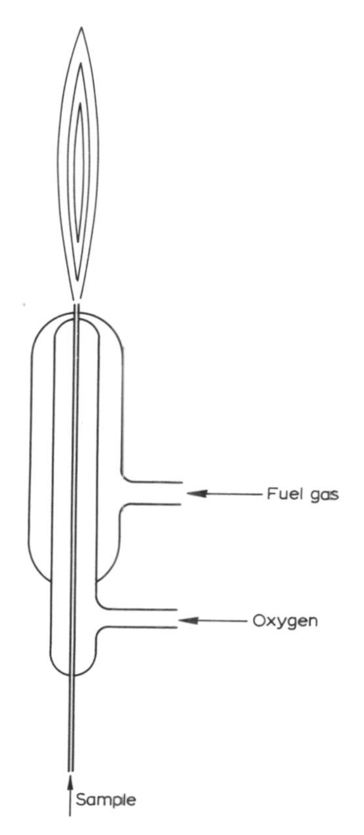

Fig.46. Direct injection burner (diagrammatic).

case of the air or oxygen as this is used to atomize the sample in addition to supporting combustion and a change in pressure will give rise to a change in the rate of atomization in addition to changing the flame characteristics.

The optical unit

Here the light from the flame is collected and the required wavelengths selected. The best way of achieving this is by means of a monochromator. Both prism and grating instruments have been used in the visible and ultraviolet regions, but the expense is not always justified when, say, sodium and potassium are to be determined and many instruments utilizing optical filters have been developed. Whichever is used, however, the function of the unit is to isolate the required radiation from all the other light emitted from the flame. The total radiation from the flame comprises a general background emission of continuous radiation and also the lines or bands produced by all the components in the sample. Inevitably the range of wavelength transmitted through a filter is higher than that through the exit slit of a monochromator and it becomes obvious that a filter will be a very inefficient way of isolating any one line from a number which are closely spaced. In addition, if the level of continuous radiation is high then a very much larger portion of this background radiation will be transmitted through a filter than through a monochromator. This, in itself, imposes severe sensitivity limitations on the equipment as a whole. For these reasons instruments which employ filters usually have cool flames and are used only for simple samples which contain a few elements, e.g., the determination of sodium in natural waters.

The detector

This comprises a photocell of some kind and a means of reading its current output. On the simplest instruments which utilize filters a barrier-layer photocell is often used and its output fed directly to a sensitive galvanometer. On the more exotic equipments, having a much lower light gathering capacity, a photomultiplier is usually employed and its output amplified and fed to a meter or recorder. On some of these instruments the wavelength range can be scanned by means of a motorized drive and the output of the equipment recorded so that a chart is produced showing light intensity against wavelength. Fig.47 is such a recording of the nickel spectrum.

It can be seen that a large number of variations are possi-

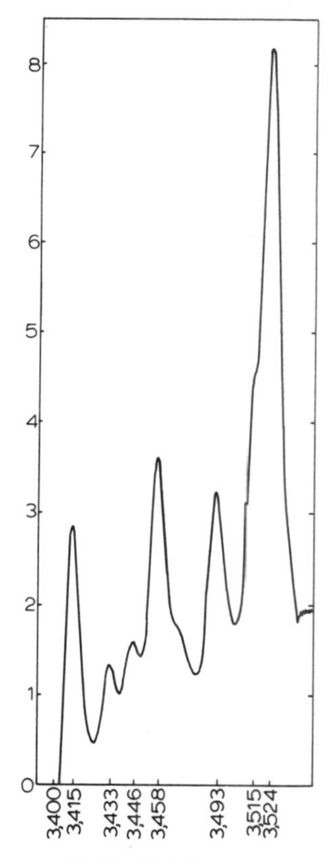

Fig.47. Nickel spectrum.

ble and many commercial instruments have been produced. The choice depends entirely on the type of work which is to be done with it.

SCOPE OF THE METHOD

Although 52 elements are excited in the oxy–acetylene flame and smaller numbers in cooler flames many of them are insensitive. Most equipment manufacturers publish tables of the lowest concentrations of elements in solution which can be detected with their instruments. DEAN (1960) prepared a list of the sensitivity limits of 48 elements. Such information is primarily useful in giving information regarding the relative qualitative sensitivities of elements but it must be remembered that flame photometry is almost always used as a quantitative procedure and the smallest concentration which can be determined will be higher than those values given in the lists quoted above. This lower quantitative limit is very much dependent on instrumental conditions and the following factors are significant:

(*a*) The temperature of the flame.

(*b*) The intensity of the radiation from the flame above ("flame background") at the wavelength of the line or band being examined.

(*c*) The rate of atomization or more exactly the mass of the element which is available for excitation per unit time.

(*d*) The light-gathering capacity of the instrument.

(*e*) The stability of the flame and electronics.

TABLE XVI

LOWER ORDER OF CONCENTRATION OF SOME ELEMENTS FOR QUANTITATIVE DETERMINATION

Element	*Wavelength (Å)*	*Lower concentration (p.p.m.)*
Calcium	4,227	0.5
Cobalt	3,454	10
Copper	3,247	5
Iron	3,720	10
Lithium	6,708	0.1
Manganese	4,033	0.5
Nickel	3,524	10
Potassium	7,670	0.1
Silver	3,383	5
Sodium	5,890	0.01
Strontium	4,607	0.5

(*f*) The effect of interfering substances in the sample.

It becomes immediately obvious that any specification of sensitivity can apply only to one set of operating conditions on one instrument and the interpretation should be made with caution. There is no substitute for investigating each case individually. Table XVI contains a list of elements of interest to the geochemist with an indication only of the lower order of concentration which can be determined with reasonable accuracy without recourse to special techniques.

ACCURACY OF THE METHOD

The accuracy obtainable in an analysis depends very much on the composition of the solution being analysed. With very simple solutions, where standards can be made up with exactly similar components in the same ratio as in the sample, the limiting factor will be stability of the instrument and the steadiness of the final reading. The accuracy will be the same as the reproducibility and can be increased by replicate readings and other devices. However, it is often not possible or expedient to produce standards of similar composition to samples. Reasons for this are:

(*1*) One may not have available a complete analysis of the sample, geochemical samples may be an example of this.

(*2*) The technique is often used for determination of impurities. In order to produce standards similar to the samples it is necessary to have large quantities of highly purified salts of the major constituents and these may be very difficult to produce. Therefore, in these cases, the accuracy will depend on how far the readings for the element being determined are affected by variable quantities of other ions in the sample solution. If there is some effect then replicate readings will not improve accuracy.

Interferences

There are many types of interferences and these can be placed into main groups as follows.

Spray rate variations

The quantity of a sample introduced into a flame depends greatly on the characteristics of the process of atomization. Certain constituents in a sample can alter these characteristics by changing the viscosity and/or surface-tension of the solution, resulting often not only in a change in the total spray rate, but also on the drop-size distribution. With any type of source unit, if the drop size is significantly reduced or the total spray rate increased then a larger quantity of sample will be introduced into the flame, thus resulting in an enhancement of the emitted radiation. The effect can be overcome usually by diluting the sample until the dissolved ions have no effect. EGGERTON et al. (1951) advocated the addition of a small quantity of non-ionic wetting agent and this practice is sometimes of use in particularly difficult cases.

In the case of the separate atomizer and burner where the sample is sprayed into an expansion chamber the rate of evaporation of solvent from the drops may be affected by ions in the solution thus altering the effective particle size distribution and this can give rise to a change in a number of drops leaving the chamber to the burner. The temperature of the sample is particularly important in this respect; as the temperature rises, the evaporation rate increases and more of the sample reaches the flame.

Spectral interferences

Any flame emits its own spectrum the shape of which depends on the gas mixture being used and on flame conditions. As can be seen in Fig. 48, the hydrogen flame has low background radiation in the region 3,500–6,000 Å whereas when hydrocarbons are used as the fuel the background is

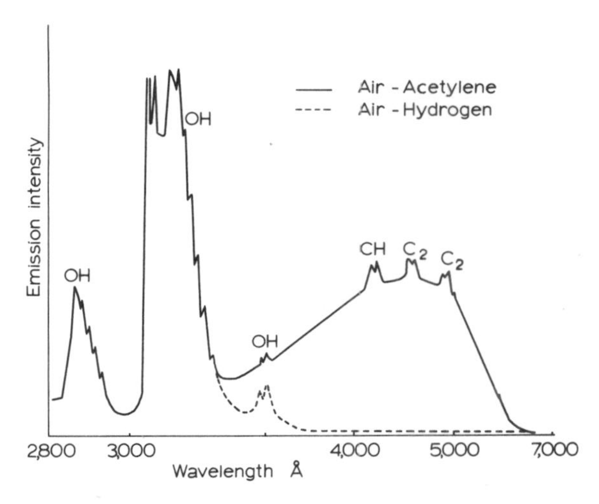

Fig.48. Background radiation from hydrogen and hydrocarbon flames.

quite high. When a solution is sprayed into this flame, then the lines and bands, due to components in the sample, are added to the background and when a particular line is selected the light passing the monochromator or filter will comprise the sum of the background, plus the line. If now any component in a sample causes a change in the background level of radiation then the intensity of the total light accepted changes and this can easily be misinterpreted as a change in the intensity of the line. Apparent enhancement can also arise from the occurrence of other metals in the sample. When present at high concentrations these can cause a change in the general background level by emitting continuous radiation, can give rise to scattered light in a monochromator or can interfere directly when their spectral line is adjacent to that of the element being determined. This is particularly important where filters are being used. The use of a monochromator largely removes these errors since it is possible to read the intensity of the total radiation over a wavelength range around the line. In this way the background level can be established.

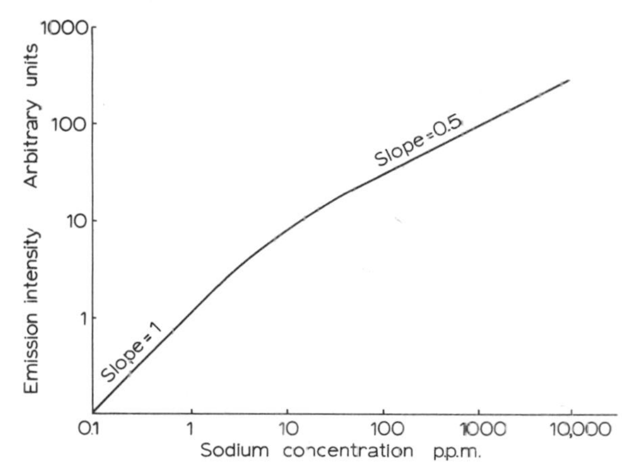

Fig.49. Emission intensity of sodium over wide concentration range. (Note logarithmic coordinates.)

Self absorption

The radiant energy produced in the centre or at the back of the flame is subject to absorption by atoms of its own kind in the ground state which exist in abundance around the cool edge of the flame. This gives rise to a diminution in intensity of the spectrum line which becomes more severe as the concentration of atoms in the flame increases. This is illustrated in Fig.49 where it is seen that, at low concentrations, a linear relationship exists between the sodium level and its emission intensity. At higher levels, however, a parabolic curve is obtained and the intensity increases as the square root of the concentration. Obviously the remedy is to work at low concentrations so that the linear response curve is applicable. If, as sometimes occurs, it is more convenient to use solutions at higher concentrations an ion line can be used. Ion lines do not suffer from self absorption in flames as the ions cannot exist in the cool outer fringes of the flame.

Ionisation

The hotter flames used in flame photometry possess sufficient energy to ionise some elements, principally the alkalis and to some extent the alkaline earths. When this occurs, for a given total concentration of an element, the number of atoms in the ground state is reduced by the number of ions produced. The intensity of the resonance line, therefore, diminishes as the intensity of the low line increases. The degree of ionization is dependent on flame temperature at low concentrations so cooler flames are always indicated for the determination of easily ionized elements. At higher element concentrations the proportion of ions to atoms decreases and this results in a calibration curve which is concave upwards. Fig.50 and 51 illustrate this effect and the advantage of using a cool flame for the determination of potassium at low concentrations is clearly shown. (These curves also illustrate the self-absorption effect at higher concentrations.)

The reason for this non-linear response is that an equilibrium will always exist between the number of ions and

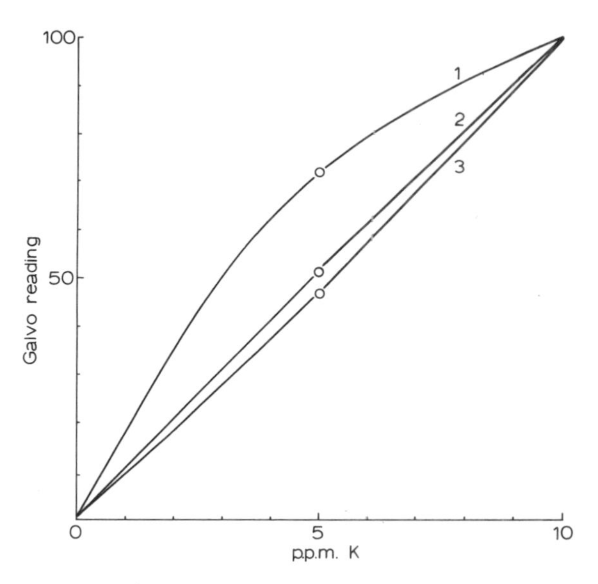

Fig.50. Potassium calibration curves using air–propane flame (λ = 7,665 Å). Curve *1*: 0–1,000 p.p.m. K; curve *2*: 0–10 p.p.m. K; curve *3*: 0–1 p.p.m. K.

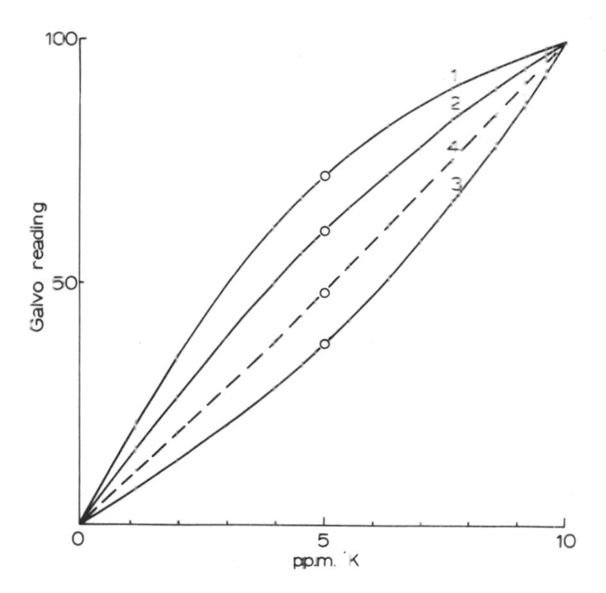

Fig.51. Potassium calibration curves using air–acetylene flame (λ = 7,665 Å). Curve *1*: 0–1,000 p.p.m. K; curve *2*: 0–100 p.p.m. K; curve *3*: 0–1 p.p.m. K; curve *4*: 0–1 p.p.m. K + 100 p.p.m. Na.

the number of free electrons in the flame. The degree of ionization is reduced as the concentration of free electrons increases. It becomes apparent, therefore, that the introduction of a second element, which can provide free electrons, will result in a reduction of the number of ions produced from the first. Fig.51 illustrates this, and it can be seen that when sodium is added to standard potassium solutions a linear calibration is obtained under the conditions which resulted in a curve when sodium was absent.

It can be seen from the examples already given that the requirements for avoiding different kinds of interferences are sometimes antagonistic. In order to avoid self-absorption a low sample concentration should be used; this, however, increases the probability of ionization. Another element may be added to reduce ionization but this may cause spectral interference. Usually a suitable compromise can be found and in the case of the alkali metals a cool flame such as air–coal gas, air–propane or oil–hydrogen flames are most satisfactory.

Anion effects

When a solution enters the flame, the solvent is first evaporated, then the liberated salts are vaporized, then dissociated before the metal element can be excited. With some salts, at low concentrations, this procedure is accomplished quantitatively and there is no difference, for example, in the emission from a given concentration of sodium whether it is present in the solution as a chloride or as a nitrate. Some salts are less easily vaporized and dissociated so that the intensity of the line is depressed considerably. The best known example of this is calcium phosphate. Table XVII gives the intensity readings obtained from solutions all containing 5 p.p.m. calcium but with different anions present. It can be seen that the depression is less marked in the hotter air–acetylene flame and this is a common feature in this work, the hotter the flame, the less marked is the depression. The various methods of overcoming this difficulty will be discussed later.

TABLE XVII

RELATIVE INTENSITIES OF CALCIUM SALTS

Solution	*Instrument response*[1]	
	Air–propane flame	*Air–acetylene flame*
5 p.p.m. Ca as chloride	100	100
5 p.p.m. Ca as sulphate	76	91
5 p.p.m. Ca as phosphate	39	88

[1]For each flame the instrument sensitivity adjusted to give a reading of 100 with the chloride solution.

ANALYTICAL TECHNIQUES

Several basic quantitative procedures are available, but whichever one is used for a particular determination a number of standardized solutions are required to calibrate the equipment. Since this is a vital pre-requisite particular care must be taken in the preparation of solutions and the following precautions observed:

(*1*) Spectrographically pure salts must be used for the components present in major amounts in the standards. The purity of the salt used to provide the element being determined is not important so long as the concentration of that element in the final standard can be guaranteed. For example, if one is making up a standard solution containing, say, 5 p.p.m. sodium and 500 p.p.m. calcium, and sodium is the element being determined, then a trace of impurity in the sodium salt used is not important, but the slightest trace of sodium in the calcium salt used will render the standard solution useless.

(*2*) Standard solutions must be stored in chemically inert containers, such as polyethylene bottles. When these have been thoroughly cleaned by allowing them to stand with 10% hydrochloric acid in them for several days they have proved to be very satisfactory. Glass bottles are highly suspect owing to their possible contribution of traces of alkalis to standard solutions from the glass.

(*3*) Great care should be exercised in handling all solutions, particularly diluted ones. Dust should always be excluded and contact with the hand avoided.

It is convenient to prepare stock solutions of about 1% or 1,000 p.p.m. concentration and to dilute these as required. Table XVIII gives details for preparing some suitable stock solutions each containing 1,000 p.p.m. of the appropriate element.

Many analytical techniques are available, the choice generally depending on the complexity of the sample. The follow-

TABLE XVIII

DIRECTIONS FOR PREPARATION OF STOCK SOLUTIONS

Element	*Directions for 1 l*
Calcium	slurry 2.497 g $CaCO_3$ with water[1] and add HCl until just dissolved; warm gently to eliminate CO_2 then dilute to volume with water
Cobalt	dissolve 4.7676 g $CoSO_4 . 7H_2O$ in water
Copper	dissolve 3.942 g $CuSO_4 . 5H_2O$ in water
Iron	dissolve 1.000 g iron wire in HCl, simmer to drive off excess acid and dilute to volume with water
Lithium	slurry 5.3233 g Li_2CO_3 and treat as under Ca
Manganese	dissolve 1.5828 g MnO_2 in HCl and proceed as under Fe
Nickel	dissolve 1.000 Ni metal in HNO_3 and proceed as under Fe
Potassium	dissolve 1.9070 g KCl in water
Silver	dissolve 1.000 g silver in HNO_3 and proceed as under Fe
Sodium	dissolve 2.5416 g NaCl in water
Strontium	slurry 1.9991 g $SrCO_3$ in water and proceed as under Ca

[1]Where water is quoted, distilled or deionized water must be used.

ing methods are included to give a wide indication of the state of the subject, but not all are suitable for the analysis of geochemical specimens. The degree of the usefulness of each method is indicated.

Direct method from calibration curve

In this method a number of standards are used to calibrate the equipment and to prepare a curve showing the concentration–intensity relationship. In the simplest case where the sample contains nothing which will affect the intensity of the element being determined a range of standards is prepared to cover a suitable concentration range. The zero of the instrument is set with distilled water and the sensitivity adjusted so that the strongest standard gives a full scale or any other suitable deflection of the recorder. The intermediate standards are then used to obtain readings which are plotted against concentration. The curve obtained, by averaging replicate readings, is sometimes reproducible from day to day providing the gas flows to the burner can be set accurately. This must never be taken for granted and experience must be built up with every particular instrument.

When the sample contains components which interfere with the intensity of the element being determined the above method cannot be used. There are cases, however, when the concentration of the interfering substances is known and is constant from sample to sample. A "blank" and a set of standards is then prepared, each containing the interfering salts. The instrument is set to read "zero" using the blank and then procedure is as before.

The method has severe limitations; although it is useful for any kind of instrument a great deal must be known about the samples. They must either contain no interfering ions, or the concentrations must be known accurately. In the latter case it must be possible to make up standards virtually of the same composition in all respects save that of the concentration of the element being determined and this may prove difficult in that the components must be completely free of that element. For these reasons the technique is virtually limited to the determination of the alkali metals in complex samples such as geochemical specimens.

Background correction

In some cases the interfering materials in the sample can be shown to have no effect on the line intensity but do alter the general background level. If the concentration of these substances alters from sample to sample or is unknown in any particular sample then the first method above cannot conveniently be applied. With an instrument which employs a monochromator, however, the difficulty can be easily overcome. Instead of working at a fixed wavelength the background level can be established for both samples and standards and the intensity of the line above this level measured for quantitative purposes. Two methods are available for doing this.

In the first, the line, plus background intensity, is measured at the wavelength corresponding to the peak of the line. The wavelength is then changed until a minimum reading is obtained to one or both sides of the line. This reading is then subtracted from the first one obtained and the difference used for quantitative assessment. If the background is sloping then the readings on the two sides of the line will be different and an interpolated value must be used for correction.

Much the better way of accomplishing the required correction is to use a recording instrument where the intensity can be recorded against wavelength and the background estimated by drawing a line across from the minimum levels ar each side of the line (see Fig.52). The background, due to the sodium slopes, is variable but the intensity of the CaO band at about 5,540 Å does not vary with increasing sodium content. In this case, therefore, it is possible to calibrate the instrument with standards containing a known amount of calcium and no sodium and then obtain a good estimation of calcium on samples containing large amounts of sodium providing other materials causing true enhancement or inhibition are absent.

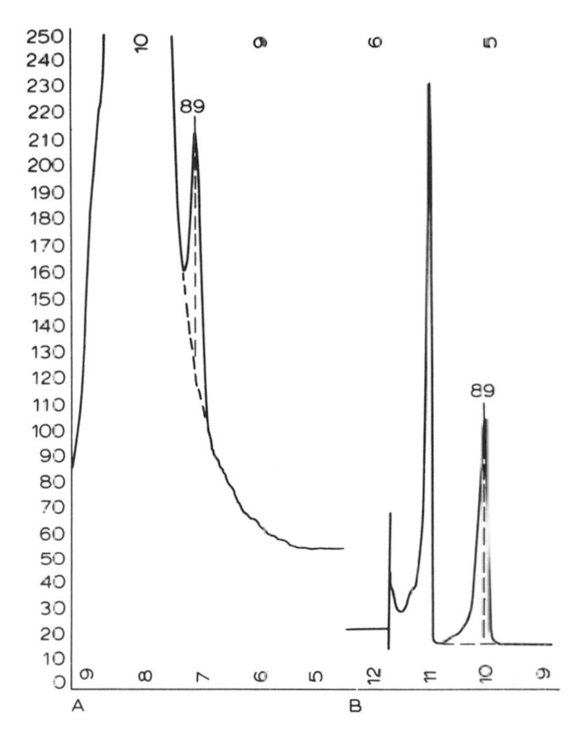

Fig. 52. Background effect due to sodium and calcium determination in an oxy–hydrogen flame. A. 1,000 p.p.m. Na and 5 p.p.m. Ca. B. No added Na and 5 p.p.m. Ca.

Dilution method

This technique, which was proposed by GILBERT et al. (1950) can be used on samples where interferences are suspected and it is known that the intensity of the analytical line is directly proportional to concentration, that is, the calibration curve of the element is linear. The instrument is first calibrated with solutions containing no interfering substances and the apparent concentration of the element is determined in the sample solution. The sample solution is then diluted to half concentration and the element again determined. The second

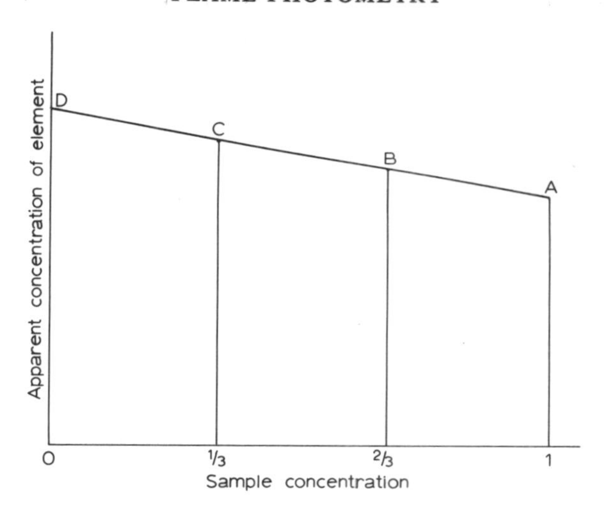

Fig.53. The dilution technique. (*A, B, C, D*: see text.)

value will be half the first if there is no interference, but if this is not the case then the whole dilution procedure can sometimes be repeated over a lower concentration range until the interference is eliminated. In certain cases, however, the interference is directly proportional to concentration of interfering materials and it is impossible to eliminate it by dilution, in this case an extrapolation procedure to zero-concentration of interfering substances can give a good approximation. In Fig.53 the sample itself gives apparent concentration *A*. After dilution to 2/3 concentration the apparent element content is multiplied by 3/2 and becomes *B*. After dilution to 1/3 concentration, the apparent content is multiplied by 3 and becomes point *C*. Back-extrapolation to zero concentration gives point *D* and this is the real concentration of the element in the sample. It must be emphasized that the method is applicable only if two conditions are satisfied.

(*1*) The intensity of the element line is directly proportional to concentration over the range used.

(*2*) The degree of interference is directly proportional to concentration of interfering substances.

The standard addition technique

This method is more generally useful than those described above since it is applicable to those samples which are believed to contain interfering materials at unknown concentrations. It has been applied by CHOW and THOMPSON (1955 a,b) to the determination of strontium and calcium in sea water and by BREALEY et al. (1952) to the determination of lithium in a range of samples which contained large excesses of sodium. There are several methods of applying the technique, the choice would depend on personal preference and on the knowledge one had of the sample. The one which has been found to yield the most reliable results on a wide variety of samples is as follows.

Four equal aliquots of a solution of the sample are measured into volumetric flasks and labelled *A, B, C* and *D*. Different known quantities of the element being determined are added to *B, C* and *D* and all four solutions are diluted to similar volumes. The intensity of the element line is then determined for each solution using the recording technique so that the background effect is eliminated. In Fig.54 the method is shown and the intensities are called I_s, I_1, I_2 and I_3 respectively. Referring now to Fig.55 the values obtained are plotted against the concentration of added element in the solution, in the diagram:

Solution *A* has no added elements and has intensity I_s
Solution *B* has y p.p.m. added elements and intensity I_1
Solution *C* has $2y$ p.p.m. added elements and intensity I_2
Solution *D* has $3y$ p.p.m. added elements and intensity I_3

This line is extrapolated back, shown dotted, and its interception with the concentration axis gives the amount of element in the solution (x p.p.m.).

The only requirement in this method is that a linear calibration is obtained as shown and this version of the technique ensures that this is so. If a curve is obtained the determination should be repeated at a lower concentration range.

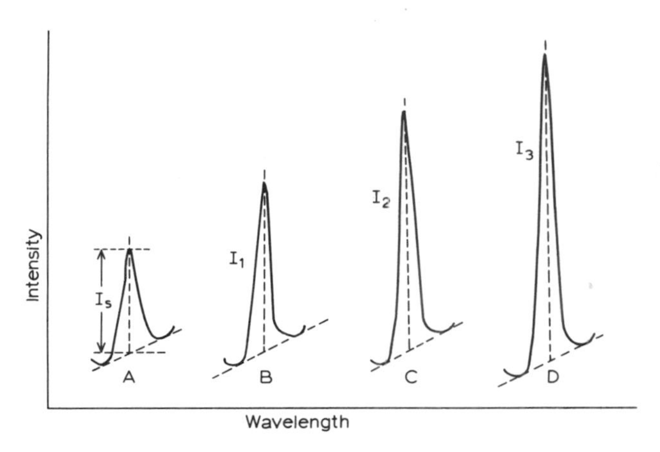

Fig.54. Recordings from solutions used in standard addition technique. (For the meaning of symbols, see text.)

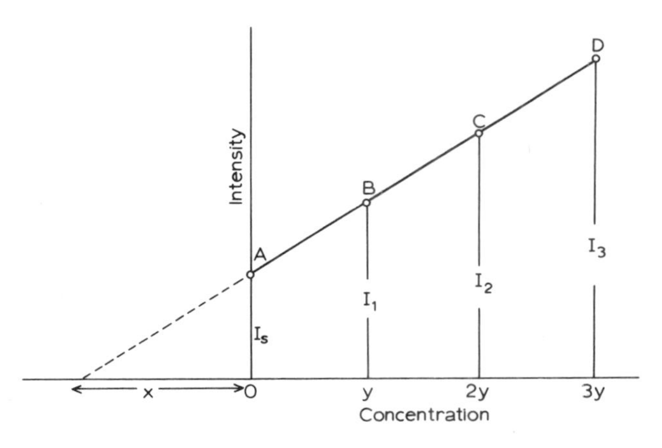

Fig.55. The standard addition technique. (For the meaning of symbols, see text.)

The internal standard method

Internal standard methods are used extensively in other branches of spectrography and they have been advocated in flame photometry; indeed some instruments have been designed specially for the purpose. The principle is to add to all standards and samples a fixed quantity of an element not present in the sample. The emission intensities of the sample and standard line are measured simultaneously on a two channel instrument or the spectral region is scanned on a recording instrument and the ratio of the intensities of the sample and standard lines are used for analytical purposes.

If the application of the technique is to be successful the internal standard line must be chosen with the same kind of care as is given to the selection in arc or spark spectrography. The requirements are outlined in that section on p.234. It can be seen that the selection will prove difficult because of the limited number of lines excited in a flame and in fact, except for some specialized cases, the technique has not been widely applied in flame photometry. In the case of geochemical specimens it would usually prove very difficult indeed to find a suitable internal standard element which was entirely absent in the sample.

The use of spectroscopic buffers

Anionic interference is particularly troublesome in many applications and the alkaline earths are particularly affected. The depression of the calcium emission in the presence of phosphate has been known for a long time. The phenomenon was first reported by GOUY (1879) and it has been investigated extensively. As phosphate is added to a solution of calcium the intensity of the calcium lines and bands decreases progressively until a certain concentration is reached beyond which further additions of phosphate have no further effect. Fig.56 illustrates this quite clearly. Advantage can be taken of this property by adding excess phosphate to all samples and standards so that the effect of any present in the sample is swamped. This is a simple application of a buffer but the method is not always applicable.

In the presence of aluminium calcium lines or bands can

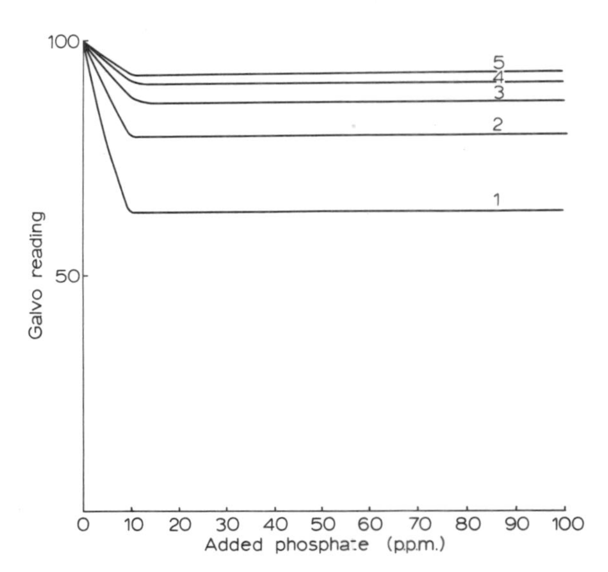

Fig.56. The emission from solutions containing 5 p.p.m. Ca + phosphate at different levels in an air–acetylene flame (λ = 5,500 Å). The base of slit above burner top is for curve *1*: level; curve *2*: 1 cm; curve *3*: 2 cm; curve *4*: 3 cm; curve *5*: 4–5 cm.

disappear completely. This is thought to be due to the fact that in the flame a highly refractory aluminate is produced which is not dissociated at the temperatures involved. The phenomenon was first reported in detail by MITCHELL and ROBERTSON (1936). Similar depressions have been encountered in the presence of silicate (DIAMOND, 1955), chromate and iron (SERVIGNE and DE MONTGAREUIL, 1954), beryllium (PINTA, 1951) and sulphate (IKEDA, 1956). As in the case of phosphate the calcium line does not always disappear completely and the degree of the depression depends on flame temperature, the lower the temperature the greater the depression. It is also noticed that the effect is not the same all through the flame but decreases towards the top (see Fig. 56). All of these observations support the contention that refractory salts or combinations are produced.

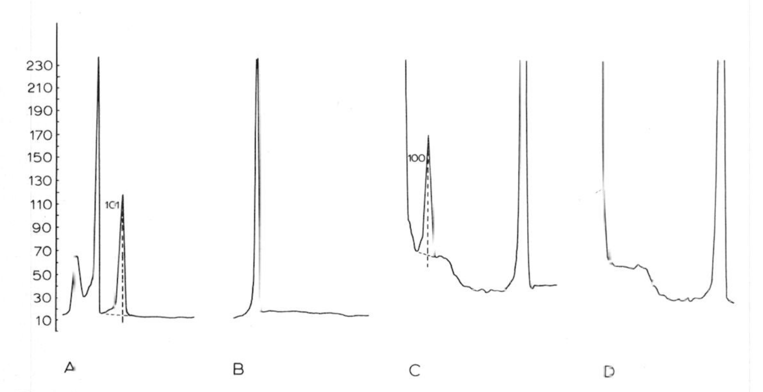

Fig.57. Restoration of calcium emission in presence of aluminium on adding strontium, in an oxy–hydrogen flame. A. 5 p.p.m. Ca; B. 5 p.p.m. Ca, 100 p.p.m. Al; C. 5 p.p.m. Ca, 100 p.p.m. Al, 2,000 p.p.m. Sr; D. 2,000 p.p.m. Sr.

To overcome the difficulty use may be made of the principle of mass action. The addition of an excess of another element which also forms an equally, or more refractory salt with the interfering ion is added to the sample. An example is illustrated in Fig. 57. Here it can be seen that the addition of strontium to samples containing calcium and aluminium result in the theoretical intensity of the calcium being restored. This technique is widely applicable and it is suggested that it would be useful for geochemical samples, particularly when a recording instrument is available.

Other buffers have been advocated for the determination of the alkaline earths. For example, the addition of dextrose was shown by PRO and MATHERS (1954) to restore to some extent the calcium intensity in the presence of phosphate. In a similar problem WIRTSCHAFTER (1957) found that the addition of EDTA (ethylene diamine tetra-acetic acid-Na salt) was effective. It is difficult to imagine, however, by what mechanism these organic materials could have an effect in a flame unless the presence of free carbon derived from these compounds is significant. If this is the case it might also explain the report by Knutson that in determining magnesium, anion nterference is decreased as the proportion of acetylene in an oxy-acetylene flame is increased.

A study of the determination of alkaline earths, particularly calcium, leads to the opinion that the use of the cationic buffer is most widely applicable to geochemical samples.

Solvent extraction

Solvent extraction techniques have been introduced into flame photometry, largely as a result of the work of Dean and his co-workers. Selective extraction by an organic solvent, with or without a complexing agent, can be used to accomplish the separation of a given element from a mixture and the solvent solution aspirated directly into the flame using a direct injection burner. In addition to separating an element from a mixture which might contain interfering components a greatly enhanced intensity of the line of the element being

TABLE XIX

SOLVENT EXTRACTION PROCEDURES

Element	*Extraction procedure*	*Reference*
Iron	acetylacetone from strongly acid solution	DEAN and LADY (1955)
Iron	4-methyl-2-pentanone from 5 *M* acid	DEAN (1960)
Chromium	4-methyl-2-pentanone from 1 *M* HCl	BRYAN and DEAN (1957)
Lanthanum	4-methyl-2-pentanone from solution at pH5	MENIS et al. (1959)
Aluminium	cupferron in 4-methyl-2-pentanone from solution at pH 2.5	ESHELMAN et al. (1959)
Copper	salicylaldoxime in chloroform	DEAN and LADY (1956)
Copper, Nickel, Manganese	diethyldithiocarbamate in chloroform	DEAN and CAIN (1957)

determined usually results. The reason for this is twofold. Firstly, the solvent is completely vaporized in the flame so that the whole of the sprayed sample becomes available for excitation and, secondly, the temperature of the flame is often increased as a result of the combustion of the organic solvent vapour.

The technique has been shown to have great potential and is being increasingly applied to a wide variety of samples. Table XIX lists a number of extraction procedures which have been reported by Dean and his co-workers.

Ion exchange methods

Ion exchange has been used to separate certain metals from an interfering environment. For example, calcium has been absorbed on a cation exchange column and then eluted with nitric or hydrochloric acid. Alternatively, phosphate can be eliminated from a solution by passing the solution through anion exchange resin in the chloride form. GEHRKE et al. (1955) used Ambertlite IR-413 for the purpose.

The technique is most useful for repetitive work where a large number of samples merit the effort required to prove the efficiency of the separation and where the necessary columns can be assembled and maintained as a long term fixture. Ion exchange techniques are less useful when samples are only examined occasionally.

CHAPTER 7

X-Ray Spectrography

INTRODUCTION

Soon after his discovery of X-rays in 1895, Röntgen made every effort to disperse a beam into its spectrum but his experiments were unsuccessful because his prisms had refractive indices so nearly unity at the short wavelengths of X-rays that dispersion was not possible. The work of VON LAUE et al. (1913), which showed that salt crystals could be used as diffraction gratings, enabled BRAGG (1912), MOSELEY (1913, 1914), and others to resolve spectra and show that the emission lines were characteristic of the material used as a target in the X-ray tube. It was early realized that this phenomenon was potentially useful for analysis but the laborious process of mounting a sample on the target of an X-ray tube and evacuating the tube militated against its widespread use.

It was then found that when materials were irradiated with X-rays, a secondary emission was excited and the spectrum produced was identical with that obtained by electron bombardment. VON HEVESY (1932), COSTER and VON HEVESY (1923) and others investigated the possibilities of fluorescent X-ray spectroscopy as a means of qualitative and quantitative analysis, but at that time there were insuperable difficulties which arose from the extremely low intensities of the fluorescent X-rays. The only method of recording available was a photographic process and very long exposures were required. From that time the technique was known as being potentially useful but virtually impossible from a practical point of view.

Since 1946 X-rays tubes have been produced, having output energies several orders of magnitude higher than those available to the early workers. In addition highly sensitive detectors have been produced and the technique has become established as an analytical tool of some importance. It has been mostly used for metal analysis but geochemical specimens in powder, solution or fused forms can be analysed and a great deal of work has been done, particularly in the U.S.A., to accumulate standard samples and develop methods of analysis.

The main attraction of the technique lies in the fact that under favourable circumstances some elements can be determined with an accuracy as high as that attainable by the best chemical methods. Sensitivity is generally much lower than that given by optical spectrography and so X-ray spectrography is most useful for the determination of elements at concentrations of parts per thousand or higher.

Until about 1959 the most serious limitation was that elements of atomic number less than 20 could not readily be determined and this excluded many of great interest to the geochemist. Recently commercial equipment has been available with which elements down to atomic number 12 can be determined and it is anticipated that range will be extended even further.

THE ORIGIN OF X-RAY SPECTRA

When an element receives energy which results in changes of energy level of electrons in the outer orbits the nett result is the emission of radiation having wavelengths between 2,000 Å and 8,000 Å. When the available energy is such that electrons from the inner orbits are involved the emitted radiation is in the X-ray region of the spectrum, i.e., 0.1–15 Å.

The Bohr theory of the atom postulates that the electrons surrounding the nucleus occupy shells which are named K, L, M, N, etc. Two electrons can be contained in the K shell, eight in the L shell, eighteen in the M shell, etc. The electrons in the K shell possess the same energy but there are three sub-levels in the L shell named L_I, L_{II} and L_{III} and there are *t* sub-levels in the M shell, M_I to M_V (see Fig.58).

If the energy received by the atom is such that an electron from the K shell is removed from the influence of the nucleus then the atom is in an excited state and will revert to the ground state by an electron from an outer shell falling into the vacancy. This is accompanied by the emission of radiation, the wavelength of which depends on the transition involved. If the electron movements are L→ K then the lines emitted are known as the $K\alpha$ lines. The possible transitions are $L_{II} \rightarrow K$ and $L_{III} \rightarrow K$ ($L_I \rightarrow K$ is "forbidden") and the two lines produced are $K\alpha_1$ and $K\alpha_2$. As the sub-levels in the L shell are separated by a very small difference in energy the

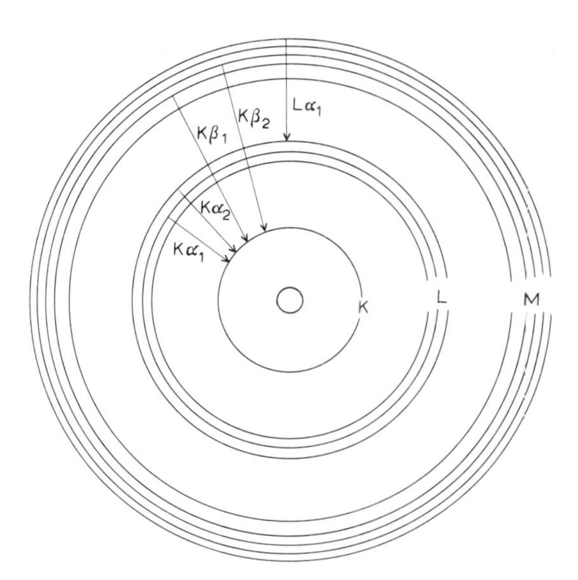

Fig.58. The origin of X-ray lines. (For the meaning of symbols, see text.)

two lines produced are Kα_1 and Kα_2. As the sub-levels in the L shell are separated by a very small difference in energy the two lines are close together and appear in the spectrum as a closely spaced doublet. When the transition is M→ K then the Kβ spectrum is produced, in this case there are two possible transitions (three are forbidden) and, therefore, two lines are produced.

The same principles apply to the origin of the L spectrum but since more sub-levels of energy are involved there are many more possible transitions and the spectrum is more complex. The energy change associated with an electron falling into the L shell is less than that of one falling into the K shell and, therefore, for a given element the L spectrum is at a

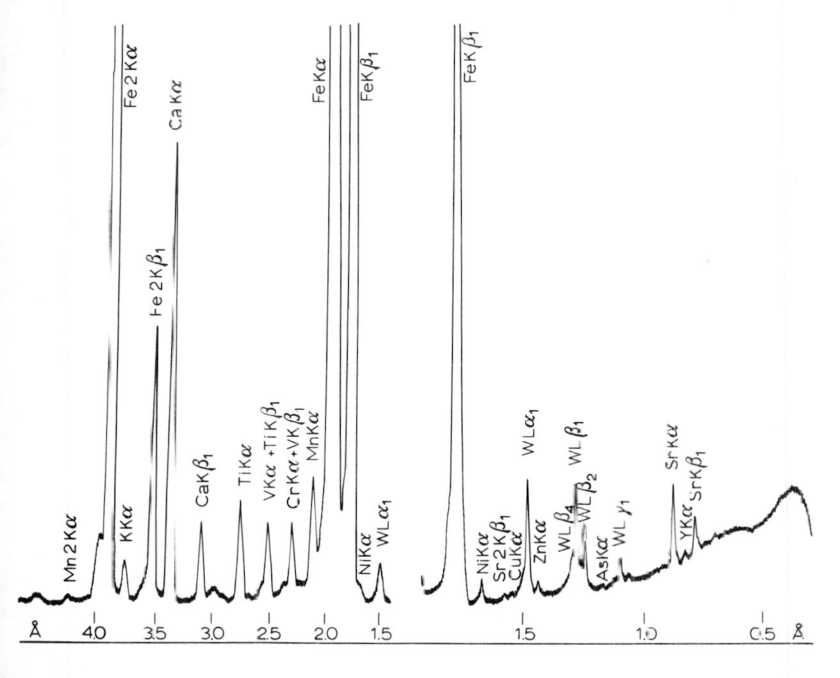

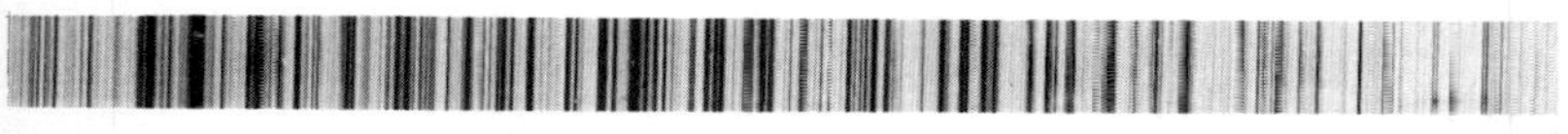

Fig.59. X-ray and emission spectra of ironstone.

longer wavelength than the K spectrum. The M and N spectra are at still longer wavelengths and are rarely used in analysis.

It is seen that since the number of possible electron transitions involving the inner orbits is small, X-ray spectra are relatively simple (see Fig.59) and in 1913 Moseley showed that the wavelength of a given line and the wave number of the element bore a systematic relationship according to the equation:

$$\frac{1}{\lambda} = c\,(Z-\sigma)^2$$

where λ = wavelength of line in Å, c and σ are constants, and Z = atomic number.

Using this law and classifying the elements in order of atomic number led to the identification of hafnium and rhenium by their X-ray spectra.

THE EXCITATION OF SPECTRA

The oldest method of generating X-rays is by accelerating a beam of electrons to a high velocity and causing it to fall onto a target of some kind. This is accomplished in an evacuated tube, the electrons are commonly emitted from an electrically heated filament which is held at a high negative potential with respect to the target. Most of the available energy is converted to heat and so the target is usually water cooled.

The line spectrum produced in this manner is superimposed on a background of continuous radiation which takes the form shown in Fig.60 and has this same general shape for all target materials. This continuum is the result of a continuous rather than discrete loss of energy of the electrons in the target material. The emission is cut off very sharply at a minimum wavelength at *A* and the position of this point depends on the energy of the electron beam. This minimum

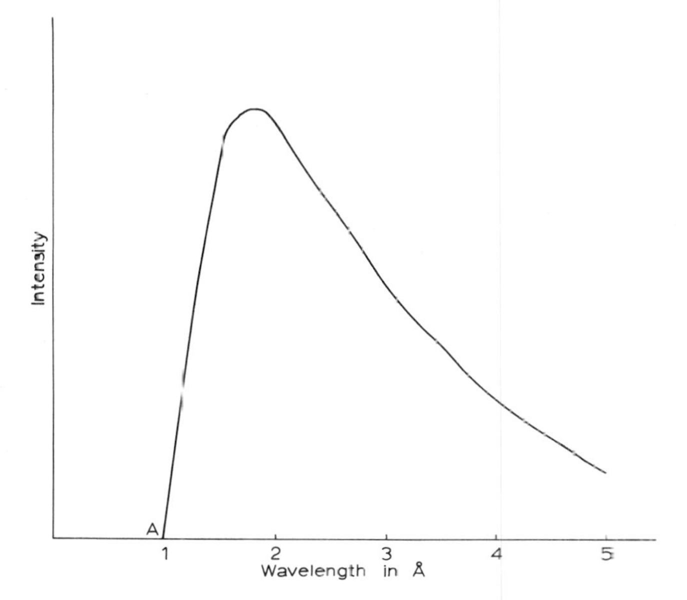

Fig.60. Continuous radiation from X-ray tube. A = cut off of emission.

wavelength may be calculated approximately from the expression:

$$\lambda_{\text{min.}} \simeq \frac{12{,}350}{V}$$

where λ is in Ångtrom units (Å), and V= voltage between filament and target.

This expression is derived from the equation:

$$eV = \frac{\text{hc}}{\lambda_{\text{min.}}}$$

where e = charge on an electron, V = voltage applied, h = Planck's constant, c = velocity of light, and $\lambda_{\text{min.}}$ = minimum wavelength emitted.

With modern equipment X-rays down to wavelengths

TABLE XX

CRITICAL EXCITATION POTENTIALS

Atomic no.	*Element*	*Excitation potential (kV)*			
		K edge	*L edges*		
			I	*II*	*III*
12	Mg	1.3			
16	S	2.5			
20	Ca	4.0			
26	Fe	7.1			
40	Zr	18.0	2.5	2.3	2.3
48	Cd	26.7	4.0	3.7	3.5
56	Ba	37.4	6.0	5.6	5.2
74	W	69.5	12.1	11.5	10.2
92	U	115.6	21.8	20.9	17.2

of about 0.01 Å can be generated by applying a sufficiently high voltage to the tube.

Providing the energy available is high enough the characteristic line spectrum of the target material is excited and is superimposed on the background radiation. The voltage through which an electron must be accelerated in order to give it sufficient energy to remove an electron from an inner orbit is known as the critical excitation potential. In Table XX the values are given for the inner shells of some elements. It is seen from these figures that it is possible to excite the L spectrum of an element without exciting the K spectrum by a suitable choice of applied voltage on the X-ray tube. When an electron is removed from the K shell it may be replaced by one from the L shells with the emission of a Kα line. This leaves the L shell short of one electron and an L line is consequently emitted when a further electron fills the gap. The process is repeated so that it is not possible to excite a K spectrum without also exciting the L, M, N, etc., spectra.

Fluorescent spectra

If X-rays of sufficiently high energy and intensity are directed at a target, then a secondary radiation of the X-ray spectrum of the target material may be produced. This phenomenon is analogous to fluorescence in the visible region and is of considerable importance in the routine practice of X-ray spectrography.

The same energy is required to remove an electron from a given shell whether the source of the energy is an electron beam or an X ray beam. In the latter case the energy level can be defined in terms of a wavelength which will be somewhat shorter than the wavelength of the line being excited. This required wavelength coincides with the appropriate "absorption edge" of the element concerned. It has been shown previously that the shortest wavelength of the continuous radiation emitted from an X-ray tube can be determined from the applied voltage across the tube and so if the absorption edge for a given shell is known the tube conditions required to excite spectra can be defined. The formula used is:

$$V = \frac{12{,}350}{\lambda}$$

where λ is the wavelength of the absorption edge.

The intensity of the fluorescent spectrum increases rapidly as the minimum voltage across the tube supplying the exciting beam is exceeded and so in practice the voltage used is well in excess of that required to excite all the required elements in a sample. The fluorescent spectrum is normally excited by the continuous radiation but on occasion a line from the target metal might coincide with the absorption edge of the shell of an element in the sample and thus increase the intensity of the appropiate line or lines in its spectrum. It is not often possible to make use of this feature but it can be useful occasionally when the maximum possible sensitivity is required. Examples are the excitation of elements between arsenic and iron when using a tungsten target X-ray tube for excitation and phosphorus and silicon when using a molybdenum tube.

When a spectrum is excited by fluorescence no continuous radiation is produced similar to that which occurs when electrons are used because photons cannot lose energy in a continuous manner. A low intensity background is produced by two scattering phenomena.

(*1*) Unmodified scattering of the primary radiation by the sample. Here photons suffer elastic collisions with the atoms in the sample and are scattered without undergoing a wavelength change. Thus the general shape of the background is the same as that of the primary beam, but at a much lower intensity.

(*2*) Compton scattering, caused by inelastic collisions with atoms of low atomic number such as oxygen, carbon and particularly hydrogen. In this case the primary photons lose some energy in the collision and, therefore, are scattered as radiation of a longer wavelength than the primary beam. This effect is not important when metallurgical samples are being examined but powder samples, such as silicates, and particularly solution samples, produce this type of scattering (see Fig.61).

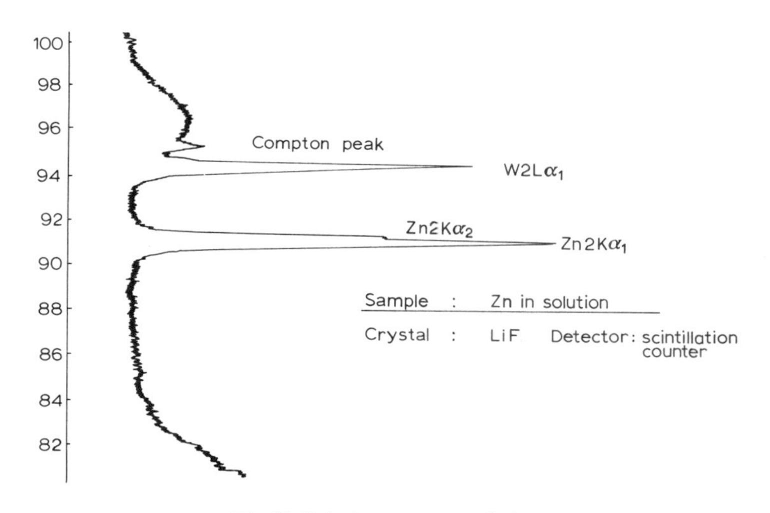

Fig.61. Solution spectrum of zinc.

A further source of scattered radiation can arise in the monochromator.

INSTRUMENTATION

The equipment required for the practice of the technique comprises a source unit, spectrometer and detecting and recording apparatus. Originally the source unit was a demountable X-ray tube in which the sample was mounted as the anode and the direct emission used for analysis. Photographic detection was used and since it was not very sensitive a high initial intensity was required or long exposures needed. In order to provide high intensity the electron beam in the X-ray tube had also to be of high intensity and consequently the sample was often overheated and decomposed. This, coupled with the fact that, at that time, high vacuum technology was not far advanced made the examination of samples a lengthy and cumbersome procedure and quantitative methods were impractical.

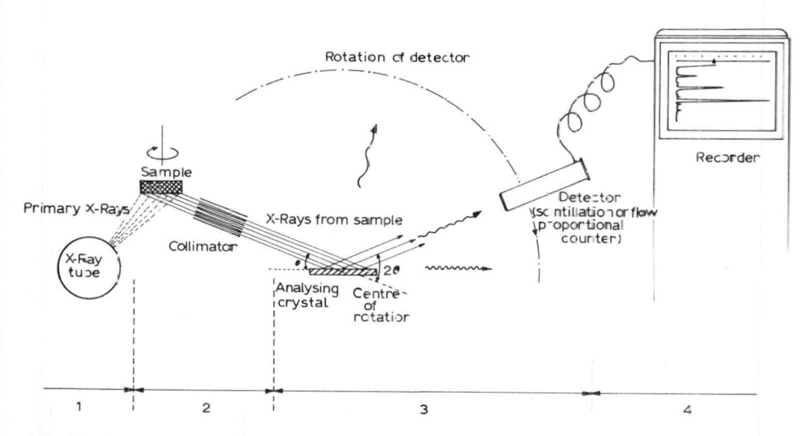

Fig.62. X-ray fluorescence spectrograph (schematic). *1* = X-ray generator; *2* = X-ray spectrograph attachment; *3* = wide range goniometer; *4* = electronic circuit panel with recorder.

Fluorescent methods of excitation were therefore introduced when detectors of sufficient sensitivity were developed and the method now forms the basis of modern equipment. A schematic diagram of a typical unit is given in Fig. 62.

The source unit

Sealed off X-ray tubes are commonly used to generate the primary X-ray beam. The window of the tube is usually made from beryllium and the target is usually tungsten, molybdenum, chromium or gold. As mentioned earlier the line spectrum of the target material may be superimposed at low intensity on the fluorescent spectrum. This is easily recognizable and becomes important only when small concentrations of the same element are being determined in the sample. For example, an X-ray tube with a tungsten target would not be used for determining small concentrations of tungsten in a sample although it could be used for high concentrations as the intensity of the scattered radiation would be insignificant compared with the intensity of the sample spectrum.

In order to obtain as high an intensity of primary radiation onto the sample as possible the distances between target, window and sample are kept to a minimum. When elements of atomic number less than 22 are being determined the wavelengths of both exciting and emitted light are in the range 2.7 Å and over, and air absorption becomes significant. The radiation paths must, therefore, be evacuated or filled with either hydrogen or helium to eliminate this attenuation.

The window of the X-ray tube also absorbs the primary radiation. At wavelengths less than 3 Å this is not highly significant but at 4.5 Å only 1% of the primary beam is transmitted by a beryllium window 0.030 inch thick and the transmission intensity falls to 0.1% at 5.18 Å and 0.01% at 5.7 Å. The use of a thinner window may be a partial answer to this problem in certain cases but since the lighter elements are of great interest this phenomenon is the cause of one of the principle limitations of fluorescent X-ray spectrography. BIRKS (1959) has suggested that the sample should be mounted inside a demountable, continuously evacuated X-ray tube and excited by fluorescence. The tube could be separated from an evacuated spectrometer by a thin plastic window since the pressure difference would be small. The scheme would be almost a reversion to the earliest techniques of VON HEVESY (1932) but would be more practical now in that high vacuum techniques have improved vastly. It would be no more of a problem in fact than that encountered in electron microscopy where the sample must be mounted in an evacuable space.

The spectrometer

Having excited the spectrum some form of dispersion is necessary in order to distinguish the lines of various wavelengths. Alternatively the wavelengths can be distinguished by taking advantage of the energy levels of the different lines. The latter method does not give very good resolution and is not normally used for the type of analytical equipment under consideration.

Various types of crystals are used as diffracting elements.

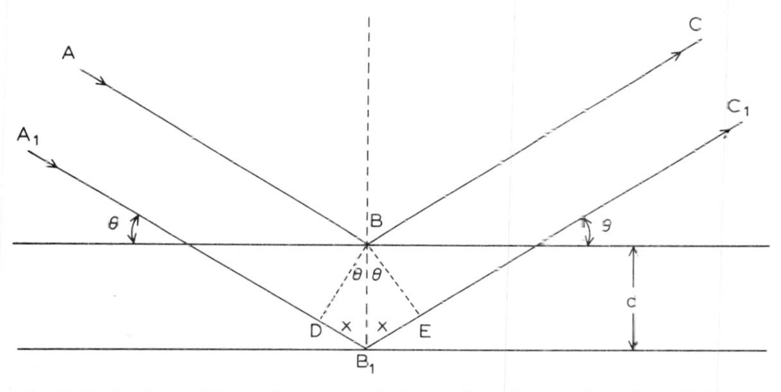

Fig.63. Reflection of X-rays from crystal planes. (For the meaning of symbols, see text.)

In Fig.63 the atom planes in a suitable crystalline material are represented by the lines $X\,Y$ and $X_1\;Y_1$, distance d apart.

$A\;B\;C$ and $A_1\;B_1\;C_1$ represent a parallel beam of X-rays impinging on the crystal at an angle θ to the normal and reflecting from the planes at the same angle. If $B\;D_1$ and $B\;E$ are constructed normal to the appropriate parallel beams it is seen that the distance travelled by $A_1B_1C_1$ exceeds that travelled by $A\;B\;C$ by $D\,B_1 + B_1E$.

Now $D\,B_1 + B_1E = 2\,d \sin\theta = 2x$.

If $2x$ is equal to a whole multiple of the wavelength of the incident beam then the intensity of the beam is reinforced. At other wavelengths interference attenuates the beam intensity. Thus when the beam comprises polychromatic radiation the wavelengths which are selectively reflected are given by the Bragg law:

$$n\,\lambda = 2\,d \sin\theta$$

where n is a whole integer. When $n = 1$ the reflected beam is said to be a first order diffraction, when $n = 2$ it is second order, and so on.

It can be seen that the maximum wavelength which may be diffracted is $2\,d$ when $\sin\theta = 1$ and since the planar distances of many crystals is of the same order as the wavelength of X-ray radiation a careful choice of crystal material must be made. For example, the planar distance of LiF is 2.01 Å so this material cannot diffract wavelengths greater than 4.02 Å. The K lines of potassium are at 3.7 Å so the spectra of elements of atomic number less than potassium cannot be resolved by crystals of LiF. By a similar argument quartz crystals can be used down to phosphorus and ethylenediamine ditartrate (EDDT) to aluminium. Obviously crystals of wider spacing are capable of diffracting shorter wavelengths and it may be thought that they would be suitable for dispersing the spectra of the heavier elements. By differentiating Bragg's law the following expression is obtained:

$$\frac{d\theta}{d\lambda} = \frac{n}{2\,d\cos\theta}$$

From this it is seen that the dispersion decreases as $2d$ increases and so the best separation of neighbouring wavelength is obtained by using a crystal which has the minimum possible spacing.

The simplest type of spectrometer utilising a flat crystal is shown diagrammatically in Fig.64.

The primary X-ray beam A strikes the sample S and the fluorescent spectrum is emitted in all directions. The collimator C which is normally a bundle of fine tubes or plates allows an almost parallel beam to fall onto the flat crystal E. Only one first order wavelength will be diffracted according to Bragg's law and this radiation is received by the detector placed as shown. The mechanism is arranged such that the angular movement of the detector is twice that of the crystal so that the whole spectrum can be scanned and the wavelength received by the detector is always known from its angular position. The wavelength range is covered when θ varies from 0° to 90°.

The resolving power of the arrangement depends largely on the geometry of the collimator. The greatest efficiency is obtained when the tubes or plates are closely spaced and the

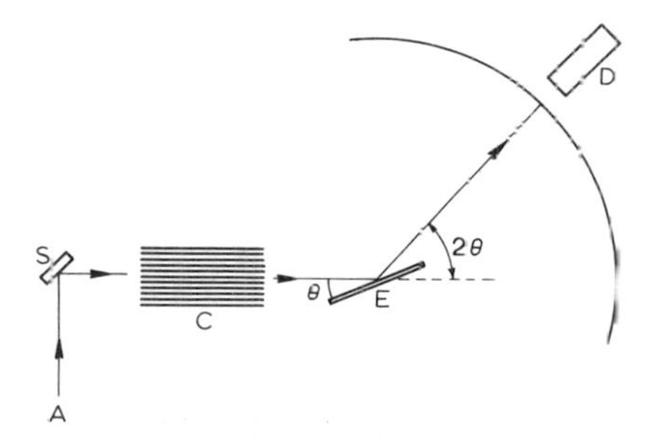

Fig.64. Geometry of plane crystal spectrometer. (For the meaning of symbols, see text.)

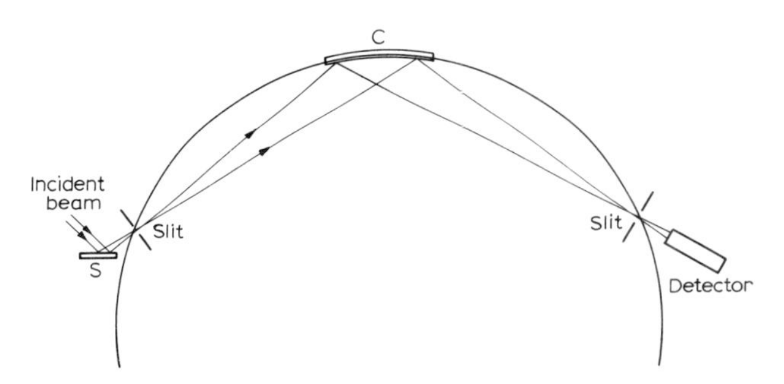

Fig.65. Curved crystal spectrometer. (For the meaning of symbols, see text.)

bundle as long as possible. The plate type is called a Soller Slit and this is usually about 4 inches long and the blades placed 0.010 inch apart. The cross-sectional area of the collimated beam is normally about 0.5 sq. inches.

The second type of monochromator in common use utilises a cylindrically curved crystal. The geometry of the arrangement is shown in Fig.65.

The fluorescent spectrum emitted by sample S is passed through a slit and the divergent beam strikes the curved crystal C. This crystal is bent and cut in such a way that all parts of the beam meet it at the Bragg angle and all radiation of the appropriate wavelength converges on the exit slit. The crystal acts in a manner analogous to a concave grating in an optical spectrograph. It must not be thought though that the beam is focused optically but it is diffracted in a manner such that one wavelength converges to one point. Fig.66 can be used to explain how this happens and also show the manner in which the crystal must be bent and cut.

The crystal is considered as a series of planar segments distributed on the circumference of a circle, C and C_1 are two such segments. Monochromatic radiation diverges from point A on the same circumference and two rays strike the segments C and C_1 at the Bragg angle θ when the planes of the segments are normal to the lines C_1D and $C D$. After diffrac-

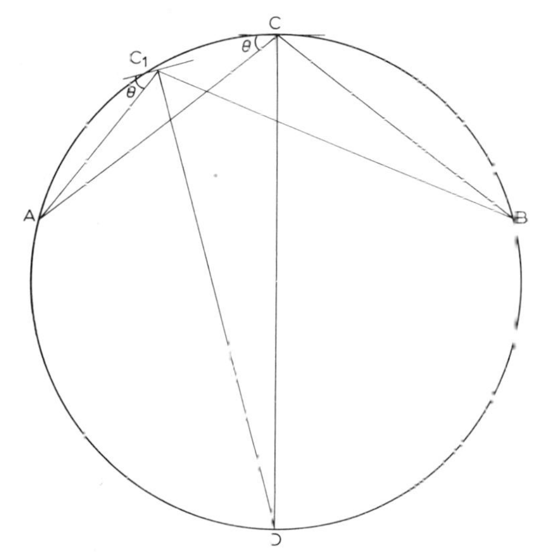

Fig.66. Geometry of curved crystal spectrometer. (For the meaning of symbols, see text.)

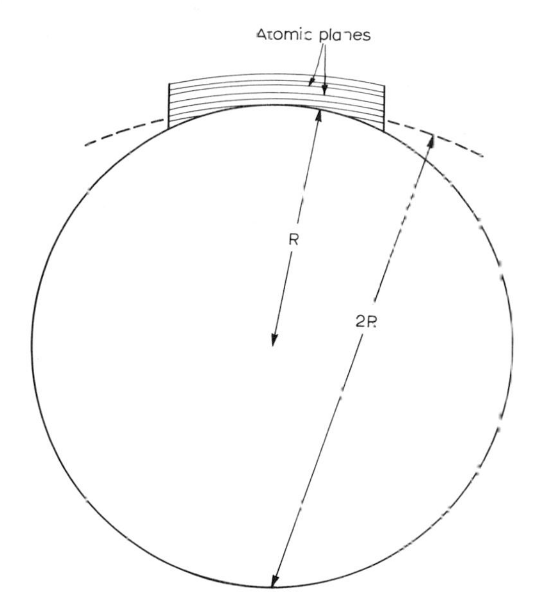

Fig.67. Form of curved crystal.

tion the two rays converge on B. Since B and D are also on the circumference $\angle A C_1 D = \angle A C D = \angle D C B = \angle D C_1 B = 90° - \theta$. Thus it is seen that the ideal conditions are obtained when the planes in the segments are tangents to circles drawn on D as centre and having radii of $C D$ and $C_1 D$ respectively. This requirement is satisfied if a crystal is bent to a radius $C D$ and the surface is then ground down to a radius $C D/Z$. It will then have the form shown in Fig. 67.

The detector

The detectors commonly used are Geiger, proportional, and scintillation counters.

Geiger and proportional counters consist essentially of a vessel filled with a noble gas such as argon, krypton or xenon. A wire is stretched down the centre of the vessel and is maintained at a fixed positive potential with respect to the walls of the vessel itself. Radiation enters through a thin beryllium or mica window. When a photon enters, it causes atoms of the gas to ionize. The released electrons are accelerated towards the anode and by impact may ionize other atoms thus releasing more electrons. This process is repeated and gives rise to an avalanche of electrons so that an appreciable current flows for each photon entering the detector. The magnitude of the current pulse depends on the applied voltage (see Fig.68).

At voltages higher than the threshold at A, the current rises almost linearly with applied voltage until at B the curve flattens and a plateau occurs between B and C. Here the amplitude is independent of the applied voltage and also of the energy of the photon initiating the discharge. The Geiger counter is operated in this condition, the applied voltage being chosen to lie about midway between B and C. The plateau usually starts at about 500–1,000 V and extends over a range 100–200 V, depending on the design of the tube.

The proportional counter is operated in the region between

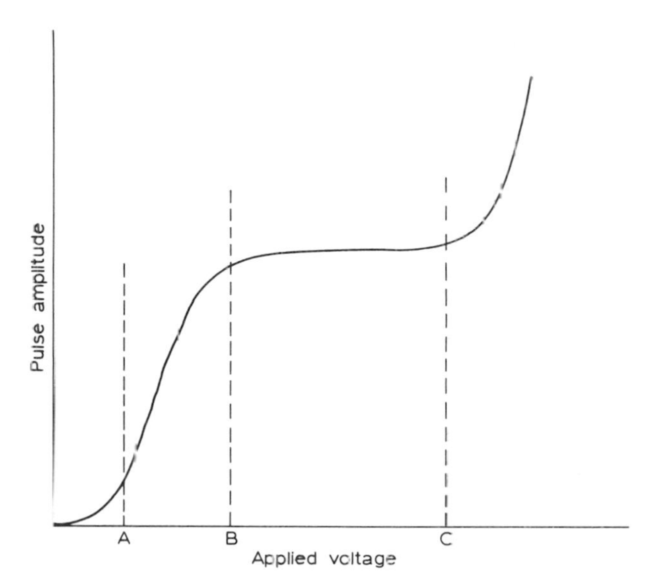

Fig.68. Effect of applied voltage on pulse amplitude from a Geiger counter. (*A, B, C*: see text.)

A and *B*, where the pulse amplitude is dependent both on applied voltage and on the energy of the photon initiating the discharge.

Scintillation counters depend on the fact that when an X-ray photon falls on some materials such as sodium iodide containing a trace of thallium, a flash of visible light is produced. This can be detected with a photomultiplier. As with the proportional counter the amplitude of the pulse obtained is proportional to the energy of the incident photon.

The Geiger counter is not much used nowadays despite the fact that it gives a high amplitude pulse of about 1–10 V which requires relatively little amplification. Its great disadvantage is a large "dead time" which becomes serious at high counting rates. The dead time is that period during which a photon is received and the avalanche of electrons is formed and collected. During this period any other photon entering the cavity is not detected. If N counts per second are recorded and t is the dead time in seconds then Nt counts are lost and the corrected count is given by the expression:

$$N(\text{corr}) = \frac{N}{1 - Nt}$$

The dead time of a typical Geiger counter is about 150–300 μsec and this gives rise to appreciable count losses even at low counting rates.

The proportional counter on the other hand produces much smaller pulses but has a dead time only of the order of 0.2 μsec since the electron avalanches are confined to the region where the photon enters and the remainder of the tube can detect a further photon. The scintillation counter has a dead time of the same order and with either detector this becomes important only at counting rates of 10^7 or higher.

The choice between the last two detectors is usually determined by the wavelength of the line being measured. The scintillation counter has a much better response at wavelengths up to about 1.0–1.5 Å but above this loses its advantage. At wavelengths above about 2.5Å the proportional counter has the better response. In addition the scintillation counter is at a disadvantage at very low count rates due to its "dark current" and so would not be chosen, other factors being equal, for the detection of trace constituents.

Measurement of line intensity

Monochromatic radiation is emitted from the sample in the form of photons the energy of each of which is a function of the wavelength. The "intensity" of the line is the rate of photon emission. The detector converts each photon into an electrical pulse, within the limits already specified so the measurement of line intensity resolves itself into the measurement of the rate at which pulses are received from the detector. This can be done in several ways of which the following two are most commonly used.

The first method and the one usually employed for qualitative, semi-quantitative and quantitative work where the greatest possible accuracy is not required is to pass the pulses through a ratemeter. This gives a current output dependent on the pulse rate. The spectrum is scanned and the pulse rate fed to a recorder so that a continuous trace of the spectrum is obtained as shown in Fig.69. The line intensity is obtained from the height of the peaks.

The second technique is to set the detector on the required

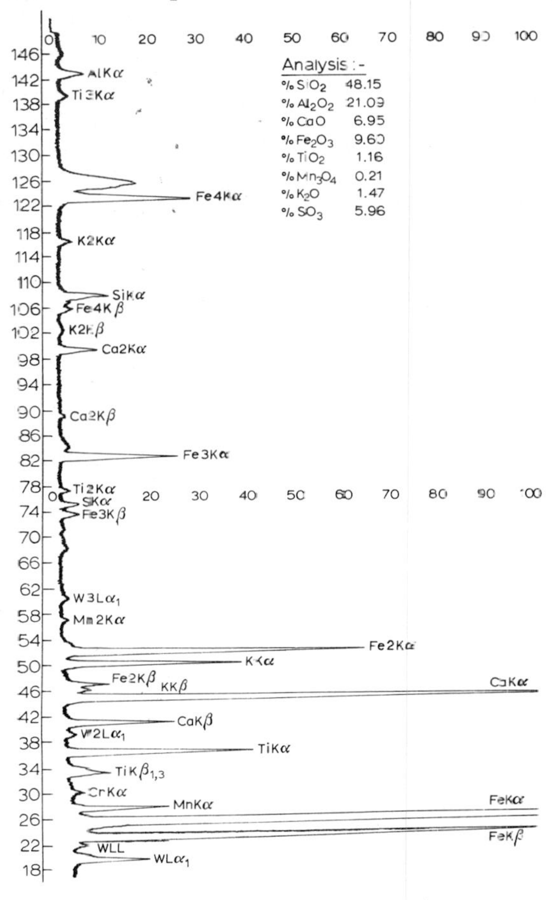

Fig.69. X-ray spectrum of coal-ash.

Fig.70. The Philips X-ray spectrograph.

line and either count the pulses in a fixed time or accurately time the interval during which a predetermined number of pulses is received.

The X-ray spectrograph has now been developed as a complete unit of equipment for routine geochemical analysis, and one such type is illustrated by Fig.70. The close-up view of the spectrometer (Fig.71) shows the ease with which the samples can be processed for analysis using the holders illustrated by Fig.72.

Fig.71. Close-up view of the spectrometer.

Fig.72. Sample holders and typical samples used with the Philips instrument. a = pressed disc; b = metal specimen; c = fused bead.

THE PREPARATION OF SAMPLES FOR ANALYSIS

Geochemical samples may be examined as solids, powders, compressed briquettes, fused masses or solutions. For greatest sensitivity and precision in quantitative work the sample surface irradiated by the primary beam should be fairly large and normally the full area of the sample holder is filled. This is normally greater than the cross-sectional area of the primary beam. The effective penetration of the X-rays into the sample is small and in practice a critical thickness is defined as that depth which gives a fluorescent intensity of 99% of the intensity obtained from a sample of infinite thickness. KOH and CAUGHERTY (1952) measured the critical thickness of iron, chromium and nickel metals and found it to be about 0.003 cm. With powders, of course, the value is greater but it is very important to ensure that the surface of the sample is representative of the whole.

Solid samples

Only rarely can solid geochemical samples be examined particularly in quantitative work. It might be expedient occasionally to examine small specimens qualitatively but in this case the primary beam must be reduced in area and loss in sensitivity will result. The only reason for using the method at all in this manner would be for the non-destructive examination of valuable specimens. If this is not necessary optical spectroscopy is preferred.

Powder samples

For greatest accuracy the sample should be powdered to a particle size much less than the critical depth. When this is done the manner of packing into the specimen holder does not much affect the results according to SHALGOSKY (1960). Modern instruments provide the facility to rotate the specimens and this is always done to avoid directional scattering of the X-rays from crystalline particles. This should not be serious if the particle size is small enough.

Briquetted samples

ADLER and AXELROD (1954) found that improvised reproducibility was obtained when powdered samples were compressed into briquettes, using an equal quantity of aluminium or corn starch as a binder. The effect of the binder is to reduce slightly the intensity of the spectrum but with the suggested materials this reduction does not exceed 20% . Briquetted specimens are easily handled and can be stored more conveniently than powders.

Fused samples

Any inhomogeneity of samples can be eliminated by fusing into a bead with a suitable mixture. CLAISSE (1956) used borax for this purpose, a known weight was mixed with a weighed amount of sample and fused in a platinum crucible over a Mekker burner. A small crucible furnace at 1,100°C has been found to give more reproducible results. This melt is poured into a hot mould and allowed to cool slowly to form flat discs of reproducible size. These are easily handled and stored.

Solution samples

One of the first applications of the techniques was the determination of lead in petrols by BIRKS et al. (1950). Aqueous solutions can also be examined in a special sample holder having a thin plastic window. One of the main problems is the background due to "Compton scattering", which arises from the hydrogen and oxygen in the water. This reduces the sensitivity of the method. The detection limits of a number of elements in solution have been determined and are given in Table XXI.

TABLE XXI

APPROXIMATE LIMITS OF DETECTION IN SOLUTION

Atomic no.	*Element*	*Limit of detection (p.p.m.)*
19	K	80
20	Ca	38
22	Ti	150
24	Cr	5
25	Mn	5
26	Fe	4
28	Ni	3
30	Zn	3
33	As	4
38	Sr	3
41	Cd	11
82	Pb	12

QUALITATIVE ANALYSIS

Because X-ray spectra are so simple qualitative analysis is easy and rapid. The sample can be used in almost any form and with the counter and crystal rotated by a motor and the output from the detector feeding a ratemeter and recorder a complete spectrum can be obtained in about 45 min. Fig.59 shows such a spectrum obtained from a powder sample of an iron ore and all the lines are identified. To illustrate the relative simplicity of the method a portion of the ultraviolet spectrum of the same sample is shown for comparison. Even complex samples seldom give more than 100 lines in their spectra.

QUANTITATIVE ANALYSIS

To obtain quantitative results it is necessary to relate line intensity and the amount of the element in the sample. Since the intensity depends on the voltage applied to the X-ray tube supplying the primary radiation it is essential to standardize this parameter for a given analysis and to maintain a constant voltage and current during the exposure time. Modern instruments, therefore, utilize stabilized power supplies and provide the facility of setting the tube voltage accurately.

The limiting factor in practical work is the "matrix effect", that is, the effect of the major constituents in a sample on the line intensities of other components. Either line enhancement or depression can occur. The latter occurs because some emission occurs below the surface of the sample and, therefore, has to pass through some matrix material and can be absorbed. Enhancement is caused when fluorescent lines of the matrix material are of a sufficiently short wavelength to excite the spectrum of the element being determined. This can happen when the matrix consists of elements higher in the periodic table than that of interest.

The most obvious method of eliminating the difficulties is to calibrate the equipment with standards of known composition or samples which have been analysed by other methods. Whilst this technique might be of use on an occasion it is usually too laborious for the analysis of geochemical specimens particularly when only a few samples of a particular type are to be examined.

Internal standard techniques can be very useful for the determination of minor constituents. If well chosen the matrix effect can be virtually eliminated and thus relatively simple standards can be used for calibration. It must be stressed that the method is most useful for minor constituents since the best results are obtained when the internal standard is added at about the same level of concentration as the element being determined and this may be an unreasonable requirement for major constituents, particularly when rare and rather expensive materials are used for the purpose. That this may often be the case becomes obvious from a consideration of the choice of internal standard.

(*a*) It must not be present in the sample.

(*b*) It must be close to the required element in the periodic table, the ideal choice may be an element separated by only one or two atomic numbers from the required element.

(*c*) It must not cause enhancement or depression of the element spectrum.

(*d*) It must be incorporated into the sample in a homogeneous manner so that if a powder or briquetted sample is being used the internal standard material must be reduced to the same particle size and thoroughly mixed. The problem is very much simpler for solutions and fused beads.

It is seen immediately that the choice of material for internal standard may be a difficult one and rarely will the situation arise when the same internal standard is suitable for the determination of more than one element.

If the requirements for choosing an internal standard cannot be satisfied then the method of standard addition can be used and this is probably more generally useful since it eliminates the dependency on standards of any kind. To our aliquot of the sample is added a known amount of the element being determined and the intensity of the line obtained for the sample and the sample with the addition. The concentration of the element in the sample may then be computed from the following equation:

$$\frac{I}{I_A} = \frac{C}{C + x}$$

where I = intensity of the line in the sample, I_A = intensity of the line in the sample with known addition, C = concentration of element in sample, and x = amount of element added.

The one requirement for the successful application of this technique is that the concentration–intensity relationship should be linear. If this is in doubt then two or more additions at different concentrations will soon establish the situation.

It is considered that both the above methods are most useful for the determination of elements up to 10% concentrations. For higher concentrations the sample may be diluted with some suitable material which effectively alters the matrix to that material and makes the preparation of standards a simple operation. In this category might be included the use of the fused borax button.

APPLICATIONS

The determination of hafnium is of particular interest since the element was discovered by X-ray spectrography and was

TABLE XXII

SELECTION OF REFERENCES

Element	*Sample*	*Reference*	
Hafnium	baddeleyite	SHALGOSKY	(1960)
Niobium and tantalum	ores	BIRKS and BROOKS	(1950)
Niobium and tantalum	ores	CARL and CAMPBELL	(1955)
Niobium and tantalum	rare-earth ores	MORTIMORE et al.	(1954)
Niobium and tantalum	bauxite	BRISSEY	(1952)
Niobium and tantalum	bauxite	ADLER and AXELROD	(1953)
Niobium and tantalum	ores	STEVENSON	(1954)
Niobium and tantalum	titanium dioxide	PATRICK	(1952)
Iron	pyrophyllite	PATRICK	(1952)
Iron, manganese	amphibolite rocks and hornblendes	CHODOS and ENGEL	(1960)
Barium	barite	GULBRANSEN	(1955)
Thorium	ores and solutions	PFEIFFER and ZEMANY	(1954)
Caesium and rubidium	mica, beryl, feldspar	ADLER and AXELROD	(1957)
Zinc and lead	ore	DESPUJOIS	(1952)
Calcium	wolframite concentrates	CAMPBELL and THATCHER	(1958)
Titanium, calcium, magnesium	rocks and ores	WEBBER	(1957)
Uranium	ores	WATLING	(1961)
Uranium	ores	WILSON and WHEELER	(1958)
Scandium	ores	HEIDEL and FASSEL	(1961)

one of the first elements to be determined by the pioneers FRIEDMAN and BIRKS (1948). SHALGOSKY (1960) has described in some detail its determination in baddeleyite, ZrO_2.

Other work is summarized in Table XXII from which it is seen that a fair variety has been reported. The concentration range over which determinations have been made are from around 0.3% hafnium by SHALGOSKY (1960) to 25–100% barium in barite by GULBRANSEN (1955).

In addition to the references in Table XXII a great deal of work has been done in the U.S.A. by the Bureau of Mines reported by CARL and CAMPBELL (1955) and the U.S. Geological Survey reported by ADLER and AXELROD (1955) in establishing standards and examining methods of analysis for elements from aluminium, silicon and magnesium to thorium and uranium.

THE ELECTRON PROBE MICROANALYSER

As mentioned earlier a fairly large sample is normally used in X-ray spectrography although smaller areas can be examined if some loss in sensitivity is acceptable. ADLER and AXEL ROD (1957b) examined inclusions in the surface of mineral specimens by masking the primary beam to an area of about 0.5 mm diameter and by moving the specimen about were able to map the concentrations of iron, cobalt and selenium in a pyrite ore.

At an earlier date CASTAING (1951) designed an instrument in which the sample was mounted inside an X-ray tube and the electron beam falling on it was brought to a very fine focus. The primary emission from the sample was then passed through a monochromator in the usual way so that an analysis could be obtained from a very small sample area.

The equipment has since been developed and is now available commercially. In its usual form the electron spot which impinges on the target is about 1 μ in diameter and is made to scan the sample in a pattern similar to that used in television. The emitted radiation is passed through a monochromator and the detector set on the line of a chosen element. The output is amplified and controls the intensity of a spot on a cathode ray tube. The spot is made to scan the tube in a motion synchronized to the scan of the electron beam on the sample so that an enlarged picture of the scanned area appears on the screen in the form of light and dark areas which correspond to the concentration of the chosen element in the sample.

The technique is of obvious importance in geochemistry and one example from the work of BIRKS and BROOKS (1957) may serve to illustrate its use. In chalcopyrite streaks of a birefringent material had often been seen. The optical properties had suggested that they might be composed of valleriite which contains iron and copper in the ratio 2 : 1, but since they were only about 10–15 μ wide it had not been possible to remove enough material for analysis. An examination of a specially prepared surface by the electron probe microanalyser showed that the streaks contained no copper but were probably FeS.

The instrument is potentially of extreme importance. At this time commercial models are very expensive and not many laboratories can justify the cost of installation but it is certain to be widely used in the future for it is probably the most convenient technique for detecting elements down to an absolute level of about 10^{-13} g and analyse such small areas. Certainly it is the only technique capable of doing this non-destructively. It is further suggested that this type of microanalysis, in which the changes in composition of samples can be determined semi-quantitatively, may prove of much greater use and interest than the accurate average analysis of a relatively large specimen.

CHAPTER 8

X-Ray Diffraction

INTRODUCTION

It was shown in Chapter 7 that the wavelength range of X-rays is of the same order of magnitude as the inter-atomic and inter-molecular distances in solid materials and that crystals, in which the atoms or molecules are oriented in a geometrically regular pattern, can be used as diffraction gratings to disperse polychromatic beams of X-rays in monochromators. This phenomenon can also be utilized to furnish information on crystalline substances which cannot be obtained in any other way.

It was BRAGG (1912) who first visualized the process and showed that X-rays were diffracted according to the law:

$$n\lambda = 2\,d \sin\theta \text{ (see p.286)}$$

This condition is fulfilled when a monochromatic, unidirectional X-ray beam is incident to a number of geometrically similar, parallel planes in a crystal at an angle such that the reflected beams from each layer reinforce each other and an intense diffracted beam results. At other angles the reflection from each layer interferes with that from each other layer and any resultant beam is of feeble intensity. It should be noted that whereas it can be said that a beam is "reflected" from any one plane in a crystal, the process as a whole is known as "diffraction."

It can readily be seen that if the wavelength λ of the incident X-ray beam is known and the angle of diffraction θ can be measured then the interplane distance d may be calculated. This in fact can be done with considerable accuracy, a precision of 1% is easily reached and with care a precision of 0.1% can be achieved.

The ultimate application is the determination of crystal structure but the technique is also widely used in qualitative and quantitative analysis to identify substances and to determine the composition of mixtures. In the latter applications the crystalline structure of each substance is accepted and used rather than investigated.

It is immediately obvious that here is a tool of great importance to the geochemist particularly when used in conjunction with other analytical techniques. Not only has it been used to determine the structures of minerals such as kaolinite and montmorillonite but the concentration of such minerals in clays can be estimated. Whereas the other analytical techniques described in this book may be used to determine the total concentration of, say, aluminium or silicon in a clay, by means of X-ray diffraction the different minerals and impurities making up the clay may be identified and often their concentration determined.

METHODS OF X-RAY CRYSTAL ANALYSIS

In nature crystals usually occur as an agglomerate although in some cases single crystals of fairly large size are found. Large crystals can also be produced artificially. Several different methods of X-ray examination have been devised each specializing in yielding information of a particular kind.

Laue method

This first X-ray method used was developed from a suggestion made by Von Laue and differs from all other subsequent techniques in that continuous radiation is employed. The X-ray tube used as a source usually employs a tungsten target and is operated with a voltage sufficiently low as not to excite the tungsten K radiation (see p.278). A slit system is set up in front of the window of the tube and a narrow collimated beam thus produced passes onto a fixed specimen which consists of a small single crystal (see Fig.73). A flat photographic plate is positioned on the remote side of the crystal to receive and record the diffracted X-rays. The distance between the X-ray tube target and the specimen is usually 15–20 cm and specimen to film is 5 cm, this latter distance being very accurately adjusted. The slit employed is generally 1 mm and the thickness of the specimen crystal varies between 0.1 and 2 mm depending on the type of material being investigated.

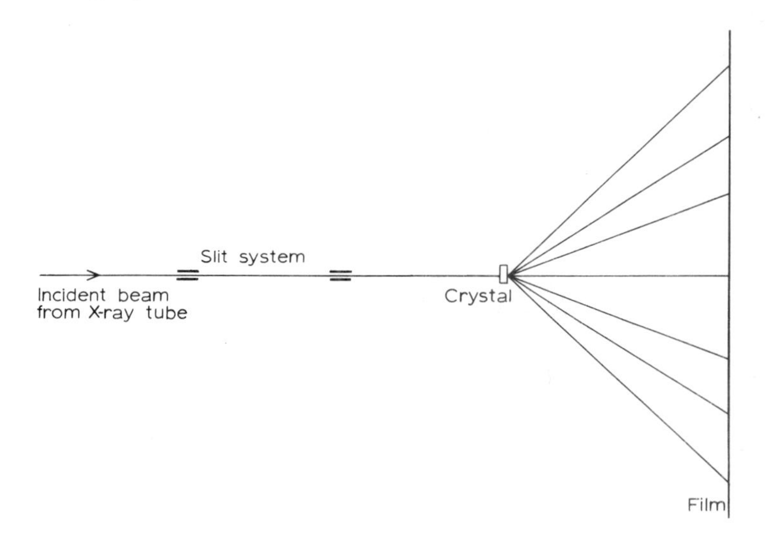

Fig.73. The Von Laue method.

Each spot on the recorded photograph is the result of reflection from a number of parallel crystal planes the basic Bragg formula $n\lambda = 2\,d \sin \theta$ having been satisfied since the numerous crystal planes each diffract that wavelength of the polychromatic beam required by the law.

When the incident X-ray beam is parallel to one of the principal axes of the crystal the resultant photographic pattern reveals the symmetry about that axis and by use of the gnomonic projection method described by WYCKOFF (1935) indices may be assigned to each spot. For a complete analysis of the crystal structure it is then necessary to know the wavelength of the radiation producing each spot and this cannot be determined until the dimensions of the unit cell are known. The latter may be estimated roughly from the Laue pattern but are more conveniently and accurately found by other X-ray diffraction techniques.

Although the method has been used with great success by several workers for the deduction of crystal structures it is a very complex process because the relative intensities of the spots produced on the plate depend not only upon the structure alone but also on the intensity of the particular wavelength in the incident beam, each wavelength having a different intensity (see Fig.73) and in addition the photographic plate has a different blackening response to the various wavelengths.

Rotating crystal methods

In this method of analysis a single crystal is rotated or oscillated about an axis perpendicular to an incident beam of monochromatic radiation. The diffracted beams are recorded on either a flat photographic plate or, more usually, on a cylindrical film with the crystal at the centre of the cylinder. When the specimen is set with an important lattice axis perpendicular to the X-ray beam the resulting photograph is in

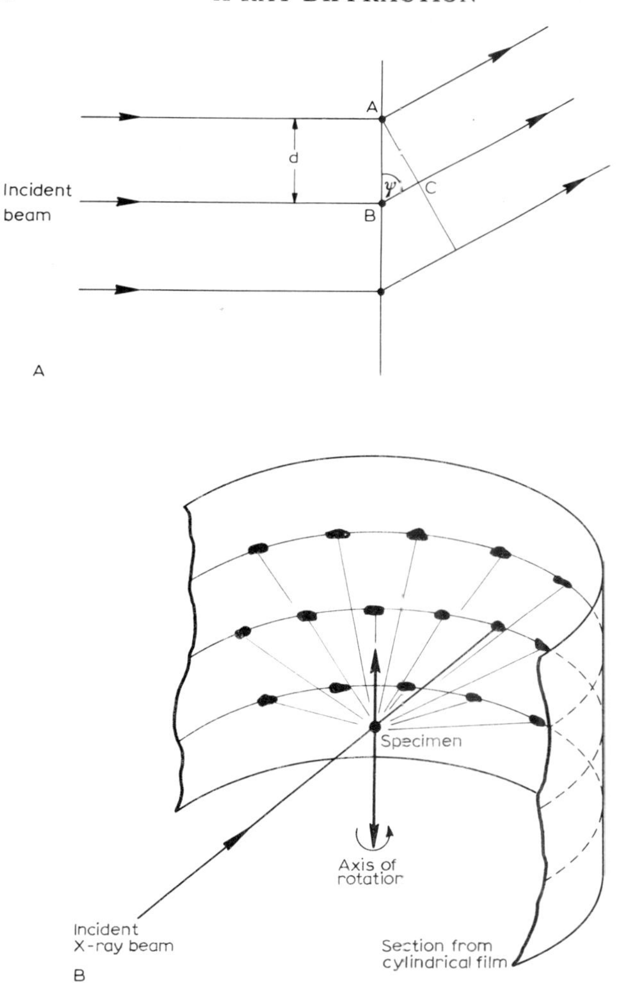

Fig.74. The rotating crystal method. A. The diffraction. (For the meaning of symbols, see text.) B. The recording on a cylindrical film.

the form of a series of spots lying on a number of circumferential lines; this effect is illustrated in Fig.74.

If the crystal is rotated such that an axis is parallel to the axis of rotation and A and B are successive lattice points along that axis then, for diffraction to occur, the difference in path length between the radiation scattered at points A and B must be equal to a whole number of wavelengths of the incident beam so that $BC = n\lambda = d \cos \psi$, where ψ is the angle of diffraction.

Thus, from the photograph, the displacement of the circumferential lines from the linear projection point of the incident beam can be measured and the angle ψ calculated. From this the lattice spacing of the crystal in a direction parallel to the axis of rotation is obtained. A number of photographs obtained with the crystal differently orientated will give the spacings along the different crystal axes and the size of the unit cell is obtained directly.

For the full elucidation of the atomic distribution in a crystal, however, each spot along a circumferential line must be assigned and this is a most difficult operation since the angular orientation of the crystal in producing a given spot is unknown. WEISSENBERG (1924) developed a method of overcoming this difficulty by moving the film in a reciprocating motion along the axis of rotation and gearing this movement to the angular rotation of the crystal. The displacement of the spots along a longitudinal line on the cylinder facilitated the assignment.

Powder methods

Since agglomerates of small crystals are more readily available than single crystals, methods based on such samples are the most widely used of those available. The amount of sample required is very small and the accuracy with which a diffraction pattern may be measured is very high.

In agglomerates the crystals have a random orientation so that some individual crystals must always be in a position to diffract a beam of monochromatic X-rays falling on the specimen. The powder sample is normally fixed at the centre of a cylindrical camera on whose circumference the photographic film is located. One exposure is sufficient to secure the reflections of the monochromatic beam from all crystal planes. This results in simplicity of technique but introduces limitations in that no direct information can be obtained with regard to the relative orientation of the incident X-ray beam and the crystal axes.

The essential features of the powder diffraction method are illustrated in Fig.75. The diffracted beams radiate from the specimen in the form of coaxial hollow cones, the common axis of which coincides with the path of the incident radiation. The impingement of these diffracted beams on the cylindrical film strip produce sets of lines as shown, and the spacings of the line pairs corresponding to a single cone can be measured and the Bragg angle calculated since the angle subtended at the specimen is 4θ.

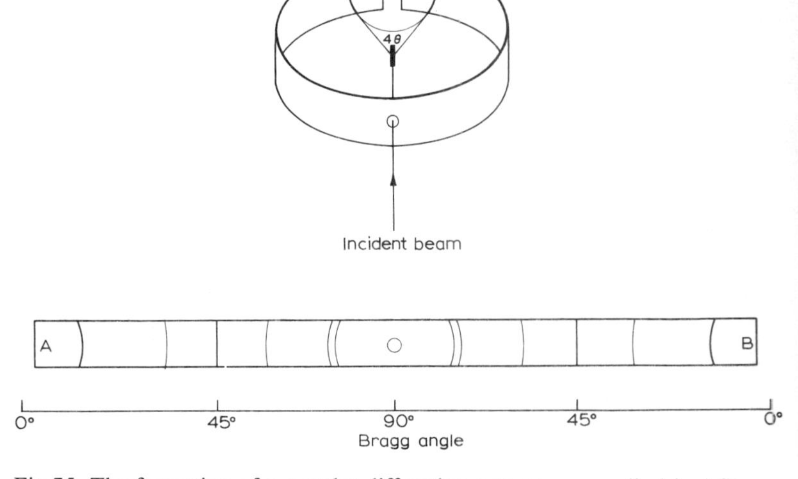

Fig.75. The formation of a powder diffraction pattern on a cylindrical film. $4\ \theta$ = Bragg angle.

THE POWDER TECHNIQUE

Equipment

X-Ray source

The principles of design and operation of X-ray tubes for spectroscopic equipment were described in Chapter 7 and the requeriments for X-ray diffraction are somewhat similar. The main differences lie in the need to produce a characteristic K radiation spectrum of the target material with a minimum of continuous radiation and to focus the electron beam onto a small area of the target so that as much of the X-ray beam as possible may be concentrated in the very narrow cones admitted by the collimator slits into the camera.

It is also necessary to use monochromatic radiation of as high an intensity as possible since the diffraction effects are feeble and exposure times are long if sufficient blackening of the photographic film is to be achieved. It is not possible to obtain pure monochromatic radiation except by the use of a monochromator and this entails severe loss in intensity. Filters are therefore used to give a radiation which is sufficiently pure for most purposes.

An examination of the Bragg formula immediately reveals two facts. Firstly it is advantageous to use X-rays of as long a wavelength as possible in order to obtain the maximum diffraction angle for a given substance. Secondly no diffraction will occur if the wavelength of the incident beam exceeds twice the interplane spacings in the crystal lattice. For optimum working conditions for a wide range of materials it is therefore necessary to have a range of target materials either by employing an X-ray tube with demountable targets or having a number of sealed tubes with different target materials.

The camera

A number of different arrangements has been used in the past but the Debye–Scherrer camera is nowadays commonly

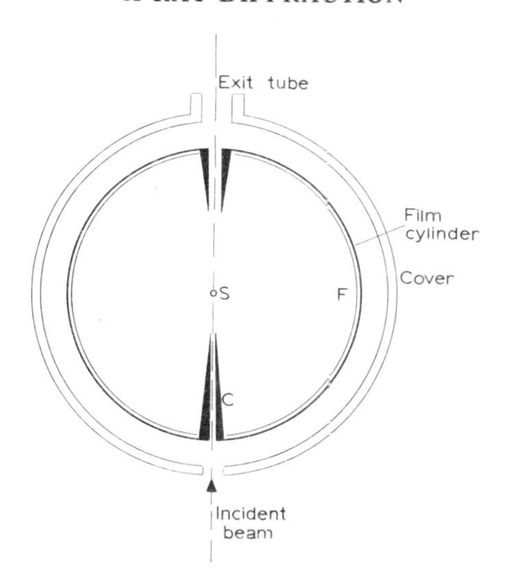

Fig. 76. A typical Debye–Scherrer camera. (*C, F, S*: see text.)

employed in Britain. Such cameras were developed by Bradley and his colleagues and the general design is the subject of a British Standard Specification (B.S.1693:1950); their main features are illustrated in Fig. 76.

The specimen *S* is prepared as a small elongated cylinder about 5–15 mm long and 0.1–1.0 mm diameter. The photographic film *F* is bent into the form of a cylinder and is held firmly in position so that its axis coincides exactly with that of the specimen. Thus, all diffracted beams from the specimen are normal to the film and travel the same distance to it. The collimator *C* allows a narrow, almost parallel ray of radiation to pass from the X-ray tube *X* onto the specimen and the geometry is such that the width of this beam is not less than the specimen diameter. Beyond the specimen there is a beam exit tube which allows the incident radiation to emerge onto a fluorescent screen and thus provides facilities to check alignment. In normal use this exit is plugged to prevent the emergence of X-rays. A number of methods are used to fix the film in the camera. The two most commonly used are the

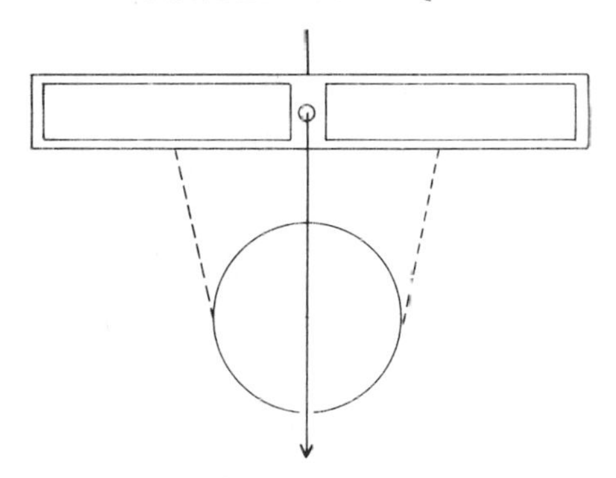

Fig.77. The Van Arkel film mounting

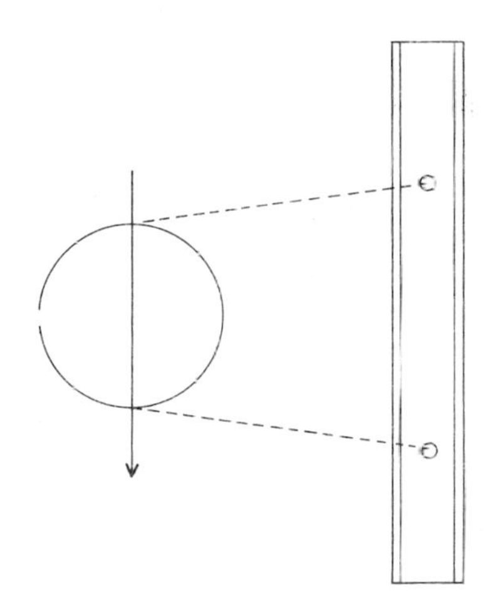

Fig.78. The Jevins and Straumanic film mounting.

Van Arkel and Jevins and Straumanic mountings, illustrated in Fig.77 and 78.

A typical camera of this type is shown in Fig.79. In this instrument the film is held in place by flexible metal straps and the dimensions are such that Bragg angles of 5–85° may be photographed on each half of the film. Changeable collimating slits of 0.5, 0.8 and 2.5 mm width may be fitted and suitable metal foil filters can be mounted at the beam inlet

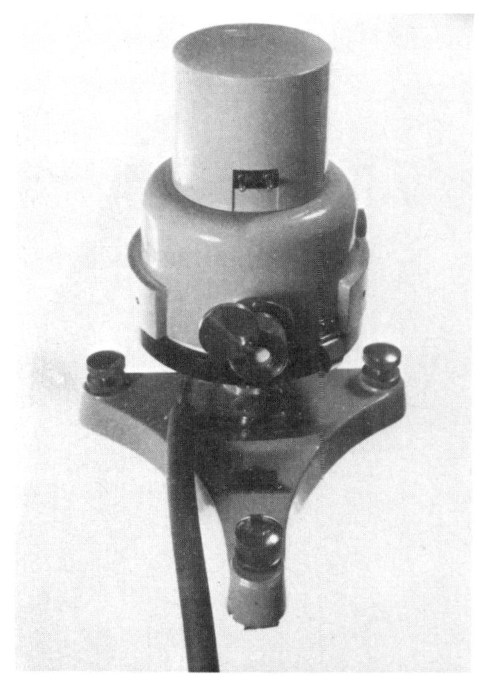

Fig.79. A typical 9 cm X-ray powder diffraction camera designed for the Van Arkel mounting of the film.

aperture. The sample is mounted in a bearing provided with two perpendicular adjusting screws so that it can be positioned accurately on the axis of rotation. Means of rotating the sample are provided, a synchronous motor giving a rotational speed of one revolution per 7½ min. The camera chamber itself is sealed by an "O" ring and is fitted with a tube so that the camera may be evacuated by means of a vacuum pump.

The camera fits onto a stand which has rotational and translational adjustments necessary for accurate alignment with the X-ray beam. A special film punch and guillotine are provided to cut and fit films accurately into the camera.

Sample preparation

A number of methods of sample preparation are in common use and selection depends both on the type of sample and on the particular radiation required for the analysis. As stated earlier, in nearly all cases the sample for irradiation is required in the form of a cylinder 1 mm or less in diameter and up to 1.5 cm long.

Rolled specimen

The most satisfactory method of sample preparation is to roll a small amount of the powder, mixed with a binder, between two microscope slide glasses. The sample is first ground carefully to pass a 270 mesh screen and then a small quantity is placed on a slide glass and mixed thoroughly with an approximately equal volume of gum tragacanth. A single drop of deionized water is added and the sample mixed to a thick paste. As it slowly dries out mixing is continued until it is of such a consistency that it may be formed into a small cylinder and gently rolled out between two slide glasses. Light rolling is continued until the sample is of the required diameter and has thoroughly hardened. It can then be cut to a suitable length and fastened on the specimen mount with a small piece of plasticine.

Capillary tubes

Specimens may also be prepared by introducing the finely ground sample into a pyrex capillary of 0.4–0.6 mm inside diameter and with walls not more than 0.1 mm thick. A small plug of plasticine or cotton wool serves to retain the sample in the capillary. This method is not suitable when long wavelength radiation, e.g., Cr Kα or Fe Kα is to be used.

Extruded specimens

KOSSENER (1955) has described a small press suitable for preparing specimens by extrusion. This method is of advantage when the use of a binder is precluded.

Fibre mounted specimens

A further technique is to mix the finely divided sample with an adhesive such as Canada balsam and to roll the mass into a fine fibrous support such as a hair or very fine glass fibre until the latter is evenly coated with the sample.

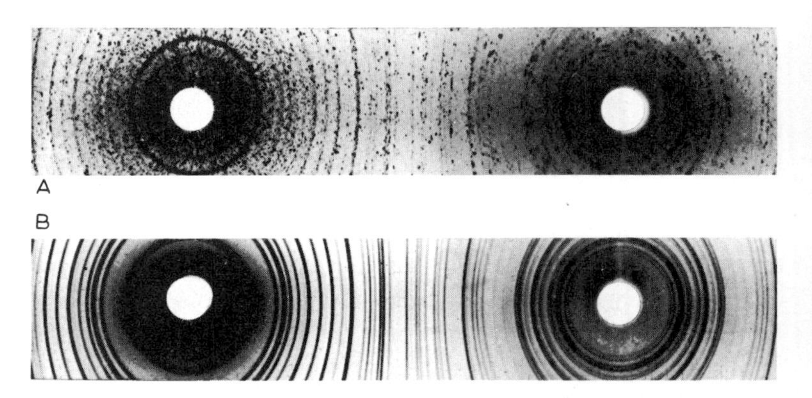

Fig.80. The effect of rotating a sample of α-quartz taken with Cu Kα radiation on a 5.75 cm camera with a Straumanic film mounting. A. Specimen stationary. B. Specimen rotated.

In all cases it is advantageous to rotate the sample during exposure in the camera since this overcomes problems of preferred orientation. It also reduces the effects due to large particle size; Fig.80 shows two diffraction patterns, one with the specimen stationary and the other with the specimen rotated. The broken appearance of the lines in the former pattern is due to the rather large particle size of the sample.

When the sample contains elements of high atomic number dilution with a filler such as cornstarch or tragacanth is necessary to reduce absorption of the radiation within the specimen. As a guide, when the mean atomic number of the sample is less than 10, no dilution is necessary but when the mean atomic number is 70 or greater, a 10:1 dilution is required. Obviously when a sample of unknown composition is being examined trial exposures are frequently required to establish the optimum conditions.

CHOICE OF CONDITIONS

Type of radiation

It has already been shown that a choice of quality of X-rays is needed if a wide range of materials is to be studied and it is necessary to measure *d* spacings over a wider range than can be covered by using one target material. The extreme limits for different target materials is given in Table XXIII.

In considering the choice of target material it is also necessary to select one which gives an incident radiation which will not give rise to fluorescent X-rays from the sample, resulting in a high background to the diffraction pattern. The phenomenon occurs when the sample contains as major constituent an element of atomic number two or three less than that of the target element. Thus an X-ray tube with a copper target should not be used to examine a specimen known to contain a high proportion of iron.

TABLE XXIII

CAMERA COVERAGE

Target material	*Wavelength of* $K\alpha_1$ *(Å)*	*Range of* d *spacings (Å)*	
		max.	*min.*
Cr	2.2896	13.1	1.149
Fe	1.9360	11.1	0.972
Co	1.7889	10.3	0.898
Cu	1.5405	8.8	0.773
W	1.4764 ($L\alpha_1$)	8.47	0.741
Mo	0.7093	4.07	0.356

Choice of filter and working voltage

The output from an X-ray tube consists of a continuous background with high intensity peaks superimposed upon it as shown in Fig.81. The main peaks are designated $K\alpha_1$, $K\alpha_2$, etc., and a similar series of "lines" called the L series is

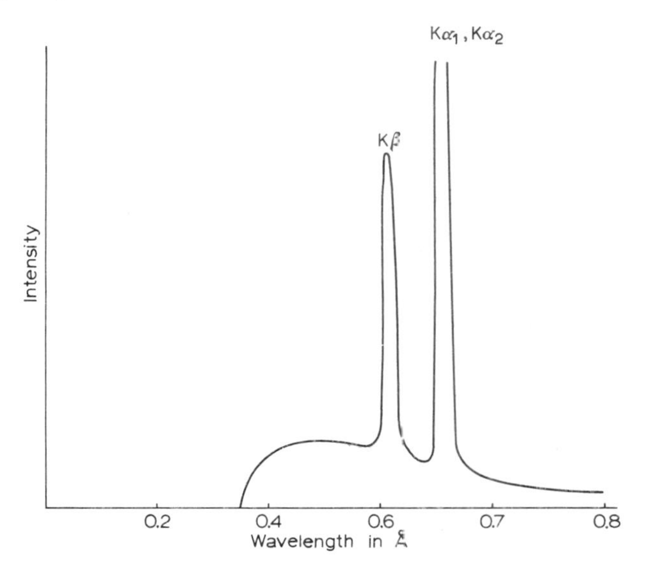

Fig.81. Radiation from a Mo target with the tube operating at 35 kV.

also excited (see p.278). To produce these series of lines the voltage applied to the X-ray tube must exceed a given level known as the excitation potential. This is lower for the L series of any given element than for the K series. It is therefore possible to excite the L series without producing the K series but the reverse is impossible. In X-ray powder diffraction, however, the L series of lines produced from the commonly used target materials are of such long wavelengths that they are easily absorbed in the tube window and by the air so that they do not reach the film.

The wavelengths of the significant lines of some target materials used in diffraction work are given in Table XXIV. From this it is seen that the Kα_1 and Kα_2 lines are of almost identical wavelength, whereas the Kβ line is of shorter wavelength. It is thus possible, by interposing in the X-ray beam a thin filter made from a material having an absorption edge between the Kα and Kβ wavelengths, to absorb the Kβ radiation almost entirely whilst the Kα radiation suffers only relatively slight attenuation. The absorption edge of an element is illustrated in Fig.82 where it is seen that at a particular wave-

TABLE XXIV

DATA FOR X-RAY TARGETS

Element	*Atomic number*	*Wavelength*			*Absorption edge* Å	*Excitation potential* (*kV*)
		$K\alpha_2$ Å	$K\alpha$ Å	$K\beta$ Å		
Ag	47	0.56381	0.55941	0.49701	0.4855	25.5
Mo	42	0.71354	0.70926	0.63225	0.6197	20.5
Cu	29	1.54434	1.54050	1.39217	1.3802	9.0
Ni	28	1.66168	1.65783	1.50008	1.4869	8.3
Co	27	1.79279	1.78890	1.62073	1.6072	7.7
Fe	26	1.93991	1.93597	1.75654	1.7429	7.1
Cr	24	2.29352	2.28962	2.08479	2.0701	6.0
		$L\alpha_2$	$L\alpha_1$	$L\beta$		
Au	79	1.28763	1.27636	1.083	1.0403	11.8
W	74	1.48739	1.47635	1.282	1.2154	10.2

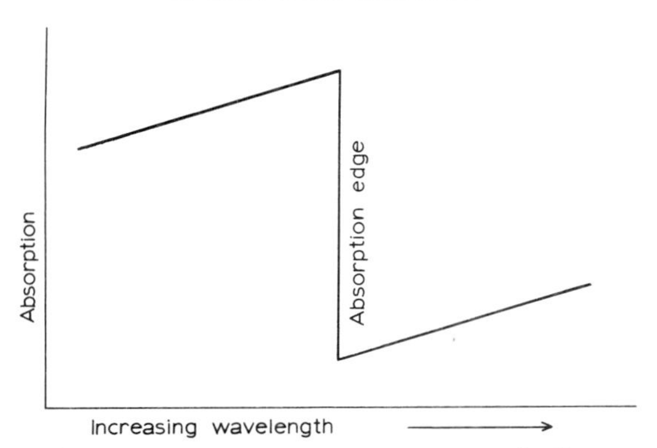

Fig.82. Illustration of the "absorption edge" effect.

length, depending on the element, the absorption of X-rays suddenly decreases. The correct filters for various target materials are given in Table XXV. For a tube with a cobalt target it will be seen that an iron filter of 0.0007 inch thickness is suitable.

The intensity ratio between characteristic and continuous radiation is dependent on the operating voltage on the X-ray

TABLE XXV

DATA FOR X-RAY TARGETS AND FILTERS

Target element	*Optimum potential (kV)*	*β-Filter*		*Thickness (inches)*
		Element	*Absorption edge (Å)*	
Ag	102	Rh	0.5341	0.0031
Mo	80	Zr	0.6888	0.0043
Au	47	Ga	1.1926	0.0018
W	41	Cu	1.3802	0.0008
Cu	36	Ni	1.4869	0.0008
Ni	33	Co	1.6072	0.0007
Co	31	Fe	1.7429	0.0007
Fe	28	Mn	1.8954	0.0006
Cr	24	V	2.2676	0.0006

tube. The most favourable ratio is obtained at a voltage about four times the excitation potential for any target material and series of lines. This optimum potential is not critical, however.

The tube current used is normally the highest possible at the given operating voltage consistent with long tube life so as to keep exposure times down to the minimum.

INTERPRETATION OF RESULTS

Measurement of d *spacings*

A typical powder photograph is shown in Fig.83. In order to identify a compound or a mixture of compounds it is necessary to translate the diffraction pattern on the film into the *d* spacings in the sample.

From a knowledge of the wavelength of the incident radiation and the diameter of the camera it is possible to calculate the *d* spacings corresponding to the various lines on the film. This procedure can be accomplished approximately by using special rulers which are obtainable for this purpose. This method is often not sufficiently accurate and a more satisfactory technique is to calibrate the camera using a sample whose *d* spacings are known. The positions of the images in relationship to the knife edges limiting the exposed portion of the film are accurately measured using a travelling microscope. The distance between the knife edge images for one half of the film is corrected to a standard distance and all line separations are corrected to the same scale. This method corrects for film shrinkage during processing.

Fig.83. A powder diffraction pattern of willemite (Zn_2SiO_4) taken with Co $K\alpha$ radiation on a Van Arkel film mounting.

Quartz or sodium chloride are suitable materials for calibration.

Measurement of relative intensities

It is necessary to know the relative intensities of the various lines in the diffraction pattern in order to identify materials giving rise to the pattern. This is required particularly when the sample comprises a mixture of materials.

The relative intensities may be judged visually but the most satisfactory method is to use a microphotometer such as that used in spectrographic analysis. A recording instrument in which the film is traversed whilst continuously recording the photocell output is of particular value in X-ray diffraction work.

A very suitable instrument has been described by TAYLOR (1951).

Use of the A.S.T.M. index

This index, which is in card form, presents data on the diffraction pattern of several thousand substances in such a manner that, in conjunction with an index book, the data on any substance may be quickly found either from the name of the substance or a knowledge of the three principal lines of the diffraction pattern.

Having located a card, or cards, corresponding to the three strongest diffraction lines in an unknown it is then necessary to check the *d* spacings given by weaker lines to make a positive identification of the unknown material. When more than one substance is present in the unknown it becomes necessary to identify one of the components of the sample by trial and error, eliminate all its lines from the diffraction pattern and then proceed in the same way to identify the other components.

The limit of detection by this technique is usually about

5 % although in some favourable cases 1 or 2% may produce a detectable pattern.

QUANTITATIVE ANALYSIS

From a mixture of crystalline substances, a diffraction pattern is obtained in which the intensities of the lines in individual patterns are proportional to the quantities of the respective ingredients. It is therefore possible to use the powder method as a means of quantitative analysis.

The simplest technique is to prepare a series of standard mixtures and to derive a correlation between relative intensities of standard lines and relative concentrations of materials in the standards. For maximum accuracy lines should be chosen which are of similar intensity to reduce calibration errors and close to each other to minimize the effects of absorption in the sample and angular factors.

The various factors which affect line densities on the photographic film may to some extent be eliminated by utilizing an internal standard technique similar to that used in optical spectrography.

THE X-RAY DIFFRACTOMETER

In X-ray cameras the diffraction pattern is recorded on photographic film. An obvious development of this technique is to scan the pattern with a suitable detector and record continuously the intensity of the radiation. This method avoids the delay in developing, washing and drying the film and then making a record of the diffraction pattern.

One type of instrument used for this purpose is illustrated in Fig.84. The incident X-ray beam, after passing through a slit A and then a collimator B, falls onto the flat specimen C. The receiving slit D defines the width of the diffracted beam

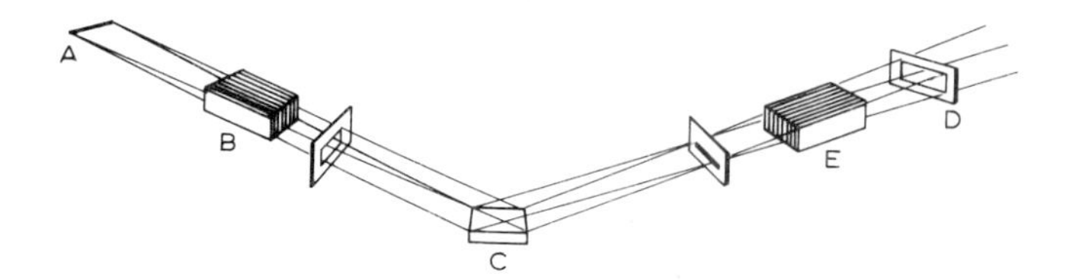

Fig.84. An illustration of a typical diffractometer. (*A, B, C, D, E*: see text.)

received by the detector. The two collimators *B* and *E* are in the form of thin parallel metal foils and limit the divergence of the beams in the plane normal to the slits. In practice, the diffraction pattern is scanned by moving the exit slit and the detector round the circumference of a circle which has the sample at the centre. At the same time the sample is rotated on the axis of the circle such that the angular movement of the sample is half that of the detector.

The arrangement is somewhat similar to that used in X-ray spectrometers and the same criteria apply to the choice of detector. For high accuracy and reproducibility in quantitative work it is essential that the output from the X-ray tube possesses both short and long term stability and the power supply is therefore stabilized electronically.

The output from the detector is amplified and the rate of counts recorded continuously, thus giving directly a chart of intensity against Bragg angle. Alternatively, the detector may be set on a particular line and a count made over an accurately measured time interval.

The preparation of the specimen follows normal practice, the finely divided sample is sieved and pressed into a holder which is usually in the form of a recess in a metal or plastic plate. Binders may be used to compact the sample but the main requirement is that the face presented to the X-ray beam is as flat as possible.

The instrument is particularly convenient for quantitative work. In order to achieve high degrees of accuracy, it is advisable to use an internal standard technique, the different methods available have been described critically by ALEXANDER and KLUG (1948). The overall accuracy of the technique is limited by sample preparation, a technique for filling the specimen holders in a reproducible manner must be developed. The samples themselves must be finely ground and there must be no preferred orientation effects.

The diffractometer can record X-ray patterns either with great speed or high accuracy. Certainly, the accuracy attainable is higher than with any other method. Recordings may often be made in less than the exposure time required by a powder camera, although at high scanning speeds it must be remembered that minor lines may easily be missed.

APPLICATIONS OF X-RAY DIFFRACTION TECHNIQUES

The identification of the main minerals present in a sample of ore, rock or sediment can be carried out with great certainty and in practice it is usually unnecessary to supplement a powder diffraction record with any other test. On the other hand, the identification of the minor mineral constituents of a sample is a very difficult and laborious task and supplementary techniques such as optical examination under polarized light and refractive index determinations are required. It is often helpful to pick out grains of mineral from a bulk sample and identify these separately but when the sample is fine grained this cannot be done. A combination of X-ray diffraction and other analytical techniques such as spectrography or flame photometry then becomes necessary.

One excellent example of the application of X-ray diffraction techniques to mineralogical research is the identification of clays. At first sight a study of clays may seem to be an unpromising subject and indeed the outstanding successes which have been achieved would not have been possible without the use of ancillary techniques such as ion-exchange capacity, chemical, physical and thermal analysis. BRINDLEY (1951) has presented a great deal of this work in his book *X-Ray Identification and Crystal Structure of Clay Minerals*.

A number of publications aids the worker in the identification of substances. The A.S.T.M. *X-Ray Powder Data File* has already been mentioned; in addition, the American Society for Testing Materials, the American Crystallographic Association and the Institute of Physics are jointly sponsoring a project at the National Bureau of Standards on the measurement and publication of standard X-ray powder patterns. BERRY and THOMPSON (1962) have published data for 295 minerals.

CHAPTER 9

Fluorimetry

INTRODUCTION

During the seventeenth century, Boyle reported that when "Lignum Nephriticum" was infused in water the resulting yellow solution appeared blue when viewed in a certain way. Much later STOKES (1852) gave the name "fluorescence" to this phenomenon and was able to explain that it was due to the absorption of light and re-emission at a different wavelength and not to scattering.

With the development of suitable apparatus fluorescence has been used for the analysis of a number of materials which exhibits the property and significant advances have been made in the subject during the past 10 years or so. It is not the intention to treat the subject intensively here as the applications in geochemistry are limited but an instrument for measuring the fluorescent intensity of solutions was developed (BREALEY and ROSS, 1957) which enabled EVANS et al. (1962) to conduct a geochemical survey of the Nottinghamshire oilfields rather more easily than would have been the case with hitherto existing equipment. Moreover, the fluorescence of minerals has recently received comprehensive treatment which at this stage cannot be improved upon. At this stage it should be emphasized that the work reported here will be concerned exclusively with fluorescence in solutions and it must not be confused with that occurring in solids such as fluorspar, from which Stokes took the name.

THE ORIGIN OF FLUORESCENCE IN SOLUTIONS

When light passes through a homogeneous solution it emerges, diminished in intensity due to scattering or reflection at the surfaces and absorption by the solvent and solutes.

Molecules possess a number of vibration frequencies characteristic of the substance. When the frequency of a light wave coincides with one possessed by the molecule, resonance occurs and light energy is absorbed. The result of this is that the molecule then exists at a higher energy level and it begins to vibrate. In this state it is unstable and it must lose this extra energy in some way. In solution this usually occurs almost immediately when the excited molecule collides with a solvent molecule and the energy is converted to heat. Alternatively, it may provide the required energy for a chemical reaction to be induced (photochemical reaction). Some molecules, however, remain in the excited state for a short time (about 10^{-8} sec) during which they lose a small portion of their energy; then they return to their original ground state with the emission of light. Since the energy emitted is less than that absorbed, the fluorescent radiation is of a longer wavelength than that which is absorbed. This explanation is of course highly simplified. A more complete description has been given by BOWEN and WOKES (1953).

Only a relatively small number of molecules possess the peculiarly stable excited states which enable them to exhibit fluorescence and it is not possible to predict which substances have the property.

QUANTITATIVE CONSIDERATIONS

All other things being equal the intensity of fluorescence is proportional to the intensity of the incident light and to absorption and the following relationship holds (SANDELL, 1950):

$$F = A\ I_0\ (1–10^{-kcl})$$

where F = fluorescent intensity, A = fraction of radiation absorbed, I_0 = intensity of incident radiation, k = constant for the material (extinction coefficient), c = concentration of the material, and l = length of incident light path in the solution.

If the solution is very dilute then kcl is small and the equation reduces to the following:

$$F = 2–3\ A\ I_0\ kcl$$

from which it will be seen that the intensity is directly proportional to the concentration. At higher concentrations this linear relationship does not hold, for linearity kcl must be less than 0.01. In addition, high absorption reduces the light intensity as it passes through the solution and the effect is as shown in Fig.85. Here, three cases are represented where solutions are irradiated, in rectangular cuvettes, with light emitted from E and observed at O. In Fig.85A the solution is dilute and the green fluorescence, weak though it may be, is uniform over the whole volume. In Fig.85B the solution is more concentrated and the fluorescent intensity is stronger where the light enters but, because of absorption of the incident light, becomes weaker as the light traverses the solution. Fig.85C is the limiting case, at the entry point the fluorescence is strong but absorption is so high that virtually none is observed at O. In all quantitative practical work the case shown in Fig.85A must be achieved.

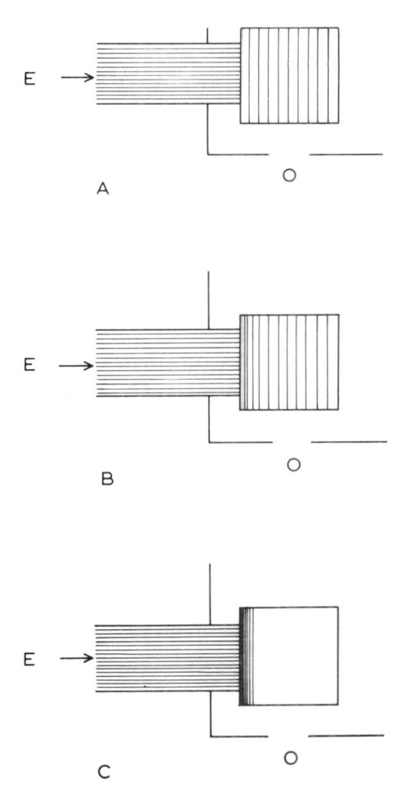

Fig.85. The effect of concentration on fluorescent intensity. A. Dilute solution. B. More concentrated solution. C. Limiting case. Light is emitted at E and observed at O.

THE MEASUREMENT OF FLUORESCENCE

The measurement of fluorescent intensity is normally achieved by irradiating the sample solution with monochromatic light and observing the emitted radiation at right angles to the incident beam. As was shown above, materials absorb selected wavelengths and the fluorescence is at a longer wave-

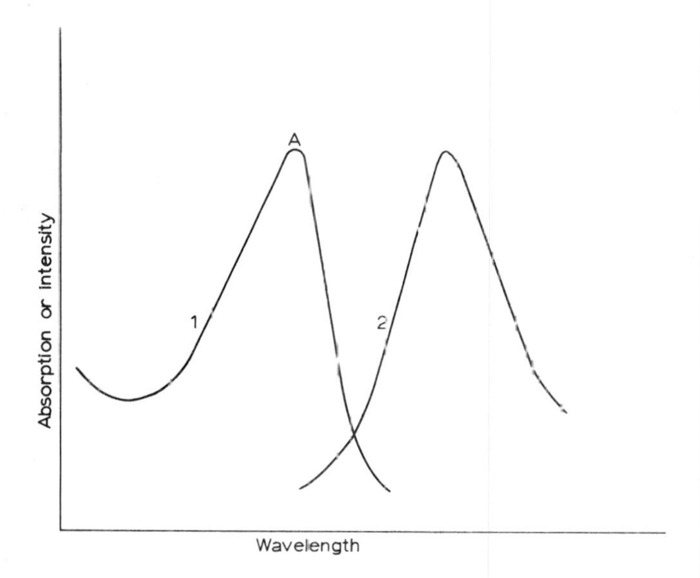

Fig.86. Curves of absorption and fluorescent intensity. Curve *1*: absorption (A = maximum of absorption); curve *2*: fluorescence.

length. In fact both absorption and emission occur in the general form shown in Fig.86 and the greatest efficiency is obtained when the incident light has a wavelength which coincides with the maximum of the absorption curve, A. The fluorescence is excited, though, by radiation at any wavelength within the absorption envelope. Because of this feature most materials which fluoresce in the visible region can be excited by the 3,650 Å line from a high pressure mercury lamp. This has the added advantage of being a source of high intrinsic brilliance.

Since the fluorescent intensity is feeble it is necessary to exclude the incident light from the detector. This is achieved by a suitable filter arrangement. One optical filter placed between the lamp and the sample isolates the 3,650 Å mercury line and another between the sample and the detector allows the fluorescent light to pass but not the incident light.

Instrumentation

Until recently all the commercially available instruments were of the null-reading type and were not very sensitive so they were unable to measure the fluorescent intensity of very dilute solutions. Consequently, some people built their own apparatus. The instrument to be described here was the most elegant model resulting from a succession of prototypes and was developed because it was required to make some 250 measurements per day over a long period, some of them on solutions which fluoresced very feebly. An instrument different in appearance but similar in mode of working was used by EVANS et al. (1962).

The basic requirements of the equipment were listed as follows:

(*1*) It should be highly sensitive.

(*2*) It should be very simple to use.

(*3*) Instrument readings should be directly proportional to fluorescent intensity so that a minimum of mathematical treatment is required.

(*4*) It should be of such a design as to permit the fitting of a run-through cuvette when large numbers of samples have to be assayed.

(*5*) It should have good long-term and short-term stability so that when samples are being measured over considerable periods of time it should not be necessary to recalibrate the equipment too frequently.

(*6*) It should be all-mains operated and highly robust.

It was obvious that the ideal instrument would be direct reading and so designed that after the initial setting up and calibrating, samples could be run through one after the other and a meter reading noted. This would at first sight seem to be a simple problem, the logical approach being a stabilized illuminating lamp together with a highly stable and sensitive measuring device. The latter is quite feasible when based upon the use of a photomultiplier but it was found impossible to stabilize completely the mercury arc which was used for illuminating the sample. Even when the applied voltage was kept constant, variations in the temperature of the lamp and the wandering of the arc around the electrodes resulted in variations in the intensity of the light. A different approach to the problem was therefore made and the instrument which was finally developed is shown diagrammatically in Fig.87.

Light from the lamp *A*, a 125 W high-pressure mercury arc type MB / D, was made reasonably parallel by lens *L* and passed through the primary filter *F1*. The sample *S* was illuminated by this light and the fluorescent output, viewed at right angles to the incident light passed through the secondary filter *F2* and fell upon the photomultiplier *P2*. The output from *P2* was amplified and read directly on the meter *M*.

The output from a photomultiplier rises sharply with an increase in applied anode voltage so that in most applications this applied voltage is highly stabilized. This feature was utilized in a different manner.

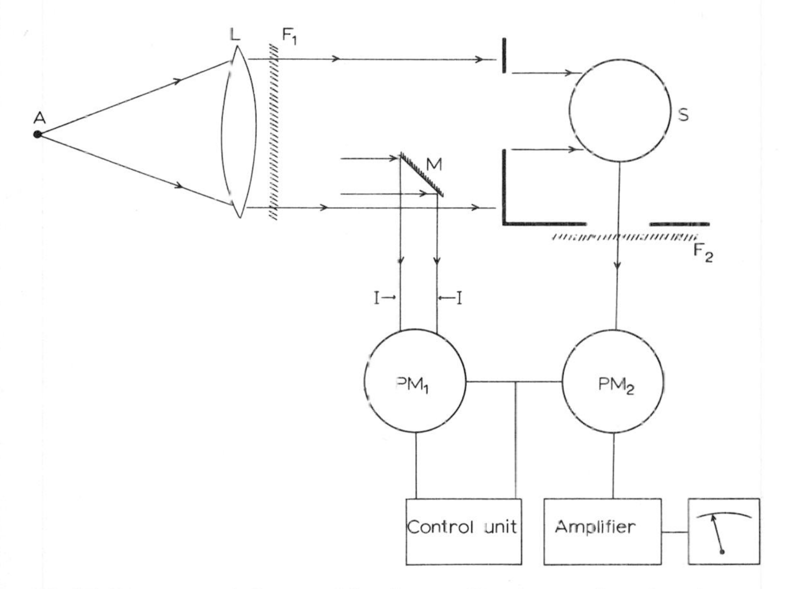

Fig.87. Diagrammatic layout of fluorimeter. (For the meaning of symbols, see text.)

A small portion of the light transmitted by the primary filter *F1* was reflected by a mirror *M* through the variable aperture *I* onto a second photomultiplier *P1*. The output from this was fed into a controlling unit which supplied the voltage applied across both photomultipliers. This voltage was made to fluctuate in such a way that the output of *P1* was kept constant, so that if the intensity of the lamp increases, the applied voltage decreases and vice-versa.

As both *P1* and *P2* were supplied from the same voltage supply and as they had similar characteristics the output from *P2* was independent of lamp intensity fluctuations and was a function only of the concentration of the sample *S*.

The amplifier

The output current from the photomultiplier monitoring the sample (*P2*) was fed through one of a series of load resistors and the voltage developed applied to the input grid of a conventional balanced cathode follower degenerative circuit. The sensitivity was controlled by means of a variable shunt across the meter and the parameters of this shunt were so arranged that full scale defection could be obtained for input voltages between 0.3 and 1.2 V. The amplifier was linear within this range. This gave a ratio of 4 : 1 on the sensitivity control and the "range" load resistors were selected to the ratio 0 : 1 : 3 : 10 : 30 : 100 : 300 : 1,000. The zero range was included so that the amplifier zero could be adjusted before commencing any readings.

The wide overall ratio of input resistors resulted in a very large sensitivity range and as the ratios were chosen to be accurate to within 1% it was possible to change the range for readings within a group of samples. If, therefore, the instrument had been standardized and a sample gave a reading over the top of the scale, then, instead of preparing another sample or diluting that in the instrument, one could switch directly to a lower range and multiply the reading by the appropiate factor.

If the solvent being used had a small fluorescence, or if the optics or filters fluoresced, or gave rise to light scattering, the unwanted constant blank could be backed off so that a zero reading was obtained. When this had been done any further adjustment of range switch or sensitivity control did not affect this zero reading. In fact, the backing off was usually performed with the range switch at the most sensitive position so that it was done with greatest accuracy.

The cuvettes

The cuvette, together with its associated optics, was housed in the instrument in a substantial brass block large enough to permit the inclusion of water circulation for thermostatic control if required. This whole block could be changed so that different types of cuvette could be used. An exploded diagram of the arrangement is shown in Fig.88. It was made so that the incident and fluorescent light were on

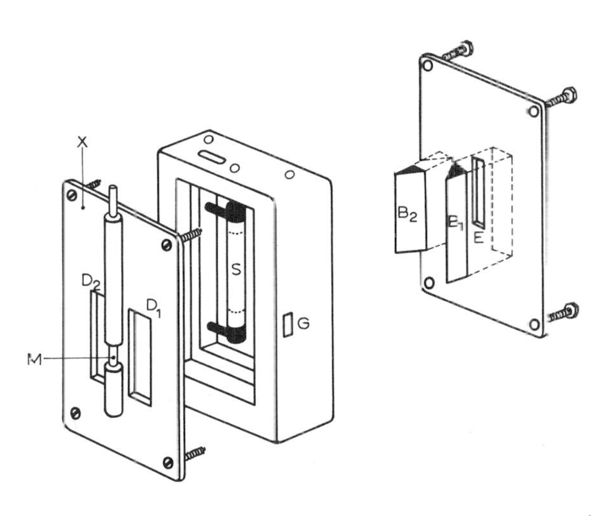

Fig.88. Cuvette housing with run-through cuvette fitted. (For the meaning of symbols, see text.)

the same optical axis. After passing through the primary filter a small portion of the incident light was reflected by mirror *M* through aperture *G* and eventually reached the photomultiplier *P1* (Fig.87). The remainder of the light fell onto plate *X* and that portion passed by apertures *D1* and *D2* was reflected by mirror *B1* and *B2* to illuminate the sample in the cuvette S. The fluorescent radiation passed through aperture *E*. The run-through cuvette was made from thin walled glass tubing about ½ inch outside diameter 1½ inch long. The central 1 inch was illuminated and to cut down scattering the ends and connecting tubing were painted black. The filling and emptying were done by vacuum—conveniently a water pump so that the discarded sample was disposed off immediately. The arrangement is shown in Fig.89. With the tap in the

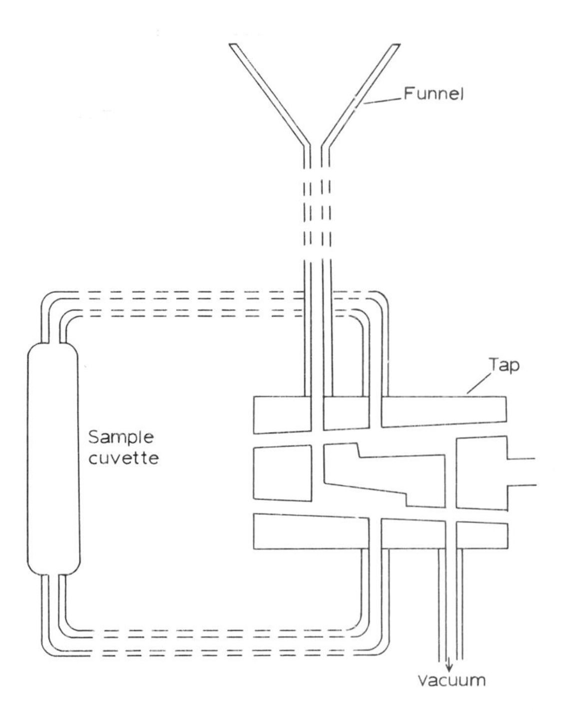

Fig.89. Filling arrangement for the cuvette with the tap in the "fill" position.

"fill" position the sample from the funnel is connected to the bottom of the cuvette and the vacuum line to the top. In the "empty" position this is reversed so that complete emptying is effected. The run-through cuvette and associated system required only 20–25 ml solution for effective washing and filling but when the sample was limited an alternative block was fitted to accommodate a replaceable tube requiring only about 4 ml of solution. The complete instrument is shown in Fig. 90.

Both short and long term stability was very good, changes in main voltage from 190 V to 270 V did not affect the reading and once calibrated the instrument could be used for long periods with only occasional checks. The sensitivity was limited only by the efficiency of the filter combination and the fluorescence of the optics.

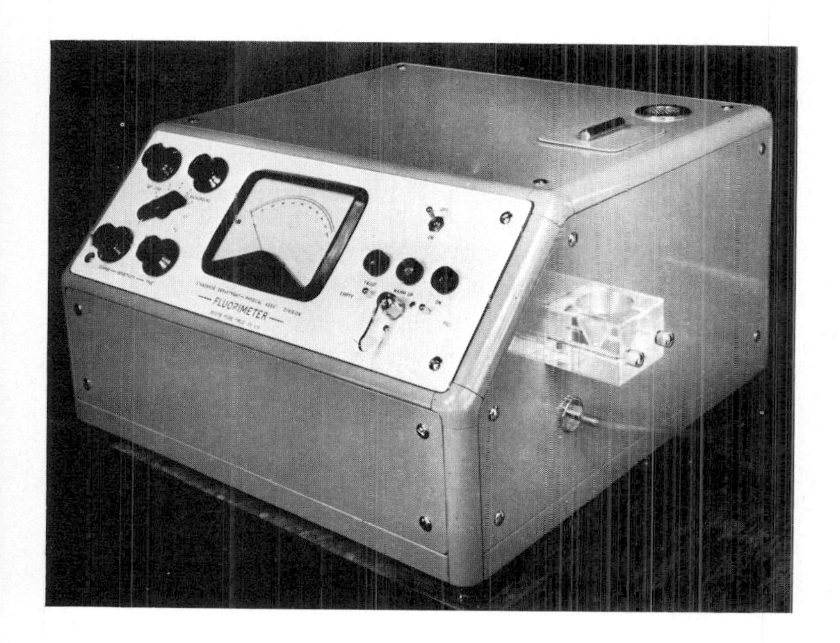

Fig.90. The complete instrument.

GEOCHEMICAL APPLICATIONS

It is rather surprising that up to now relatively little use has been made of fluorimetric methods of analysis of solutions, in spite of the fact that geologists are well aware of the importance of observing the occurrence of fluorescence in minerals. One of the best known applications of fluorimetry was devised for the determination of uranium. This is, however, not done in solution. In this case, the sample is fused with sodium fluoride into a solid button and from this the quantity of uranium is recorded in terms of fluorescence. By contrast, beryllium has been determined in silicate rocks by a solution technique (SANDELL, 1950) making use of the complex formed with morin (a pentahydroxyflavone). Likewise, traces of gallium in rocks have been determined, using the yellow fluorescence of the 8-hydroxyquinoline complex. From a study of these three examples the geochemist will readily appreciate the use of a fluorimeter in various aspects of his work as it combines simplicity and elegance with reproducibility and accuracy.

Fluorescein has long been used to trace the movement of underground waters. Normally the detection of the yellow-green fluorescence is done visually by night but by making use of the high sensitivity of a fluorimeter the method can be used with greater certainty and precision. Of special interest will be the tracking of pollution into underground water tables. Moreover, the sensitivity of this instrument has yet to be finally established. Eventually, it will clearly lie in a region capable of the direct determination of fluorine in natural water.

In a variety of ways, the distribution of fluorescent substances and, particularly, hydrocarbons found in subsoils will prove to be of value in geochemical prospecting. A typical example has been the use of the direct-reading fluorimeter in a geochemical survey of oilfields in Nottinghamshire (EVANS et al., 1962). Samples of rock and subsoil were stripped of hydrocarbons by extracting them with a "triple solvent" (equal proportions of alcohol, acetone and benzene) by means of an ultrasonic probe. In this way the hydrocarbons were dispersed in a quantitative fashion in the solvent. Using a standard solution of the principal type of petroleum derived from these oilfields, the levels of fluorescence were obtained from the solvent extracts. When plotted on a map, and linked by "isofluorescence lines", they revealed a pattern which reflected the position of petroleum at depths of over 1,500 ft. Owing to side-way diffusion of the hydrocarbons which migrate upwards from the oil-bearing anticlines, the fluorescent anomalies were somewhat diffuse. Nevertheless, this example illustrates the significance of the area as a potential oilfield and, of more general interest, the application of fluorimetry as a rapid method for surveying large areas for potential accumulations of petroleum. It is a method which can be applied with success to the survey of deserts and icefields. In the former, the sands will adsorb the more fluorescent compounds which migrate from oil-pools and allow the lighter hydrocarbons to escape. Likewise, as petroleums diffused through the ice-cap they will fractionate and yield water samples which will be identifiable at exceedingly low levels of concentrations by means of a fluorimeter as sensitive as that illustrated by Fig.87 and 90.

CHAPTER 10

Chromatography

INTRODUCTION

The term "chromatography" signifies the graphical use of colour in the identification of compounds which have been isolated by diffusion processes which are manipulated in a wide variety of different ways. At least, this was the form in which this immensely elegant and efficient method of analysis originated. Today the term chromatography is used to cover a much wider field of analytical techniques, many of which do not make use of the graphical distribution of colour values. To appreciate the background and extent to which chromatography has already assisted geochemists reference should be made to *Chromatography in Geology* by RITCHIE (1964). It is now customary to classify the various techniques into two broad groups, namely "adsorption chromatography" and "partition chromatography". In greater detail techniques are also referred to as: (*a*) paper chromatography; (*b*) thin layer chromatography; (*c*) column chromatography; and (*d*) vapour-phase or gas chromatography.

All four branches involve the use of adsorption and partition coefficients which are either employed independently or in conjunction with the process of diffusion which leads to the separation and identification of mixtures of organic and inorganic substances in a qualitative or quantitative manner. By analogy, the laboratory practice of chromatography in many ways resembles the kinetics of those geological processes which involve the migration and interaction of gases and liquids throughout the rocks of the earth's crust. For example, the fractionation and migration of petroleum from its source rocks is a function of the adsorption and partition of fugitive hydrocarbons as they diffuse through sediments of differing grain size and mineralogical composition. Likewise, in the laboratory, the solid–liquid particles used in a chromatographic column depend for their ability to fractionate and purify compounds upon the interplay of partition and adsorption coefficients. Consequently the use of chromatographic techniques provides data which can be applied to geochemical problems in the field. Moreover, many aspects of the solid–solid diffusion processes involved in metamorphic processes, like granitization, can now be described more accurately in terms of chromatographic criteria.

Adsorption is expressed as a coefficient of the ability of one substance to adhere to another. The effect of this is to decrease the total free energy of any system by the liberation of heat. Such exothermic effects will range in value from those representing the heat of condensation (5–10 kcal per mole) to those which develop at high temperatures when as much as 100 kcal per mole are developed by adsorption. Beyond this one enters the realm of chemisorption in which irreversable ionic bonding replaces the purely physical Van der Waals forces involved in adsorption.

Purely adsorptive processes are governed by a number of factors. Highly adsorptive systems develop more favourably at low temperatures mainly because the kinetic energy of the phases involved is reduced in value. The degree of ionization is also a factor which influences the coefficient of adsorption. On the other hand, the mutual repulsion of molecules reduces adsorption and likewise the nature of the surface of a solid has a profound influence on the attraction or otherwise of liquids and gases. These are important factors governing the use of so-called "adsorption chromatography". In those cases where the adsorption coefficient is to be exploited the nature of the solid surface is of overwhelming importance. Such solid surfaces may be purely catalytic in their function, as for instance the catalytic action of platinum in the manufacture of sulphuric acid or of nickel in the hydrogenation of unsaturated esters to form fatty acids. In fact this aspect of geochemistry has yet to be widely explored but it is full of intriguing possibilities especially in the field of diagenesis and the formation of meta-sediments. In this aspect of catalysis it is axiomatic that the efficacy of the catalyst is enlarged the more finely divided the particles become.

Before a reaction can take place on the surface of a solid the reactant molecules must be adsorbed. This presents a finality to such reactions, as once the surfaces of the catalyst are fully occupied by adsorbates the possibility of further reaction ceases. Therefore at high concentrations of potential reactants the rate of reaction is often zero. It is therefore important to control the quantity of material which is to be analysed by adsorption chromatography so as to maintain surfaces which can be exposed to the reactants for as long as possible.

The equations developed by Freundlich and Langmuir indicate the relationship between the quantity of substance adsorbed and the amount existing in bulk at any given temperature. These yield what are termed *adsorption isotherms*. The Freundlich equation emphasized the fact that pressure influences the degree of adsorption of gases on solids, and takes the form:

$$\frac{x}{m} = kp^{1/n}$$

in which x is the mass of gas or liquid adsorbed per m g of solid; p is the pressure of the gas or the concentration of the absorbed liquid, and k and n are respectively constants for a given gas or liquid and a given solid.

The Langmuir equation yields the so-called "Langmuir

adsorption isotherm" which includes the exothermic nature of adsorption expressed in terms of the rate of adsorption and takes the form:

$$K_1 p (1 - a) = K_2 A$$

in which p is the pressure at which the molecules are adsorbed onto the solid surface, A is the area covered by the adsorbates, $(1 - a)$ is the area not covered by adsorbates, and K_1 and K_2 are proportionality constants. The equality of $K_1 p \times (1 - a)$ to $K_2 A$ expresses the condition of equilibrium achieved by adsorptive systems in which the tendency of molecules adsorbed to leave and return to the liquid or gas phase is equated by the rate of adsorption. More generally it might be expressed as:

$$\text{rate of adsorption } [K_1 p (1 - a)] = \text{rate of evaporation } [K_2 A]$$

By combining the Freundlich and Langmuir equations it will be seen that as the pressure of gas or concentration of adsorbed liquid increases the order of reaction decreases until it becomes zero when the surface has become completely occupied by molecules. The limit to which complete coverage is attained expresses the coefficient of adsorption of gases on liquids or liquids and gases on solids. Experimental determination of such coefficients has shown a linear relationship between adsorption and pressure for any particular level of concentration maintained at specific temperatures.

Sometimes separately, but usually in combination with adsorption, use is made in chromatography of the distribution of substances between two other immiscible substances. This in essence constitutes "partition chromatography". Partition is that function of a substance which enables it to be distributed throughout other substances. In common practice partition is mainly employed in dealing with liquid phases or the distribution of gases in liquid phases. Very little is known of the partition effects of one solid as it is distributed throughout others except in the case of metals. In liquids it can be shown that the distribution of one liquid between two immiscible liquids is a constant regardless of the amount of the absolute amount of solute dissolved provided the limit of solubility is not reached in either of the two solvents. This law of partition only applies when no association or dissociation of the solute takes place in either solvent. In partition chromatography conditions can be created in which this basic law of partition can be made to function to enable mixtures of compounds to separate either on absorbant papers or powder columns. Similarly, by extending these into the realm of partial pressures gases can be manipulated to yield similar types of separations and this is the basis of gas chromatography. In its simplest form the coefficient expressing this distribution may be expressed in the following form:

$$D_A = \frac{A_1}{A_2}$$

in which D_A is a constant (the partition or distribution constant) and A_1 and A_2 the concentrations of the substance A in the two adjacent phases. For example, if a liquid is distributed over the surfaces of mineral particles its concentration on each of these particles will be determined by the adsorptive coefficient of the minerals. If an immiscible liquid is allowed to percolate through such coated particles a two-phase liquid system is produced. When the percolating liquid contains substances in solution they will partition themselves between the carrier solvent and the so-called stationary liquid phase adsorbed on the particles. In this way the partition coefficients of these substances will cause them to separate one from the other and enable them to emerge from the tube containing them in an individual fashion. In practice pure substances such as kieselguhr, quartz, cellulose powder, activated alumina, or glass beads are used to hold the stationary liquid phase in an analytical chromatographic column. The powder is saturated with the stationary liquid phase and the excess removed by filtration or by evaporation in an oven to apparent dryness. When the powder is free moving it is assumed that the particles are covered with a thin layer of the so-called "liquid phase". In the same way substances maintained in a gaseous condition can be made to exert partition coefficients when carried over these liquid stationary phases and so separate one from each other. This is the basic principle employed in vapour phase chromatography. On the other hand the stationary liquid phase can be omitted and the adsorption coefficients of the inert solid particles can be employed to separate the substances carried in the gas stream. This involves a knowledge of diffusion.

Gases have the property of uniformly filling a space and this is accomplished by the process called diffusion. Graham's law shows that the rate of diffusion of a gas is proportional to the square root of its density. Since density is proportional to molecular weight, the rate of diffusion can be expressed as being proportional to the square root of the molecular weight of the gas. For example the rate of diffusion of oxygen is 1.17 as rapid as carbon dioxide since:

$$\frac{\text{rate of diffusion of } O_2}{\text{rate of diffusion of } CO_2} = \frac{\sqrt{\text{density of } O_2}}{\sqrt{\text{density of } CO_2}} = \frac{\sqrt{44}}{\sqrt{32}} = 1.17$$

From this example it will be seen that the lower the molecular weight of a gas the more rapidly it will diffuse. The kinetics of gas diffusion must certainly be due to the rapid movement of molecules so that the pressure of a gas is due to the bombardment of the walls of its container. The larger the vessel the lower the number of bombardments per unit of time, hence the lower the pressure exerted by the gas. Likewise the higher the temperature the greater is the rate of bombardment of the sides of the vessel by accelerated molecules. This introduces the factor of partial pressures.

In a mixture of gases occupying the same volume, each gas exerts a partial pressure which is independent of the others. This partial pressure is defined as the pressure which a gas would exert if it occupied the same volume as the mixture. Thus Dalton's law of partial pressures states that the total pressure of a mixture of gases is equal to the sum of the partial pressures of the constituent gases. If for example a volume of air is at a pressure of 780 mm and the composition is assumed to be 80% nitrogen, 14% oxygen, 6% carbon dioxide then the partial pressures of each constituent would be:

$$N_2 = \frac{80}{100} \times 780 = 624 \text{ mm}$$

$$O_2 = \frac{14}{100} \times 780 = 109.2 \text{ mm}$$

$$CO_2 = \frac{6}{100} \times 780 = 46.8 \text{ mm}$$

Boyle's law states that at constant temperature the volume of a definite mass is inversely proportional to the pressure. By Charles' law, if the pressure remains constant, a given mass of gas increases in volume by 1/273 of its volume at 0°C for every degree centigrade rise in temperature so that:

$$v = v_0 (1 + t/273)$$

where v is the volume at t°C and v_0 is the volume at 0°C. This implies that if gases obeyed Charles' law at very low temperatures they would all have zero volumes at –273°C, which is called the "absolute temperature" or the basis of the Kelvin scale of temperatures (i.e., 0° Kelvin = –273°C). Deviations from the gas laws are partly due to the existence of cohesive forces between molecules and also to the fact that gas molecules occupy a finite volume. Van der Waals therefore modified the gas equations to bring them nearer to the form of real gases by the following equation:

$$(P + a/V^2)(v - b) = \mathrm{R}T$$

in which R is the gas constant, T is the absolute temperature, a and b are constants and a/V^2 the attractive force which tends to pull the molecule from the containing wall of a vessel into the bulk of the gas. Since this reduces the pressure relative to that which would be exerted by an ideal gas it means that $(P + a/V^2)$ replaces p in the formula for an ideal gas. Likewise he realized that the finite volume of the molecules decreases the volumes of the free space available for motion so that the expression for volume was reduced to $(v - b)$. R, which is the gas constant, can be calculated on the basis of the Avogadro hypothesis that 1 gmol of any gas occupies 22.4 l at 0°C and 1 atm pressure:

$$\mathrm{R} = \frac{1 \times 22.4}{273} = 0.082 \text{ l-atm/degree/gmol}$$

or:

$$\mathrm{R} = \frac{76 \times 13.6 \times 981 \times 22{,}400}{273} = 8.3 \cdot 10^7 \text{ ergs/degree/gmol}$$

or:

$$\mathrm{R} = \frac{8.3 \cdot 10^7}{4.18 \cdot 10^7} = 1.98 \text{ cal/degree/gmol}$$

If c.g.s. units are used (pressure in dynes/cm^2 and volume in cm^3) 76 cm^3 is the volume of a column of mercury 76 cm high and 1 cm^2 in cross-sectional area, 13.6 g/cm^3 is the density of mercury, and 980 cm/sec^2 is the acceleration due to gravity: 1 cal = $4.18 \cdot 10^7$ ergs.

PAPER AND THIN LAYER CHROMATOGRAPHY

The relationship between liquids moving over the fibrous structure of paper or a thin film of powder adhering to flat surfaces of glass is a function of the interplay of adsorption and partition coefficients involved in the process of diffusion. In its simplest form, the separation of the pigments in a blot of ink spreading on filter paper illustrates the control which these coefficients exercise over the formation of diffusion patterns. It is the formation of these patterns by the use of specially selected papers or thin films of adsorbents which forms the basis of this type of chromatography. Although simple, the use of paper and thin-layer chromatography has resulted in the separation and purification of compounds which otherwise had proved impossible and as a technique for the qualitative analyses of rocks and minerals it has yet to be exploited to the full.

If a strip of paper containing a spot of a mixture of compounds is dipped into a solvent (as shown in Fig.91), the liquid will ascend and elute the soluble ingredients of the spot. The further the solvent front travels through the paper the greater will become the displacement and purification of the dissolved ingredients of the spot. To enable this to take place the paper is enclosed in a container to enable the so-called "mobile liquid phase" to saturate the surrounding atmosphere. The more uniformly saturated this atmosphere becomes, the more efficient becomes the separation of the constituents of the mixture. A reduction in the pressure of this atmosphere will obviously be desirable and it will be seen from Fig.91 that this desired condition has been incorporated into the design of the apparatus here proposed for paper chromatography. As the solvent ascends after the immersion of the paper or the thin layer plate has been lowered into it, the soluble ingredients of the spot fractionate into a relative pattern controlled by their partition coefficients in the solvent

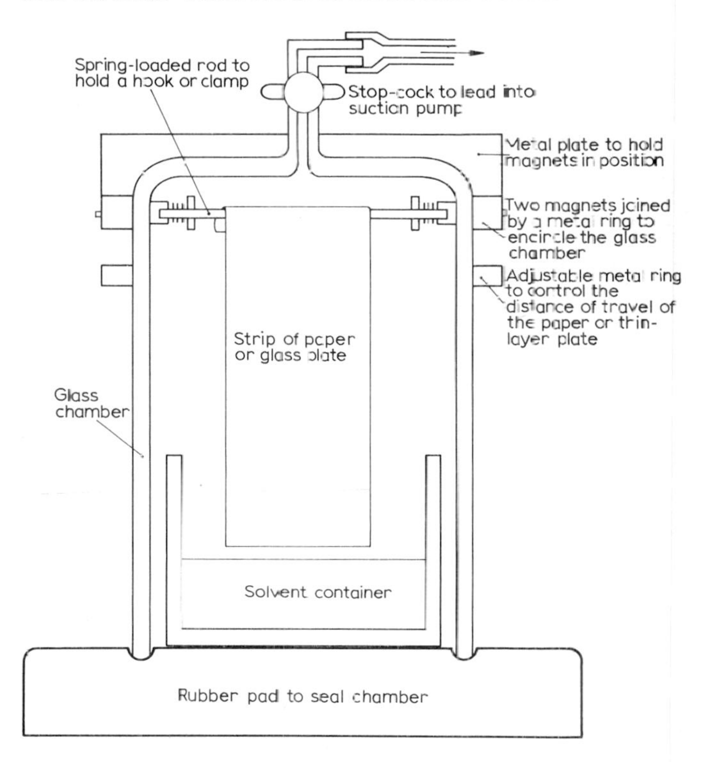

Fig.91. Glass chamber for paper or thin layer chromatography under reduced pressure.

and according to their adsorption coefficients in relation to the surface of the paper or film. Under controlled conditions this interplay of adsorption and partition results in substances achieving specific positions in relation to that of the front of the ascending solvent. This is known as the rate-of-flow, or *Rf* value, of the substance in that particular solvent. In Fig.94, the *Rf* value is given by the ratio of the distances *AB* to *AC, AD,* etc. A variation of this linear type of paper chromatography is produced by feeding the solvent into the centre of a spot of substances placed on the origin of a circular piece of paper. The radii of the circles containing the fractionated components produce the ratios which yield the *Rf* values of the substances.

When groups of compounds possess similar *Rf* values, they can be separated by the use of twodimensional paper chromatography which enables one to isolate soluble and insoluble substances with respect to a particular solvent. A simple form of apparatus for this purpose is shown in Fig.92.

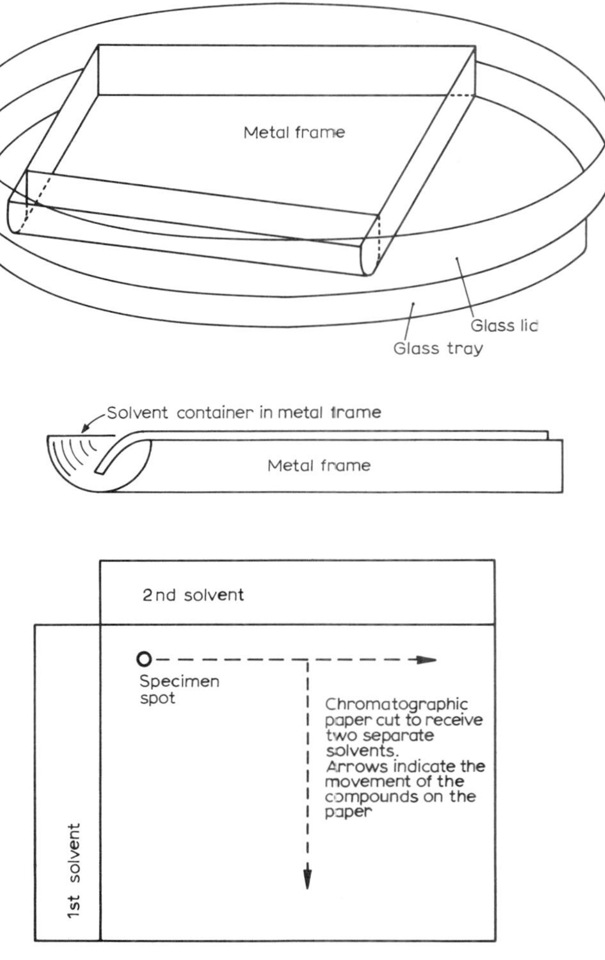

Fig.92. Apparatus for two-dimensional paper chromatography.

This was designed (unpublished) by Dr. Birks of the British Petroleum Company for the separation of the components of petroleum and such like mixtures of organic substances. A number of other methods has been devised for this purpose as they yield extremely valuable qualitative data for geochemists (see RITCHIE, 1964).

In the use of paper in one form or another for chromatography, the need to obtain an atmosphere saturated with the vapour-phase of the liquid mobile phase is paramount. To obtain paper chromatograms under low pressure conditions was an obvious way of accelerating the movement of the liquid phase and of increasing an effective interplay of the partition and adsorption coefficients of the system. This desirable state of affairs seemed difficult to achieve until the technique called "conical paper chromatography" was devised (EVANS, 1963). The equipment designed for this purpose is shown in Fig.93 and it indicates that it is the kind of apparatus which can operate in the field, since an ordinary hand-operated suction pump can be used to lower the pressure inside the vessel. As a technique, conical paper chromatography was devised originally to scan the composition of a large number of samples of natural oils from rocks and bacteriostatic substances from coals under precisely similar conditions of temperature and pressure. The apparatus shown in Fig.93 can be constructed from a sawn off cone of a glass filter funnel, to the top of which is attached a spring loaded clamp designed to hold the cone of chromatographic paper in a position out of contact with the solvent contained in a clean glass tray or Petri dish. Incorporated in the clamp is a mechanism for lowering the cone of paper into the solvent when the atmosphere inside the apparatus has been reduced to an appropriate level. By means of this technique, as many as fifteen samples can be spotted on to a 25 cm diameter paper folded into a cone. In addition to the advantage of separating the compounds under identical conditions the fifteen samples can be compared in a quantitative manner by scanning each

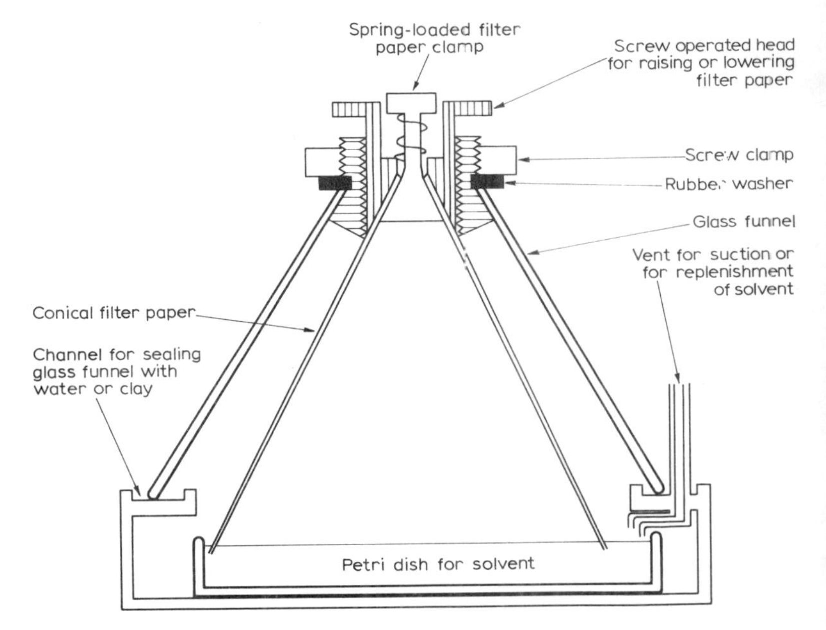

Fig.93. Equipment for conical paper chromatography.

chromatogram by means of conventional spectrophotometers or fluorimeters. Papers cut to standard size and marked at narrow intervals provide permanent records of chromatographic separations, and *Rf* values, which can be filed and preserved in polythene holders. This is of particular advantage to the field geologist using chromatographic techniques in the field.

Thin layer, or so-called thin film chromatography is a dramatic refinement of paper chromatography. As a technique it was introduced as the "chromatoplate" process (KIRCHNER et al., 1951; STAHL et al., 1956) it consists of carrying out separations in a very thin layer of adsorbents which are spread as a thin uniform film on plates of glass. The method has several advantages over paper chromatography in that these thin films accelerate the process of diffusion, accentuate the degree of separation of the components and enable the compounds to be processed by a much wider range of reagents. The thin layer is usually prepared from silica gel, alumina, kieselguhr, or various types of cellulose and ion exchange media can also be manipulated in this way. Glass plates processed in this way provide excellent chromatograms which lend themselves to photometric and fluorimetric analysis.

REAGENTS FOR INORGANIC PAPER CHROMATOGRAPHY

The following list is not intended to be a comprehensive one, but to act as a guide to the development of paper and thin layer techniques for the identification of the more important elements. Moreover, it is not intended to present an exhaustive list of the means of obtaining these elements in a concentrated condition from rocks and minerals. The preparation of the solution of the rock or mineral is a matter of choice but it is often convenient to couple this technique with conventional forms of chemical analysis and in most cases to use an aliquot of "solution A or B" for this purpose (see Chapter 3). Most solutions must be concentrated so that when spotted on to the paper or thin films the elements are retained within the smallest possible area. Batches of solutions can be rapidly concentrated at low temperatures by placing them in a container evacuated by a conventional suction pump. When the aqueous mixture has been spotted on to the chromatogram it should be dried before processing by the following solvent reagents (see RITCHIE, 1964, p.102).

Solvent 1

Butanol 50 ml, 10*N* HCl 25 ml, 40% HF 1 ml, H_2O 24 ml. Cations separate in the following relative order of *Rf* values: Au, Pt, Cr, Zn, Cd and As, Sn, Nb and Sb, Te, Pd,

Re, Fe^{3+}, Ag, Bi, Mo, Se, Te, Ti, Be, Ge, In, Hf, Zr, Mn, Mg, Li, Co, Pb, Rb, Al, K, Ca, Cr, Sr and Rn, Ba, La, Sc, Th and Ce.

Solvent 2

Butanol 50 ml, 40% hydrobromic acid 5 ml, H_2O 50 ml. Cations separate in the following relative order of *Rf* values: Zn, Au, Pt, Ru, Te and Re, Se, Ag and Cd, Sn and Hg, Sb and Cu, As, Ta, Bi, Ge, Ga and Fe^{3+}, Pb, Be, Cu, Mo, Ti and UO_2^{2+}, V and Rh, Li, Co, Mg, Al, Ni and Mn, Cr, Nb and Cs, Sc, Y, Rb and Cr, La and Ca, Ir, Th and Ce, Sr, Zr, Hf, K, Ba.

Solvent 3

Ethanol 30 ml, methanol 30 ml, 2*N* HCl 40 ml. Cations separate in the following relative order of *Rf* values: Mg, Hg and Tl, Cd, Zn, Sn, As, An and Pd, Pt, Bi, Re, Nb and Be, Cr, Ir, Ga, Sb, In, Ti and UO_2^{2+}, Sc and Al, Li, V, Cr and Fe^{3+}, Se, Te, Y and Ru, Co and Mn, La and Cu, Ni, Th and Mo, Ce, Zr, Hf, Ge, K and Te, Pb, Ca, Tl, Sr, Rb, Cs, Ba, Ag.

Solvent 4

Acetone 90 ml, 10*N* HCl 5 ml, 40% HF 1 ml, H_2O 4 ml. Cations separate in the following relative order of *Rf* values: Ta and Hg, Cr, Se, Tl, Au, Pt, Zn and Pd, Re, Sb and Fe, In and Bi, Mo, Nb and Ru, Sn, Cd, Ga, As, Ti and V, UO_2, Ag, Cu, Zr, Pb, Hf, Co, Ge, Mn, La, Be, Te, Li, Cr, Rb and K, Cs and Mg, Ir and Ni, Ca and Al, Sc and Y, Sr and Ba, Th and Ce, Te.

In order to identify the positions of these ions on the chromatogram, it is usual to apply colorimetric reagents and those listed in Tables IX and XI are suitable for spraying on to the dry chromatographic paper or thin layer material. FEIGL (1954) provides a list of suitable reagents for spot tests which might be supplemented by the handbooks of BRITISH

DRUG HOUSES (1958) and of HOPKINS and WILLIAMS (1964). Chromatographers usually prefabricate in the laboratory their own means of spraying papers and thin-layer plates with these colorimetric reagents.

COLUMN CHROMATOGRAPHY

The principal characteristic of this type of chromatography is that mixtures are placed in a glass tube packed with powder through which they are eluted by means of solvents. Consequently, two types of separation columns are applicable to the problem of separating the constituents of the mixtures. One makes use of the affinity or adsorption properties of the powders in relation to the components of the mixture. The use of adsorption columns is more effective in the separation of mixtures of organic compounds than in the dissociation and segregation of ionized solutions of inorganic compounds. Consequently, ion-exchange powders have proved more effective in the separation of inorganic mixtures than the more commonplace powders which depend upon their adsorption coefficients to isolate substances.

Partition columns more closely resemble the characteristics of paper chromatography and combine partition coefficients with the adsorption coefficients of the so-called stationary phase. A large and increasing number of particulate substances have been used as packing for partition columns. Cellulose is the most common and is particularly effective when strong eluting solvents are used to move the inorganic mixtures through the column. Silica gel, diatomaceous earth (kieselguhr), starch and artificial zeolites are also widely used. In every case, the powder is formed into a slurry with the solvent to be used to elute and partition the mixture of compounds. The slurry is poured into a glass column and allowed to settle. When sufficient sediment has been accumulated in this way the excess liquid is allowed to drain away until only a few ml remain at the top of the column. It is now ready to receive material for analysis. This is inserted in a concentrated form into the supernatant solvent and the drainage tap at the bottom of the column is opened to allow the solvent and mixture to enter the column packing. As soon as this takes place a reservoir of the pure solvent is attached to the top of the column and immediately elution of the substances to be analysed begins to take place. The partitioning effects of the ingredients result in the separation of the component compounds which are soon identifiable as differently coloured zones or even narrow bands in the solid or stationary phase of the column. As each component leaves the column it is collected and the cations and anions are determined by conventional methods of colorimetry, or flame photometry. On the other hand, the presence of elements can be identified whilst they are present in the column as they will occupy precise positions determined by their relative *Rf* values (see RITCHIE, 1964).

The use of adsorption and partition columns has been widely used in a great variety of analytical fields of investigation. Consequently a wide variety of attendant techniques have been developed which have little immediate relevance to geochemical problems. When the need for such a technique arises resort should be made to many excellent publications on this aspect of chromatography (see LEDERER and LEDERER, 1957).

GAS OR VAPOUR–PHASE CHROMATOGRAPHY

During the past 10 years gas chromatography has been developed into one of the most important analytical techniques available to science. It has provided the means of investigating many complex problems, which hitherto had defied all attempts at solution and it has also replaced other more complicated procedures which required expensive equipment. There is hardly a branch of science and technology which has not benefitted from its development and geochemistry is no exception. The analysis of petroleum products is well known but it has also been applied to the investigation of the volatile constituents in rocks, fossils, etc.; it has been used to determine the released carbon dioxide from acid solutions of minerals and the permanent and noble gases can be driven from very small samples by heating and then analysed. It can be used for the quantitative determination of all gases and for any substance which has an appreciable vapour pressure at temperatures up to about 350°C. Complete analyses can be performed on samples of a few microgrammes, and the equipment required is well within the means of small laboratories.

When MARTIN and SYNGE (1941) forecasted that a gaseous mobile phase could be used instead of a liquid in chromatography they outlined the advantages which would accrue from the use of such a system. Despite the clarity of their suggestions other workers did not follow them up and it was left to JAMES and MARTIN (1952) to develop the first practical demonstration of gas chromatography.

In gas chromatography the sample is introduced as a low volume "plug" into a moving gas stream which is flowing through a column of the stationary phase. The latter may be an active solid, or a liquid of low volatility held as a thin surface film on an inert solid. It can also be a liquid applied as a thin film to the internal surface of a capillary tube. The constituents of the sample are distributed between the two phases and when their partition coefficients differ they move through the stationary phase at different speeds emerging at the end of the column separately. The single components are then detected and the amounts determined by suitable instrumentation. This method is known as the "elution technique" and is nowadays used almost exclusively. Other techniques such as frontal analysis and displacement methods are relatively little used except for specialist problems and for this reason will not be further discussed. When a liquid stationary phase is used the technique is known as gas–liquid chromatography and when a solid is used the name given is gas–solid chromatography.

THE EQUIPMENT FOR GAS CHROMATOGRAPHY

The apparatus required for the practice of gas chromatography is essentially simple. The details vary considerably between different manufacturers and laboratories and the situation is complicated by the fact that many workers prefer to build their own apparatus. Any apparatus comprises the essential components illustrated in block form in Fig.94.

The carrier gas is supplied from a cylinder *A* fitted with a pressure reducing valve *B*. The gas first passes through a drying tube *J* and its flow rate is controlled by a precision

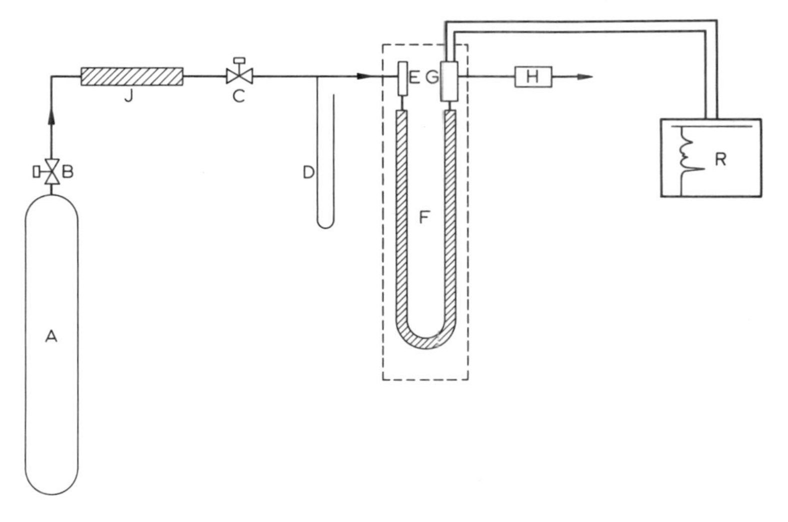

Fig.94. Block diagram of gas chromatography equipment. (For the meaning of symbols, see text.)

pressure regulator *C*. The pressure and flow conditions are monitored continuously by a manometer or pressure indicator *D* and a flowmeter *H*. The gas enters the chromatographic column *F* via a device *E*, for introducing the sample and after percolation, passes through a detector *G*. The electrical response from the detector is amplified if necessary and the resulting signal is recorded at *R*. The temperatures of the sampling system, column and detector are controlled by enclosing them in a heating jacket.

The gas supply

The gas is invariably supplied from a commercial cylinder fitted with a normal two stage pressure regulator. This in itself does not give a sufficiently precise control of pressure and it is usual to insert a precision needle valve or an additional pressure controller in series to eliminate the fluctuations which arise in the cylinder pressure regulator. The flow conditions through the chromatographic column are monitored continuously using a manometer or pressure gauge and more important, a flow-meter. The latter is shown in the gas stream emerging from the detector in Fig.94 and for the application of certain corrections which must be made in qualitative work it must be in that position. These corrections also require a knowledge of the pressure of the outlet gas, so if it is not released to atmosphere this pressure must be measured. For much routine work though, the flow-meter can be more conveniently placed in the gas stream entering the column. It is suggested by some authors that the pressure drop across the column is a sufficient indication of flow for most purposes, but the direct measurement of flow is so simple that there seems little point in omitting the flow-meter.

A very simple apparatus for the purpose is shown in Fig.95. In this simple equipment the liquid manometer measures the pressure drop across a capillary which is bent in the form of a U-tube. A series of capillaries can be calibrated for

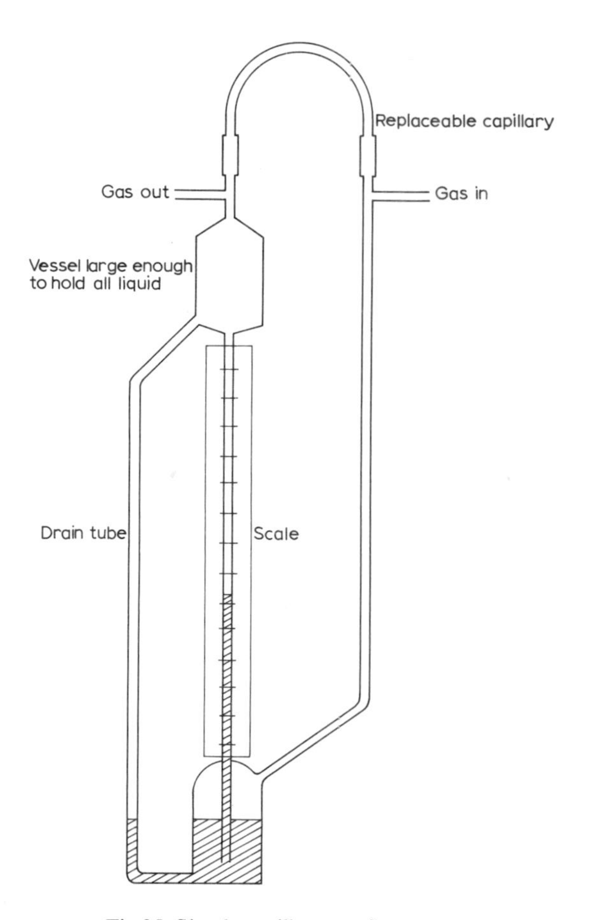

Fig.95. Simple capillary gas flow meter.

different ranges. When an accurate measurement of flow is required at the column exit or for checking the calibration of a capillary flowmeter, a simple bubble meter can be constructed from a burette as shown in Fig.96. With this apparatus a bubble is formed, by raising the level of the container *A*, and this is timed as it passes up the burette. KEULEMANS (1959) has described a slightly more elaborate system, but the apparatus above is suitable for most work. When an ac-

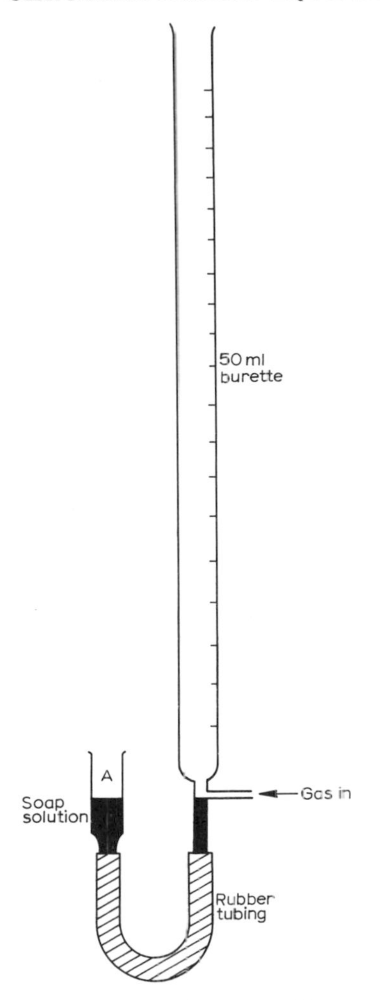

Fig.96. Simple bubble flow meter. (A: see text.)

curate estimation of gas velocity through the column is required a correction must be made for the difference in temperature of the column gas and that in the flowmeter. An allowance must also be made for the vapour pressure of water. A commercial rotameter can also be used for measuring the gas flow.

It is usually necessary to dry the gas used and a tube containing a dessicant, such as a molecular sieve or silica gel, is conveniently placed between the cylinder reducing valve and the precision valve.

Sample injection

It is necessary to introduce the sample into the gas stream in such a way that it occupies as small a volume as possible. Many methods have been used and the one preferred depends on several factors: (*a*) the amount of sample required; (*b*) the form of the sample, liquid, solution or gas; (*c*) the pressure of the gas existing at the injection point; and (*d*) the necessity or otherwise of maintaining gas flow over the column during injection.

The method which finds greatest favour for the introduction of fluids is by a hypodermic syringe through a self-sealing serum cap. The Agla syringe is suitable for this purpose but very small volume syringes designed for the purpose can be obtained and are recommended. When the column and injection point is constructed in glass, serum caps of the type shown in Fig.97 are used. They are also suitable for metal equipment but AMBROSE and AMBROSE (1961) prefer the design shown in Fig.97B. The rubber disc is compressed into the metal fitting and when the needle is withdrawn the pressure on the rubber re-seals the hole. The holes through the fitting are made slightly larger than the needle and provide a guide so that the needle tends always to go through the same hole in the rubber.

Serum caps and the discs used in the above methods are commonly made from silicone rubber which is stable up to temperatures of about 200°C. Twenty or thirty insertions can be made before the caps leak badly.

The hypodermic method is convenient for delivering liquid samples in the size range 1μl and above with reasonable accuracy, and the sample can be added very quickly, thus

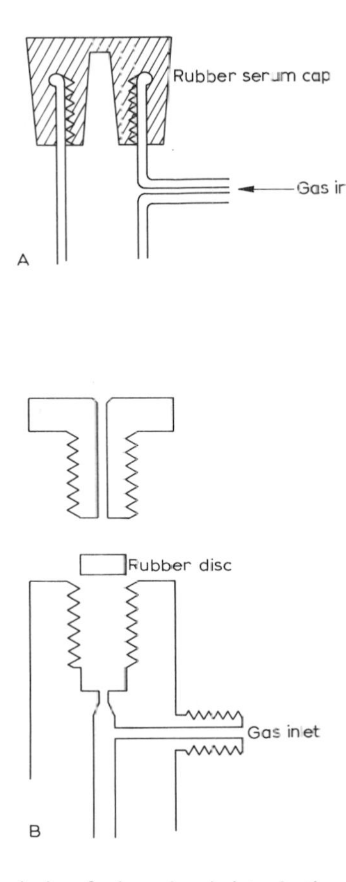

Fig.97A, B. Self sealing devices for hypodermic introduction of samples.

increasing the chances of evaporating the small volume of liquid into a low volume plug. It is essential to use a needle long enough to introduce the sample into a zone of high enough temperature to ensure that volatilization is rapid and this is often best achieved by placing the liquid directly on the top of the column packing. Occasionally it may be considered necessary to run a small portion of tubing at the entrance to the column at a somewhat higher temperature than the column itself. This "flash heater" is normally filled with glass wool to break the gas flow and to receive the sample.

Some detectors are extremely sensitive to gas flow changes and the disturbance caused by small leaks through a serum cap, through which several injections have been made, cannot be tolerated. In addition, when high gas pressures exist at the column inlet serum caps cannot be used. Many systems have been designed to overcome these problems and an excellent review of them is given by PURNELL (1962).

For much work it is not necessary to know precisely how much liquid is introduced. It is important to be able to deliver small reproducible quantities and the apparatus shown in Fig.98 has been found to accomplish this (BREALEY et al., 1959). The system is based on that described by TENNEY

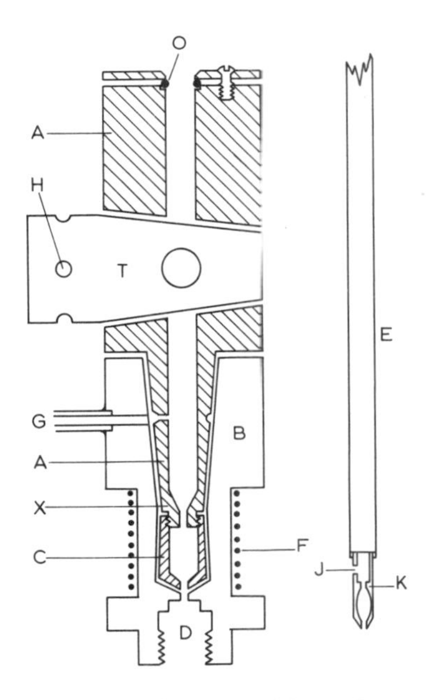

Fig.98. Pipette injection system. A = brass block; B = stainless steel block; C = cup; D = column; E = polished steel rod; F = heating element; G = gas inlet; H = handle fixing holes; J = gas inlet to pipette; K = pipette; O = O-ring; T = stainless steel tap; X = cone.

and HARRIS (1957) and the sample is added from a micro pipette that can be filled by capillary action. The pipettes *K* are made of glass of about 3/16 inch diameter and are fixed with "Araldite" cement into a polished steel rod *E*, of 1/4 inch diameter. These dimensions are such that whereas the pipette passes freely through the O-ring, the steel rod makes a gas tight seal so that the stainless steel tap *T* can then be opened to allow the pipette to be pushed down and its tip located in the cone *X*. This delivery tip is ground to seat accurately in the cone so that the gas entering the injection system at *G* is forced to pass through the hole at *J* in the pipette and thus ejects the sample very quickly into the space *C*. Once the sample has been ejected the pipette is first withdrawn sufficiently to allow the tap to be closed and is then removed. The opening and closing of the tap are facilitated by a handle that fits into one of four holes *H*. The sample is ejected from the pipette into a cup *C* which is filled with a suitable material, such as glass wool, so that any non-volatile components are trapped and prevented from reaching the column *D*. A number of pipettes of different capacity can be made to fit the equipment.

Some of the detectors used are capable of extremely high sensitivities and sample sizes in the μg range are required. It is not possible to add these quantities directly in liquid form and some method of sample splitting is used. Milligram quantities are, therefore, introduced into a small chamber and completely vapourized. The gas stream issuing from the chamber then divides in a T-tube, the majority going to waste through an orifice, and a small portion passing to the column. DESTY et al. (1959) have reported such a system and CONDON (1959) has claimed that splitting-ratios up to 5,000 : 1 can be achieved. Although as yet no other practical solution has been found to this problem, the major objection is that large quantities of gases are wasted, the higher the splitting ratio required the greater the quantity of waste.

Gases are more easily added in very small quantities than

are liquids. Whereas 1 mg of liquid occupies about 1 μl or 0.001 ml, the same weight of gas is contained in a volume of the order of 1 ml. This, and much smaller quantities, can be added accurately by hypodermic syringe but equipment designed for the analysis of gases usually incorporates some form of capillary tube sampling. One very elegant device has been described by TIMMS et al. (1958) and is illustrated in Fig.99. When the handle is in position *A* the carrier gas passes directly to the column and the sample gas sweeps through and fills the capillary. When the handle is moved to position *B* the carrier gas is diverted through the capillary and conveys the sample contained therein directly to the column.

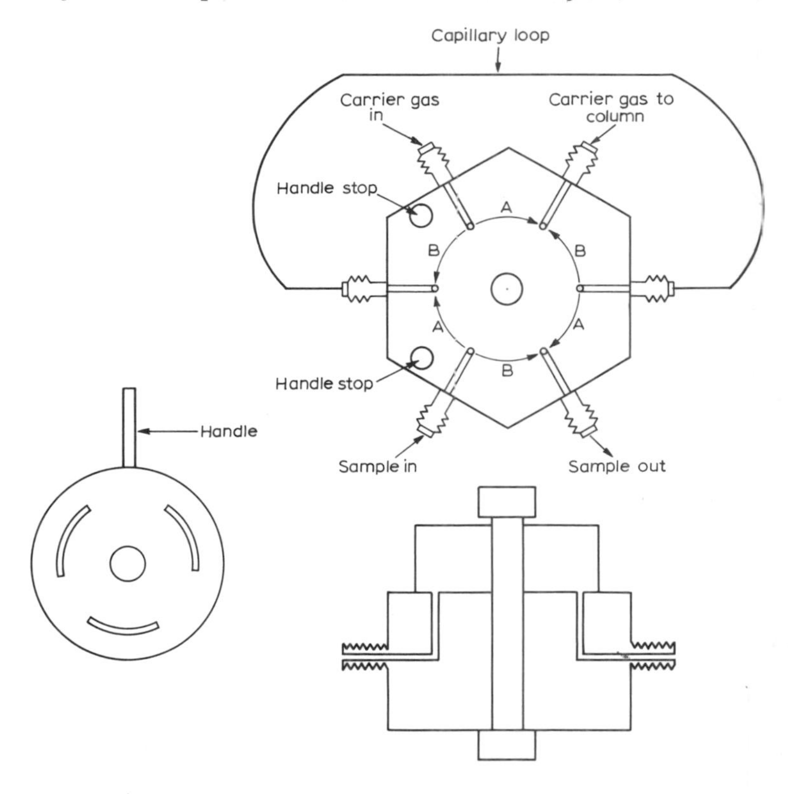

Fig.99. Gas sampling device (TIMMS et al., 1958). (*A, B*: see text.)

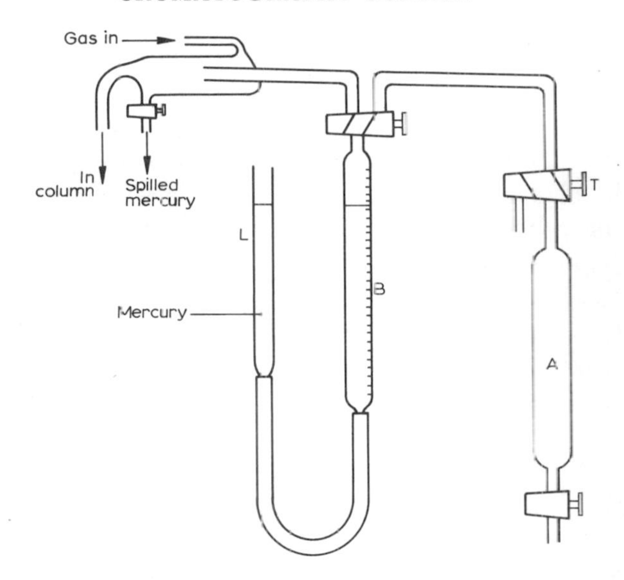

Fig.100. Gas sampling device (HARRISON, 1956). (*A, B, L, T*: see text.)

This equipment is suitable for use when a relatively large quantity of sample is available, but when the amount of sample is small some other system must be employed such as that described by HARRISON (1956) and illustrated by Fig.100. The bulb *A* contains the sample. The column *B*, and the equipment round to tap *T*, is completely filled with mercury by raising the level of *L*. By a suitable adjustment of the two taps some of the samples is withdrawn into the calibrated tube *B* and then ejected into the column.

THE CHROMATOGRAPHIC COLUMN

Two basic types of column are used in gas chromatography. These are conventional packed columns and capillaries. The former are made from tubing of any suitable material such as glass, copper, aluminium and stainless steel. Polyvinyl chloride tubing is also suitable for use at low temperatures. The internal bore may vary from 3 to 6 mm and lengths of up to 24 ft. or more have been reported, although for most purposes up to 6 or 9 ft. will usually prove adequate. The final shape of a column is purely a matter of convenience. It is usually made into the form of a U or W or a helix, the shorter ones of up to 4 ft. or thereabouts may be kept straight. The helix is most popular for the very long columns because it is obviously possible to accommodate such a form into a small space. When metal tubing is used it is important to obtain it, or convert it, into a fully annealed and softened condition so that it can be bent without kinking. It is important also to ensure the cleanliness of the internal surface by washing with solvents or detergent solutions and then drying thoroughly. Metal columns are normally filled in the straight form and then bent carefully around a mandrel after filling.

The column is packed with an inert material on to which has been adsorbed the stationary phase. The inert support is a material of high specific surface, and although a number of substances have been tried, the one most commonly employed is a form of diatomaceous earth known as "celite". The greatest separating efficiency is obtained when the celite is graded and suitable size fractions can be obtained commercially, those most commonly employed are 60–80, 80–100 and 100–120 B.S. mesh. These materials are fairly soft and easily crushed so that care must be taken at all stages of handling to avoid breaking up the particles and reducing them to a fine dust. The stationary phase is first dissolved in a low boiling solvent and then slurried with a suitable quantity of the celite. After thorough but gentle mixing the solvent is evaporated off by spreading the mixture onto a shallow tray. Up to 30% by weight of the stationary phase can be adsorbed on the celite whilst still retaining free flowing characteristics but the quantities most commonly employed are between 1 and 10% by weight.

When filling the column with this mixture it is important to achieve even packing. This is accomplished by plugging one end of the column with glass wool or silica wool and fitting a funnel to the other end by means of a short length of rubber tubing. A small quantity of the powder is then put into the funnel and persuaded to go down the column by gentle tapping. A further small quantity is then put into the funnel and the procedure repeated. The tapping may be done manually or by using an electric vibrator.

The packed column can be regarded as a series of very fine capillaries and GOLAY (1958) proposed that instead, a single length of plain capillary tubing might be used for separations. The materials used for these capillaries are stainless steel, copper, nylon and glass. The length is usually between 100 and 500 ft. and the internal diameter usually about 0.010 inch. The stationary phase is coated onto the internal surface of these capillaries by blowing a solution through the full length of the capillary at constant speed. After evaporation of the solvent, a thin film of the stationary phase is left on the wall. With these columns extremely high separating efficiencies have been achieved.

Except when gases, or very low boiling solvents, are being analysed the column is maintained at an elevated temperature. Many methods have been employed, the most common are liquid baths, vapour jackets, electrically heated air baths and metal blocks. Strict accurate temperature control is not required but it is necesary to maintain a constant temperature throughout the length of the column. Vapour jackets are favoured by many workers as they hold many advantages over other forms of heating in terms of simplicity. A typical commercial arrangement is shown in Fig.101. A suitable solvent is boiled in the flask *A*, using an electric heating mantle. The vapour rises through the jacket *J* and is refluxed back from the condenser, which may be an air condenser when high boiling solvents are used. The whole equipment is well lagged. It is important to use a fairly pure solvent for the boiling since impurities will result in fractional distillation taking place in the jacket and a temperature variation will

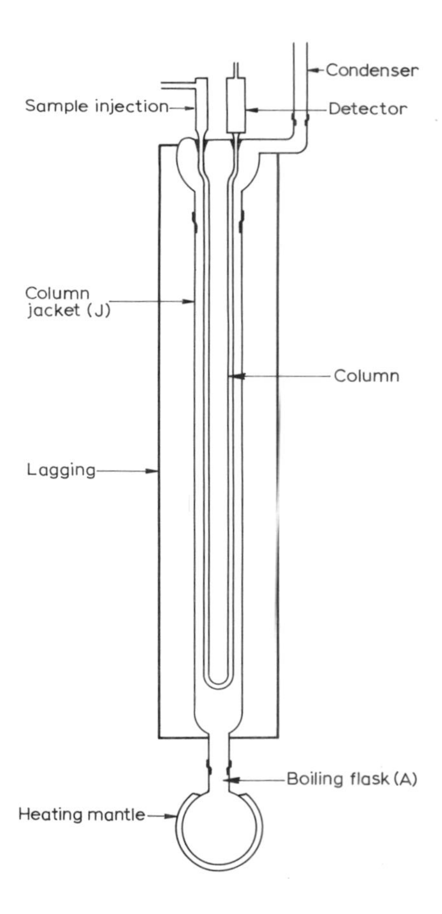

Fig.101. Vapour jacket heater.

occur over its length. Air ovens are very convenient but it must be noted that the air must be circulated around the oven in order to achieve constant temperature conditions. There can be no doubt that for the lower temperature ranges a liquid bath holds many advantages. The use is limited though, because it is very difficult to find a liquid which does not give rise to noxious fumes at temperatures above 150°C.

DETECTORS

The detection system consists of the detector itself and some means of recording the electrical output. There are two main classes of detectors:

(*a*) Integral detectors, which respond to the total quantity of material emerging from the column. The form of the chromatogram given by this type is similar to that in Fig.102A. They have distinct advantages for quantitative work but have

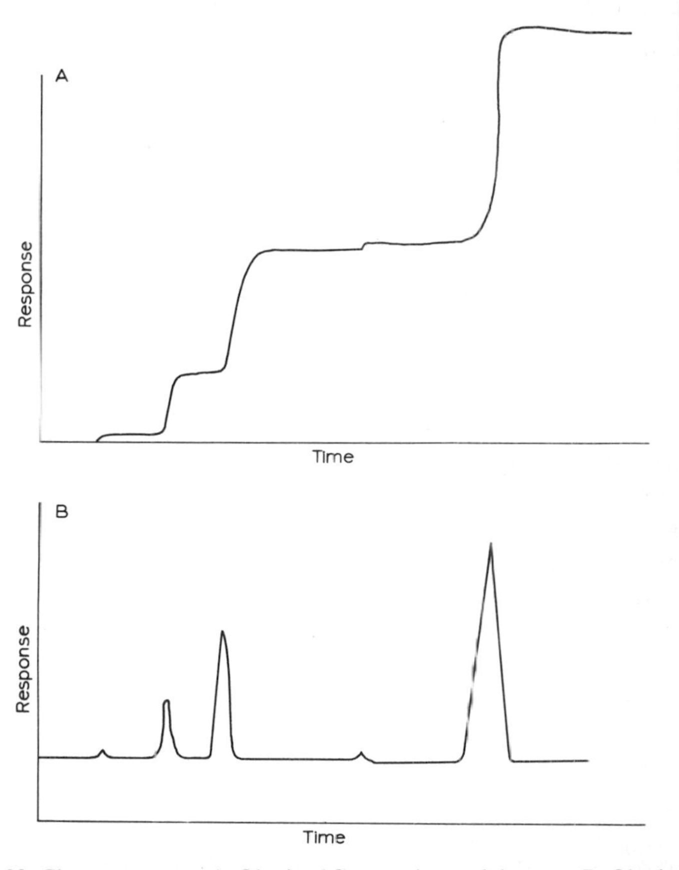

Fig.102. Chromatograms. A. Obtained from an integral detector. B. Obtained from a differential detector.

not responded to development because they suffer two disadvantages. The first is that it is difficult to determine exactly the centre of the emerging zone. Secondly, when a sample contains constituents of widely differing concentrations, the step heights may vary so much that it may be difficult to determine trace constituents.

(*b*) Differential detectors, which respond at any instant to the concentration of a vapour in the carrier gas (Fig.102B).

The second class has been developed far more fully, and some of the more important types must, therefore, be considered in greater detail. Any detector must be sensitive to small concentrations of vapour in the carrier gas and the response time to changes must be very small. The following are examples of this type of detector.

Thermal conductivity detectors (katharometers)

The name "katharometer" was introduced by SHAKESPEAR (1916) who used the principle of thermal conductivity to determine gas purities. The principle of the method is that when a hot body is located in a gas the heat conducted away from that body depends on the thermal conductivity of the gas. Katharometers consist of a fine wire or wires made from a metal having a high coefficient of resistance (platinum or tungsten) mounted in a cavity formed in a massive block of metal. The carrier gas containing the constituents from the column flows through the cavity over the wire, which is heated by passing through it a constant current. Its temperature is determined by the thermal conductivity of the gases surrounding it. As its temperature changes, so does its resistance and this latter property is measured. A differential procedure is normally adopted in which two identical cavities containing the stretched hot wires are located in the block of metal. Through one cavity flows the pure carrier gas, whilst the other receives the effluent from the column. The difference of the resistances of the two wires is measured on a Wheatstone

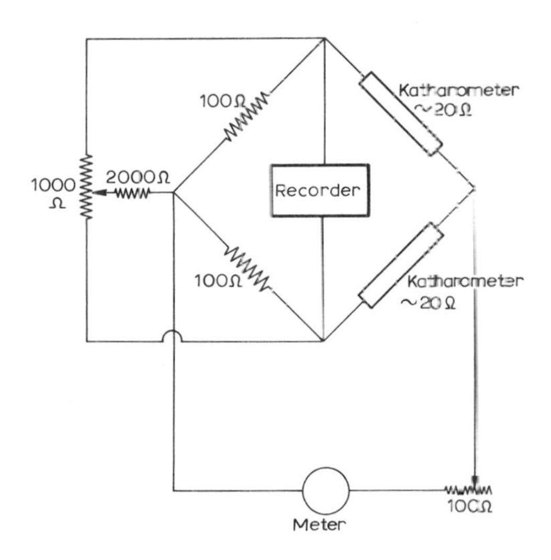

Fig.103. Wheatstone bridge circuit.

bridge, for which the circuit shown in Fig.103 has been used extensively.

When the effluent from the column is of pure carrier gas then no current flows through the meter, but when another component enters the carrier gas, the thermal conductivity in the appropriate channel changes and a current flows through the meter. The extent of this change depends upon the concentration of the impurity in the carrier gas. The relative thermal conductivities of a number of gases and vapours are given in Table XXVI.

Many types of thermal conductivity detector designs have been produced privately and commercially, but one hitherto unpublished arrangement was developed during our recent investigation. The constructional details are recorded in Fig.104 and therefore require no further description. In this katharometer, the wire is cooled not only by gas conduction, but also by convection, radiation and conduction through the electric leads. The aim in designing any instru-

TABLE XXVI

RELATIVE THERMAL CONDUCTIVITY AND DENSITY OF SOME GASES AND VAPOURS, CALCULATED FROM VALUES GIVEN IN BERL, 1951

Gas	*Relative thermal conductivity*	*Density (g/10°C at 760mm)*
Hydrogen	7.2	0.005611
Helium	6.0	0.01114
Argon	0.68	0.11135
Nitrogen	1.0	0.07807
Oxygen	1.01	0.08921
Carbon dioxide	0.60	0.12341
Methane	1.24	0.04475
Ethane	0.75	0.08469
Chloroform	0.27	
Methyl alcohol	0.59	
Acetone	0.41	

ment is to make these values as small as possible so that they are insignificant compared with the effect of gas conduction. In the correctly designed double cell these secondary effects are largely balanced out.

The sensitivity of a katharometer depends on the following factors:

(*a*) The difference in thermal conductivity between the carrier gas and the component being eluted. Table XXVI refers to pure substances, but the thermal conductivity of a mixture of two gases is not a linear function of concentration. In gas chromatography this departure is important only when a large amount of a component is being eluted but for most circumstances the relationship can be regarded as linear.

(*b*) The temperature of the wire, which is controlled by the current flowing through the wire. The higher the current the greater the temperature which, therefore, accentuates any change which takes place in the thermal conductivity of the gas.

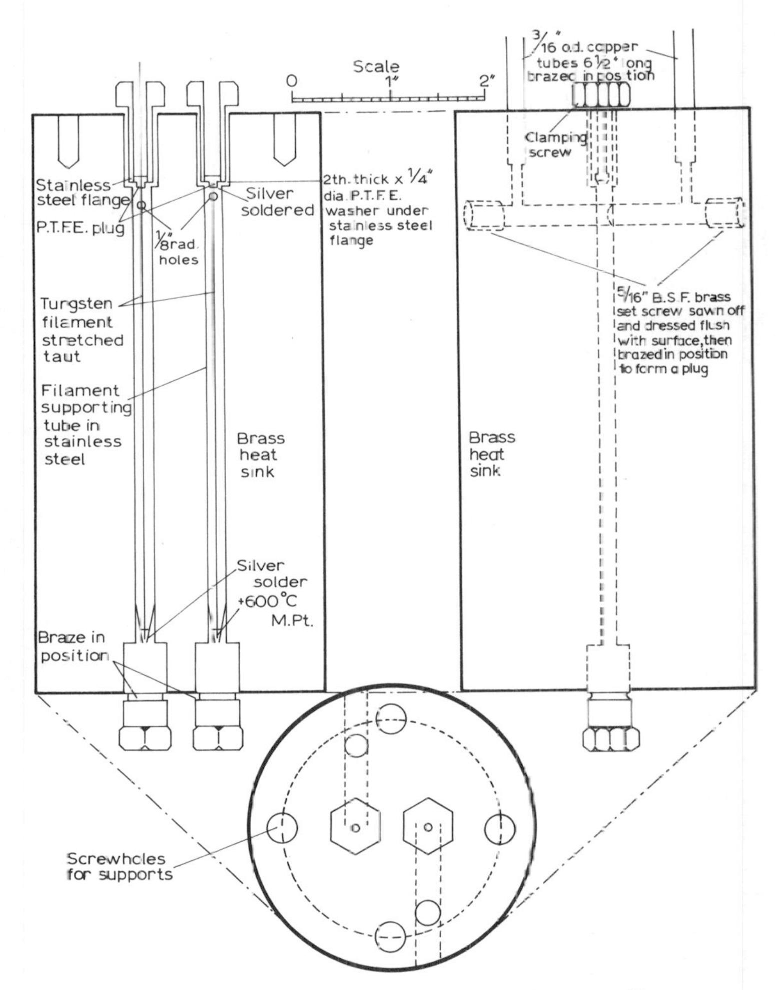

Fig.104. Details of the katharometer housing and tungsten filaments.

(*c*) The temperature of the katharometer block. A reduction in the temperature of the walls of the cells containing the wires is accompanied by an increase in sensitivity. It is, therefore, convenient sometimes to use the katharometer block at as low a temperature as possible, consistent with the avoidance of materials condensing on the walls of the cells.

(*d*) The volume of the cells directly affects the sensitivity of the katharometer. When a small amount of a component is eluted from the column it may be contained in a very low volume of carrier gas. If this is less than the effective volume of the cavity, only a portion of the wire is surrounded by the component at a given time. This in itself will diminish the response from the hot wire for any given period of time.

An examination of Table XXVI shows that hydrogen and helium have much higher relative thermal conductivities than other gases or vapours. Therefore, for analytical work requiring maximum sensitivity, they are the best choice as carrier gases. In the early days of the development of the technique, nitrogen was often used because of the better column separation which could be achieved. With a better understanding of the requirements for making more efficient columns this advantage is no longer required. Therefore, the present trend is to use hydrogen as a carrier gas when highest possible sensitivity is required. Helium is not so popular in the United Kingdom on account of its high price. When hydrogen is used, its high thermal conductivity results in the katharometer wire losing its heat more quickly than with other gases. It is normal, therefore, to increase the bridge current and thus restore the conditions for high wire-temperature. On the other hand, with hydrogen, care has to be taken that components in the sample do not react with it under the conditions prevailing in the detector. For example, a hot platinum wire may act as a catalyst to hydrogenation reactions and spurious signals may be obtained. Under such circumstances helium or nitrogen must be used to avoid damage to the detector and the production of misleading results.

Thermistors have been used instead of wires in thermal conductivity detectors. Their use provides somewhat better sensitivity at low temperatures but their use has been restricted by the difficulty of obtaining matched pairs of these units. In a matched circuit, such as that used with hot wire katharometers, it is necessary to obtain a pair of thermistors, the resistances of which are within 1.5% of each other. Even so, they are liable to decompose fairly rapidly in hydrogen, even when coated with glass or ceramic. The use of such a coating on the thermistors impairs the speed of their temperature response which reduces their efficiency.

Gas density meters

This detector, invented by MARTIN and JAMES (1956), has not received the attention it merits, probably because of the difficulty of construction. As the name suggests, it is a device for comparing the densities of the pure carrier gas with those of the components eluted from the chromatographic column. The merit of such a detector is that its response bears a simple relationship to the molecular weight of the eluted components. Moreover, it is a detector which is completely insensitive to changes in the rate of flow of the carrier gas. Consequently, the flow can be interrupted at any time to place a sample directly on the column. For details of construction and mode of operation reference should be made to the original paper (MARTIN and JAMES, 1956), from which it will be seen that this is a very complicated unit to construct. Recently a simpler model has become available commercially and this may result in an increase in the popularity of this method of detection. Likewise, we have designed a very simple form of gas density balance which has proved to be surprisingly sensitive to the changes in gas flow caused by components of differing molecular weights eluted from the column (see Fig.105 for constructional details). In essence this, like all other gas density meters, depends upon the movement upwards or downwards of gases of different densities to the carrier gas. Each light or heavy fraction respectively accelerates or retards the streams of carrier gas flowing across the upper or lower thermistor (or katharometer wires). In this way, no contact with the detector elements is made by the compounds eluted from the column. Thus the technique is peculiarly suit-

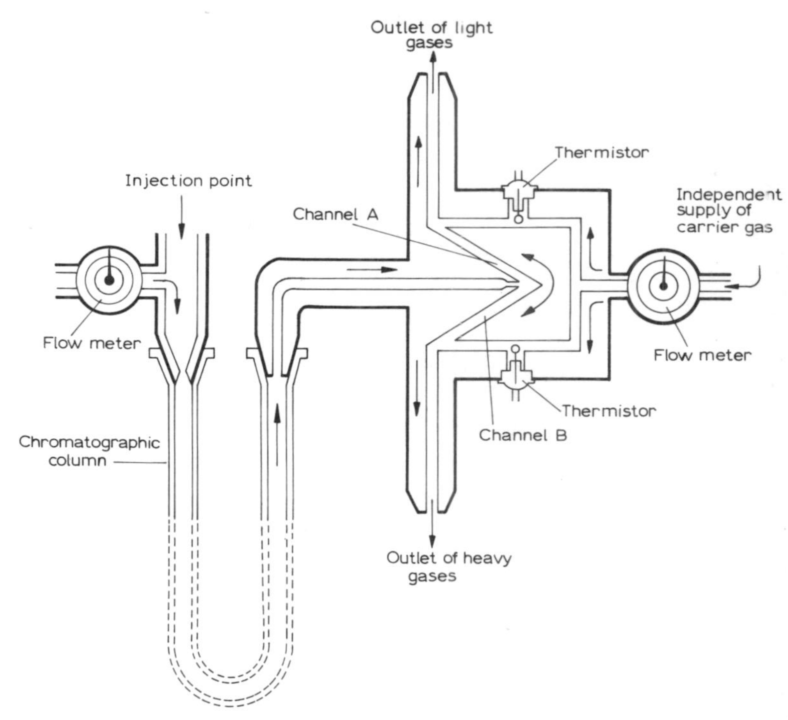

Fig.105. Details of a gas density balance designed by W.D. Evans.

ed to those forms of chromatography in which water cannot be expelled from the system. When it has been fully developed, it will be of enormous value to the geochemist in his exploration of organic matter in rocks, minerals and fossils. It is also being developed as a more elegant means of determining the water content and the carbon dioxide content of rocks and minerals. One special feature of this design is the introduction of a variable length given to the channel receiving the more dense components (Fig.105, channel *B*). By increasing its length the component travelling along channel *B* takes longer to affect the temperature of the hot wire or thermistor inserted into its particular stream of carrier gas. In this way, the chances of collision in the response created by a light component on its particular hot wire or thermistor, with that of a more dense component is minimised. In passing, it should be mentioned that this device is also being exploited to widen the time factor between the emergence of compounds from all types of columns and their entry into any particular form of detector.

Flame temperature detectors

This detector was invented by SCOTT (1955) and it is probably one of the most simple types to construct. In the original version, hydrogen or a mixture of hydrogen and nitrogen is used as carrier gas and after passage through the column it is burned at a small jet in a draught-free enclosure. A fine wire thermocouple is placed above the flame and its temperature increases when a component is eluted from the column. The output from the thermocouple is fed to a potentiometer recorder via a circuit which is used to back-off the signal obtained when pure carrier gas is being burned. The main disadvantage of this system is its extreme sensitivity to the rate of flow of the carrier gas. Obviously, if this alters, then the size of the flame will change and lead to an alteration in the response of the thermocouple. For this reason, when using most types of injection systems, a very large response is obtained when the sample is injected. WIRTH (1957) made a modification which removed some of these difficulties. Nitrogen was used as carrier gas and, after elution from the column, it was mixed at a T-junction with hydrogen, and the mixture burned as before. As a consequence, variations in the rate of flow of the column have no effect on the size of the flame.

Flame ionization detectors

These detectors possess the attractive characteristic of possessing a low degree of background interference. The detectors already described produce a small signal superimposed upon a large background which creates instability and limits the sensitivity which can be achieved. To break away from this situation, McWILLIAM and DEWAR (1958) devised the flame ionization detector. To achieve this they exploited the fact that carrier gases commonly employed in chromatography have much higher ionization potentials than organic molecules. Consequently, they were able to achieve a condition in which the sample eluted from a column becomes ionized whilst the carrier gas remained unaffected. Making use of the idea behind the flame-temperature detector (SCOTT, 1955), they replaced the thermocouple with a grid. Between it and the tip of the burner they applied a d.c. voltage and from this they recorded changes in the electrical resistance of the flame. When pure carrier gases are burned it is arranged that practically no ionization takes place. Consequently, the resistance between the grid and the burner tip is very high and virtually no current flows between them. When an organic material is eluted from the column, ions are produced in the flame, causing a flow of current which can be amplified and recorded.

The simplest electrical arrangement which gives a sensitivity equal to that of the best katharometer is shown in Fig.106A. For very much higher sensitivities the resistance R can be increased, and the voltage across it fed into a suitable amplifier, the output from which can be recorded (Fig.106B). McWILLIAM and DEWAR (1958) used a resistor of up to $1 \cdot 10^{10}\ \Omega$ in this position and obtained an appreciable signal from $1 \cdot 10^{-11}$ g of petroleum ether.

This detector is insensitive to changes in flow rates of hydrogen and nitrogen and gives no signal for permanent gases, carbon dioxide or water. It is, therefore, very useful for the examination of aqueous solutions of organic materials. Although not yet fully exploited, it has obvious uses for studies of the oil–water boundaries in oilfields and even more important is its application to samples obtained for the geochemical exploration of the sea floor.

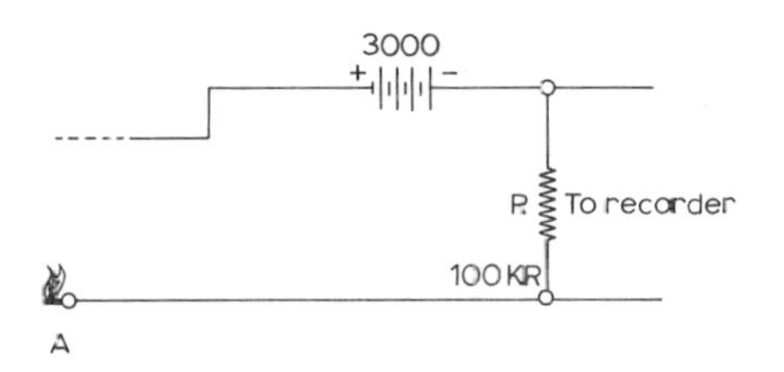

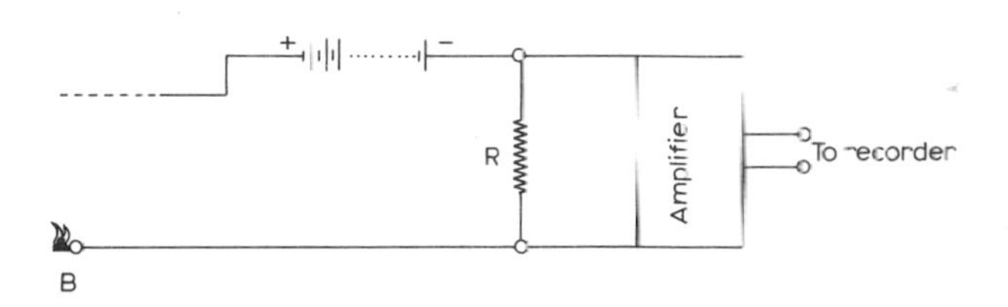

Fig.106A, B. Circuits for use with flame ionization detector.

Argon ionization detectors

Detectors depending on the ionization of components in a carrier gas by β-rays have been described by HARTLEY and PRETORIUS (1958) and by RICE and BRYCE (1957). Hydrogen or nitrogen was used as the carrier gas and the sensitivities obtained were several orders of magnitude higher than could be achieved using katharometers. LOVELOCK (1958) was able to obtain very much higher sensitivity by substituting argon as the carrier gas.

When argon or any other of the rare gases absorbs ionizing radiation, atoms of the gas are excited to a metastable state which has a relatively long life. In a pure gas these metastable atoms decay to the normal ground state emitting radiation in the process. In the presence of small traces of other gases the metastable argon atoms can donate their energy by collision. Since the molecules of most organic compounds have ionization potentials lower than the excitation potential of argon these molecules are readily ionized.

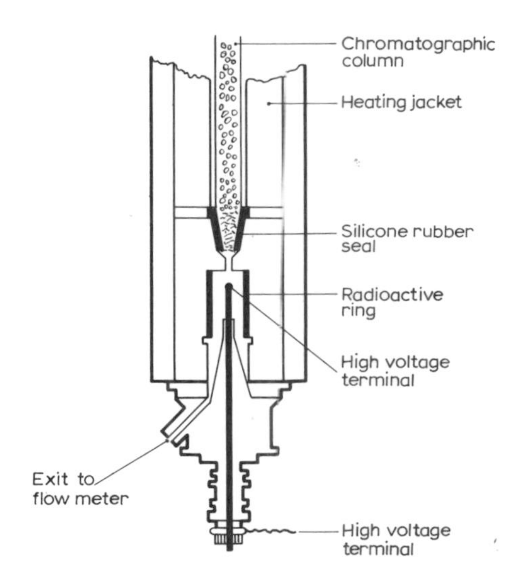

Fig.107. Diagram of the argon ionization detector.

The construction of the detector utilising this principle is shown in its simplest form in Fig.107. The cell itself is a metal cylinder about 1 inch long and $^{1}/_{2}$–$^{3}/_{4}$ inch diameter. A radioactive foil is placed on the inner wall. In Lovelock's original design this was a 10 mc source of strontium ninety. An insulated electrode comprising a 3 mm diameter rod of brass with a rounded end, is mounted in the cell. In normal usage a potential difference of about 300–1,000 V is maintained between the electrode and the body, the electrode being the anode. For certain purposes voltages up to about 2,000 V are employed.

When the cell contains pure argon a small current flows, due to some ionization of the argon, but when an organic substance enters from the column this current increases greatly owing to the induced ionization of these molecules. This current is amplified and recorded. Such an ionization detector is virtually insensitive to gas flow and pressure and the sensitivity claimed by Lovelock corresponds to the detection of a sample of about $5 \cdot 10^{-9}$ g of organic material. The cell is relatively insensitive to permanent gases and water, although the latter does give rise, during elution, to reduced sensitivity of other materials appearing at the same time. It is, therefore, necessary to remove all traces of water from the argon.

Modifications towards miniaturization have been developed more recently, which has resulted in increased sensitivity and this has made the cell more suitable for use with capillary columns.

THE COMPLETE EQUIPMENT

As can be seen from the foregoing sections, despite the fact that the technique is relatively new, a bewildering choice of equipment is available to the worker. The final assembly will depend very much on the type of work that it is required to accomplish. When complex samples are to be examined, containing very large numbers of components which require separation, a long column will normally be required and the best separations will be achieved by utilizing very small samples. This necessitates the use of one of the sensitive ionization detectors which are nowadays used in most of the commercial equipments. When mixtures of gases are being analysed as opposed to organic samples the katharometer is still extremely useful and is popular with workers who build their own equipment.

Many types of recorders have been used. When the direct output from a katharometer is being utilized a sensitive potentiometer recorder is recommended, preferably one having a full-scale deflection for an input signal of 1 mV. When the output from the detector is amplified, the recorder used will depend upon the output characteristics of the amplifier.

The present trend in commercial equipment is towards apparatus of high versatility permitting the use of either packed columns or capillary columns together with detectors of almost any description.

QUALITATIVE ANALYSIS
CHROMATOGRAPHIC ANALYSIS

The identification of materials depends on the fact that under strictly standardized conditions a substance is always eluted at a fixed time after the addition of the sample or after the passage of a fixed volume of carrier gas. These parameters are known as "retention time" and "retention volume" respectively.

The retention time is easily calculated from a knowledge of the recorder chart speed and its use gives a rapid method of comparing qualitatively the composition of mixtures of organic compounds. Whilst this is useful in a laboratory where routine work is being undertaken and the worker generally knows roughly the composition of his samples, it is of very limited value for exchanging information between laboratories. This is due to factors such as the length of time taken for recording data as this depends on gas flow rate, the temperature of the column, the total quantity of stationary phase on the column, the dead space volume and the pressure drop across the column. It is difficult to duplicate these conditions accurately between one laboratory and another. These variables can be eliminated, and a parameter of greater value obtained, by calculating the retention volume corrected for pressure drop across the column and the weight of stationary phase on the column.

Fig.108 depicts a chromatogram of a sample inserted at *A*.

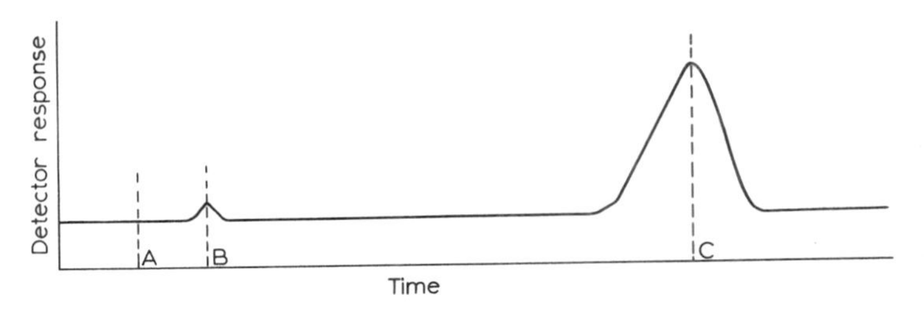

Fig.108. A typical chromatogram. (*A, B, C*: see text.)

After a short time-interval the chromatogram shows a small peak at *B*, due to the small amount of air which almost always enters with the sample. In gas liquid chromatography the air proceeds through the column unimpeded so that the volume represented by the time interval *AB* can be taken as the dead volume of the column. The component of analytical interest emerges with its elution peak at *C*.

The volume *V*, represented by *BC*, is first corrected for the effect of the pressure drop across the column:

$$V_p = V \cdot \frac{3}{2} \; \frac{(p_1/p_0)^2 - 1}{(p_1/p_0)^3 - 1}$$

where p_0 = pressure at column exit, and p_1 = pressure at column inlet. Then:

$$V_{\text{corr.}} = V_p / W_L$$

where W_L = weight of stationary phase in column.

The corrected retention volume ($V_{\text{corr.}}$) is thus independent of some of the variables and is quite useful for record purposes. Otherwise, the best method to use, is to express corrected retention volumes relative to an internal standard. A suitable material is added to the sample and from the single chromatogram the corrected retention values for sample and standard are obtained as above. The ratio of the two is then calculated and noted together with full details of the column and its working conditions. In all of this work the gas volumes must be measured at the column exit. The following compounds were recommended as suitable internal standards by the Committee of the Gas Chromatography Discussion Group and published in 1956: 1-n-butane, 2,2,4-trimethylpentane, benzene, paraxylene, naphthalene, methylethyl ketone, cyclohexanone, cyclohexanol.

Certain relationships have been found to exist between members of an homologous series of compounds which can sometimes aid the identification of an unknown material. If the log retention volume is plotted against the number of CH_2-groups in paraffins, esters, alcohols, ketones, acids, etc., linear relationships are obtained for each group when suitable stationary phases are used for obtaining the chromatograms. To help in identifying specific groups, JAMES et al. (1952) found that when the retention volumes obtained on one stationary phase were plotted linearly against those for another, straight lines through the origins were obtained for an homologous series, the slopes of the lines being characteristic of the functional group.

These methods, so far described, are suitable for identification purposes only when the worker has a fair idea of what his sample contains. Other means can be employed for completely unknown materials. If the column effluent is run through a mass spectrometer, the mass spectrum of the component can be obtained. WITHAM (1957) used this technique to analyse a petroleum fraction in the boiling range 170–260°C. Another method of direct identification is to collect the individual components in a cold trap and examine them by infra-red absorption methods. It is necessary to use a relatively large sample when this technique is employed.

QUANTITATIVE ANALYSIS

When differential detectors have been used, quantitative analysis depends on the evaluation of the peaks on a chromatogram. The more fundamental measurement is the area under the peak but the height can often be used successfully. Several methods of measuring the areas have been used. The trend nowadays is towards electronic integration, obtained whilst the chromatogram is being recorded. This is simple and accurate, provided a very stable base line is obtainable. Another method which has been used, is to cut out the peak from the recording and weight the paper. The third method is to compute an approximation of the area by measuring the peak height and the width at half peak height and multiply the values together. This procedure is reasonably satisfactory for fairly wide peaks, which can be produced by running the recorder chart at a suitable speed. If, however, this method is attempted on a narrow peak then the inaccuracy inherent in measuring a very small width is proportionally large and peak height alone would probably be more accurate an estimate of the quantity of the component it represents. Having obtained a peak height, or area, for a component this must now be related to a known quantity of that component. This is usually accomplished by running standards and preparing standard curves of the area of the peaks plotted against known concentration. In practice, the use of an internal standard is recommended. In this way a suitable substance not present in the sample-mixture is added at a constant concentration to standard mixtures and samples alike. The peaks of the resultant chromatograms can then be measured quantitatively from that of the known concentration of the compound which has been added as the internal standard.

These procedures are used when the concentrations of some of the components of a sample are being determined. When complete analyses are undertaken in a routine fashion on large numbers of samples, it may be economic to calibrate the sensitivity of the detector for the various components and compute the complete analysis, using the following equation:

$$C_x = \frac{a_x s_x}{a_1 s_1 + a_2 s_2 + \dots + a_n s_n}$$

where C = concentration of component x; a_1, a_2, a_x, a_n are the areas under individual peaks; and s_1, s_2, s_x, s_n are the calibration constants.

THE CHOICE OF STATIONARY PHASE

TUEY (1960) has compiled a list of over fifty substances which have been used for the purpose as so-called stationary phases, and gives some details for practical purposes of a representative number of substances which are available commercially in a specially prepared form. In making a choice of stationary phase the most important properties which must be considered are volatility and polarity. At the working temperature of the column, the stationary phase should have a very low vapour pressure. If it is high it will be volatilized and lost quickly so that the column would have to be renewed at frequent intervals. Of even greater importance will be the progressive entry of the stationary phase into the stream of carrier gas. This will lead to interference in the detector and its sensitivity as such would be lost. With respect to polarity, it is important to realize that polar samples are selectively retarded on a column made with a polar stationary phase. When non-polar substances, such as squalane or silicone oil, are used the non-polar components of samples emerge in ascending order of boiling point. When permanent gases are being analysed, gas–solid chromatography is normally used and the most common stationary phases are activated charcoal, silica gel and molecular sieves. The last named consist of synthetic zeolites from which the water of crystallization has been removed. The effect of the heating the sieve is to leave a uniform network of holes of molecular dimension for a given substance. In this form, molecular sieves are now available commercially as grades 4Å, 5Å and 13Å. The numbers refer to the diameter of the holes in Ångström units and this dimension determines the selective absorbing properties of the sieve. As an interesting aside, the adsorptive characteristics of these artificial zeolites are similar to that of the natural minerals. Consequently, their use in gas chromatography has shed a great deal of light on the way in which gases are contained in molten lavas and the form in which they are retained in solidified volcanic rocks.

PYROCHROMATOGRAPHY

The ability to drive volatile compounds out of rocks and minerals led to the development of a form of chromatography which has been termed pyrochromatography (EVANS et al., 1963). As the name suggests, it involves the application of heat to specimens, thus rendering its organic content volatile. For many years workers have applied pre-heating treatment to high-boiling point hydrocarbons to enable them to enter the gas chromatogram column in a volatile condition. Thereafter, the temperature and pressure conditions enabled them to pass slowly through the column as separated components which could be identified by a suitable detector. It was found that in a stream of argon a specimen could be flash-heated to very high temperatures without any appreciable pyrolytic break-down taking place. At this stage, work was commenced to produce what was virtually a micro-furnace for the distillation of small specimens of rocks and minerals. When this was achieved a whole new range of geochemical investigations commenced.

The pyrolysis unit

The first unit which was developed consisted of a coil of nichrome resistance wire welded to a glass rod which was held in a ground glass stopper. The whole was shaped to enter the upper part of a normal separation column designed to operate at temperatures above 200°C. In operation this design proved unreliable owing to the imperfect connections which could be made between the resistance wire and input current. Eventually the design illustrated by Fig.109 was developed. This proved to be reasonably satisfactory for small samples

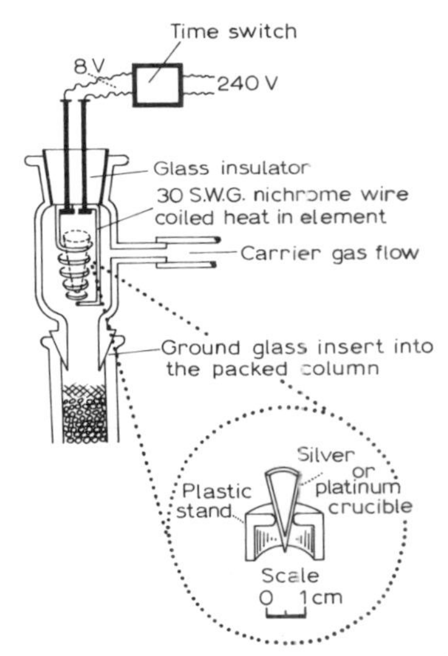

Fig.109. The heating element of the pyrolysis unit.

of rocks and minerals and such materials as coal and bitumen. Moreover, with the use of a miniature crucible of silver or platinum, the weight of the sample could be determined before and after pyrolysis. In this way it was possible to obtain a fairly accurate estimate of the quantity of volatiles expelled from a specimen and afterwards be able to examine under a microscope the effect of pyrolysis on the specimen. Using the specification quoted in Fig.109 this pyrolysis unit reaches 600°C in about 1 sec. In order to control the period of heating, a time-switch was inserted into the electrical circuit.

Using similar equipment, PARRISS and HOLLAND (1960) constructed a glass probe with a heating coil of nichrome wire (0.2 Ω). The solid or liquid specimen was placed directly in contact with the coil of nichrome wire. This excellent and simple pyrolysis unit was improved by JONES and MOYLES (1961), who developed a firing circuit to control in an accurate manner the extent of flash-heating of the specimen. In both cases the possibility of calatytic reactions taking place at these high temperatures between the specimen and the nichrome wire could not be avoided. This and the desire to determine the weight of the specimens used, made it necessary to introduce the minute conical silver sample holder shown in Fig.109. It was found that indirect heating in such a crucible inhibited calatytic reactions taking place in the pyrolysis unit.

Of cardinal importance to the development of pyrochromatography for analytical purposes, was the extent to which the substances analysed would break-down at high flash-point temperatures. PARRISS and HOLLAND (1960) applied this technique to the examination of resinous and plastic materials. They demonstrated that organic mixtures with boiling points up to about 400°C can be examined in this way without serious interference from break-down products. Remarkable chromatograms of xylenol fractions, used in the manufacture of thermo-setting varnishes, amply supported their claims for pyrolysis. Moreover, they presented convincing evidence for the use of this technique in the characterization of various types of plastics and as a control over the raw materials employed in their manufacture. Likewise, JONES and MOYLES (1961) developed pyrochromatography to investigate synthetic polymers without the use of solvent extraction techniques. They also demostrated by pyrolysing a styrene homopolymer that the chromatograms produced on a milligram scale yielded 90% of the monomer, but on a microgram scale it was practically quantitative. They also investigated the catalytic effect of the filament surface upon the specimen pyrolysed. This they state "can be demostrated at the milligram level by comparing the pyrolyses of styrene and methyl methacrylate homopolymers using nichrome and gold-plated nichrome wires; the latter surface giving slightly simpler chromatograms. However, at the microgram-level no such differences are apparent when the pyrolyses are conducted on nichrome, gold-plated platinum and platinum fila-

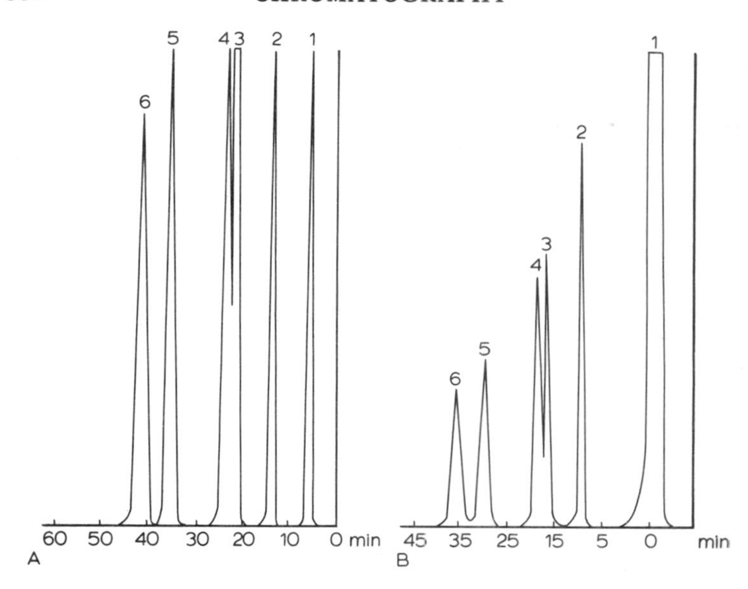

Fig.110. A. Pyrogram. B. Conventional chromatogram by liquid injection. In both cases the mixtures consist of: (*1*) xylene (b.p. = 140°C); (*2*) naphthalene (b.p. = 217°C); (*3*) methylnaphthalene (b.p. = 240°C); (*4*) 2-methylnaphthalene (b.p. = 245°C); (*5*) 2,3-dimethylnaphthalene (b.p. = 260°C); and (*6*) 2,6-dimethylnaphthalene (b.p. = 265°C).

ments"(JONES and MOYLES, 1961, p.664). In support of this, using the unit shown in Fig.109, samples of a mixture of xylene, naphthalene, 1-methylnaphthalene, 2-methylnaphthalene, 2,3-dimethyl-naphthalene and 2,6-dimethylnaphthalene were examined by normal liquid injection into a gas chromatogram. This yielded the chromatogram given in Fig.110B. The same quantity of the mixture pyrolysed in the conical silver sample holder produced a chromatogram in which all six compounds were represented by single peaks (Fig. 110A). In addition to illustrating the absence of pyrolytic breakdown, this example indicates the improvement which takes place in the separation of the compounds in the column as a result of pyrolysis.

Geochemical applications

Whilst the application of pyrochromatography is an obvious break-through into the investigation of high boiling point mixtures, its impact on organic geochemistry is even more vivid. It dispenses with the need for extracting relatively large specimens for their organic content and then losing the low-boiling ingredients in the evaporation of the solvents. The minute scale upon which pyrolysis can be undertaken, using the small conical specimen holders, has resulted in the immediate resolution of a number of mineralogical phenomena. For example, opal has always been regarded as possessing its characteristic colour from internal reflections and refractions between its constituent layers of micro-crystalline silica. Pyrochromatograms of an Austrian opal and a South American opal yielded unquestionable quantities of organic compounds (Fig.111). When the specimens of opal were removed they had lost their opalescent colours. It is, therefore, reasonable to suggest that a good deal of the opalescence of opals is due to inherent films of organic matter. Such an explanation would be more in accord with the impermanent nature of these colours which can be easily de-

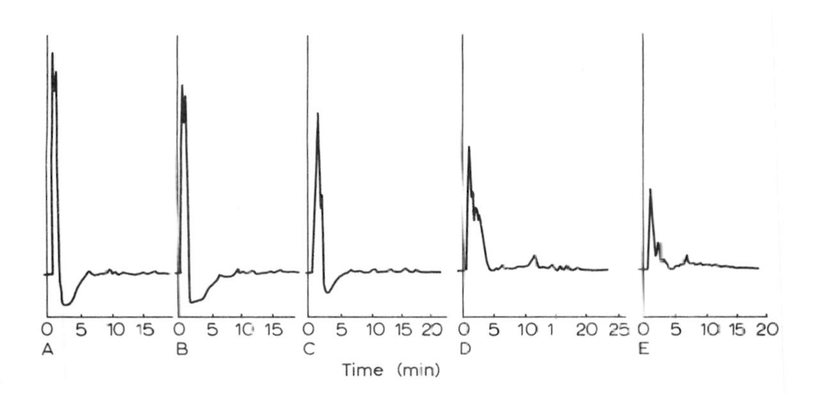

Fig.111. Pyrochromatograms. A. Brown opal. B. Multi-coloured opal. C. Moss agate. D. Authigenic quartz. E. "Blue John" fluorspar.

stroyed on heating the opal. Likewise, a specimen of moss agate yielded a similar form of pyrochromatogram (see Fig.111) and lost its colour on pyrolysis. Again it is evident that the colour of moss agates is due to organic matter present in the mineral. Another example of the use of this technique in exploring the nature of colour in minerals concerns fluorspar. It has been maintained that the colour of Blue John fluorspar is due to organic matter (FORD, 1955). "The fact that solvents failed to remove all the inclusions suggests that several types of hydrocarbons are present and that some are considerably less soluble than others. If any piece of Blue John is heated, a bituminous odour is given off" (FORD, 1955, p.45). This was an obvious subject for pyrochromatography which established the existance of ten hydrocarbons in specimens of blue fluorspar (see Fig.111 E).

In a slightly different field of research is the application of pyrochromatography to the impact of organic matter on mineralization in sediments. A typical example is the exploration of the organic content of crystals of quartz found in sediments. Authigenic crystals of quartz are fairly common in limestones. The growth in situ nature of these crystals would suggest some connections with the biogenic processes associated with the formation of limestones. A typical specimen of authigenic quartz extracted from a limestone of Carboniferous age was inserted into the pyrolysis unit. Great care was taken to avoid surface contamination of the crystal by handling it during and after extraction by means of a perfectly clean pair of forceps. Pyrolysis yielded the chromatogram shown in Fig.111D, which indicates the existence of thirteen organic compounds in the crystal. This tends to support the results obtained by the artificial production of quartz by means of humic acids and adenosine (EVANS, 1964). A somewhat similar exercise confirmed the discovery of HARINGTON (1962) that the asbestos mineral crocidolite contains 3,4-benzpyrene plus oil-wax constituents and nine amino acids. This fascinating discovery will cause a reappraisal of the origin of several other types of fibrous minerals which hitherto have been considered to be the products of metamorphism.

The application of pyrochromatography to sediments is illustrated by the analysis of the Beacon Sandstone of Antarctica (ELLIOTT and EVANS, 1963). After the ill-fated Scott expedition of 1910–13, a valuable suite of Antarctic rocks, collected by Sir Raymond Priestley, was donated to the University of Nottingham. In it was a specimen from the Beacon Sandstone series collected from Cape Royds. This was used to investigate the possibility of any part of this extensive series of sediments possessing the properties of reservoir or source rocks for oil. The Beacon Sandstones, totalling in place 3,500 ft. in thickness, extend throughout eastern Antarctica in the region of the Great Fault Scarp. The succession, consisting of nearly horizontal strata interleaved with dolerite sills, characterizes the tabular mountains of this area for more than 1,000 miles. The hand specimen from this series is a light-coloured, very coarse-grained, hard quartzite with numerous small pebbles. The sorting is rather poor and the grain size varies from 10 mm down to silt grade interstitial sediment. The majority of the pebbles are of unstrained quartz, the rest are of granite, gneiss, schist and quartzite. Very little cement has been introduced, and while a small proportion of the grains have a fringe of secondary quartz, most of them are welded along common margins. Sporadically distributed throughout are small chalky patches and also small holes often lined with chalky material. On X-ray analysis this chalky substance has proved to be albite-rich alkali felspar. Of interest and significance to this investigation are euhedra and granular aggregates of sphene. While no metamorphic significance is normally attached to this mineral, in some instances elsewhere, it is clear that it has begun to crystallize at higher temperatures than associated epidote. The minimum temperature of formation of epidote is 160°C, so that of sphene must be higher. If the recrystallization of the

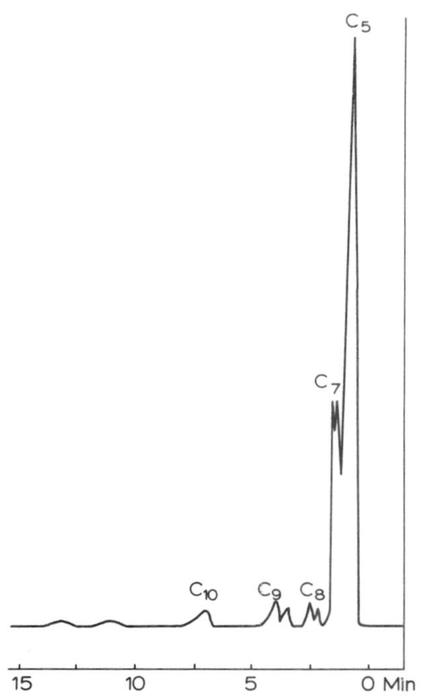

Fig.112. Pyrogram of Beacon Sandstone specimen. (For the meaning of symbols, see text.)

sphene and the quartz were at the same time, then the formation of the quartzite must have occurred above 160°C. In essence, therefore, this is a quartzite of low porosity of which the induration and association with dolerite intrusions would, at first sight, appear to rule out the possibility of the survival of hydrocarbons. Moreover, the rock was digested with hydrofluoric acid in order to scan the sediment for micro-organisms such as dinoflagellates. No traces of these or other micro-fossils were found. Nevertheless, seven powdered specimens were subjected to pyrochromatography and they all yielded identical chromatograms such as that shown in Fig.112. Of the ten compounds extracted from the specimen, five consist of paraffins ranging from C-5 to C-10; the remainder have not yet been identified. These oils differ from the indigenous oils of normal sediments not only in containing no light fraction but also in the narrow range of their boiling points, and from those of metamorphosed oil-bearing strata in containing no naphthalene or phenolic compounds. In other words, they are neither original nor residual but are probably migrants which have been fixed by condensation in these sandstones. It would be imprudent to speculate too far from one sample, but the following conditions might be envisaged. The presence of sphene suggests that during metamorphism, presumably by the dolerites, the Beacon Sandstone was subjected to a temperature in excess of 160°C. Adjacent to the intrusions, higher temperatures would develop, accompanied by hydrothermal activity derived from the inherent water content of the invaded sediments. Consequently, by a process akin to steam distillation, the volatile constituents of the sediments would migrate laterally and upwards throughout the thermal aureole and would condense as soon as they found themselves in a rock with a temperature lower than their own boiling point. Such a thermo-dynamic process would account for the narrow range of saturated hydrocarbons found in this specimen. The fact that the boiling point of the highest recorded C–10 paraffin is 174°C, which is very little above the minimum temperature indicated by the sphene, would support such a hypothesis. Moreover, such an explanation seems to indicate that this considerable column of Upper Palaeozoic and Mesozoic sediment might well contain source rocks capable of producing oil. Unfortunately, so far as we know, there are no structures capable of concentrating such oil into an oilfield, but from the nature of these deposits there might conceivably be stratigraphic traps.

The exploration of organic residues in limestones may be illustrated by an investigation of the conditions of sedimentation of the Lincolnshire Limestone Series of the Middle Jurassic extending throughout the eastern and central areas of England (GUNN and COOPER, 1963). Using the pyrolysis unit described in Fig.109, Gunn and Cooper showed that these

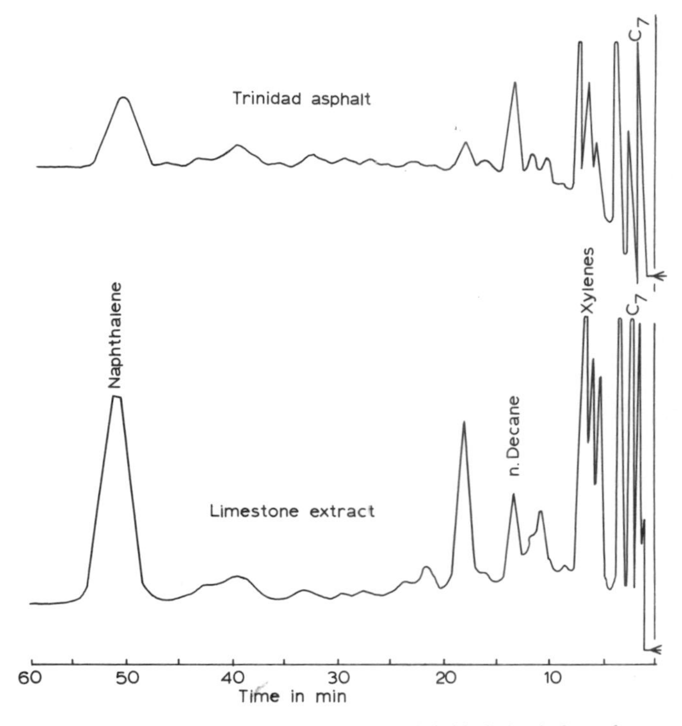

Fig.113. A pyrochromatic comparison between Trinidad Asphalt and extracts from the Lincolnshire Limestone (GUNN and COOPER, 1963). Column characteristics: 10% Apiezon M, 100°C. Argon detector at 1,750 V. (For the meaning of symbols, see text.)

limestones contained organic matter which closely resembled the asphalt of Trinidad Lake (see Fig.113). The limestones yielded hydrocarbons up to C–7, and the principal aromatics, benzene, toluene, ethyl benzene, xylene and naphthalene were identified. There were also significant amounts of oxygen-containing compounds with boiling points between 100° and 200°C. By the use of orthodox methods of solvent extraction, no free phenols were identified, but after fusion of these extracts with potassium hydroxide in the nitrogen atmosphere of the pyrolysis unit p-hydroxybenzoic acid and vanillic acid were indicated. Gunn and Cooper concluded that the organic matter in these limestones is directly associated with the clay constituents. As such they are probably chelated to the alumino-silicates as aromatic-based complexes. Previously no evidence of this type has been available, and it serves to illustrate the value of pyrochromatography in the study of calcareous sediments.

The extension of these pyrochromatographic techniques into studies of organic matter in fossils, the geochemistry of coal macerals, and the evaluation of the organic constituents of sediments offers considerable possibilities. For example, a sample of graptolite shale containing *Didymograptus murchisoni* has been used to examine the possible organic content of the fossil. Over fourteen organic compounds emerged from a small fragment of the stipe and among them the amino-acid alanine. This is the first record of amino-acids from graptolites. Likewise, the ocular area of the head-shield of the trilobite *Phacops* has yielded six hydrocarbons. Similary, two specimens of fish from the Old Red Sandstone have shown that they still contain significant quantities of biogenic matter.

Perhaps the most significant use of pyrochromatography is in the evaluation of the petrology of coal. Since it is a technique which can analyse exceedingly small specimens, it could be described as a micro-chemical probe. A typical example of its value to workers in coal petrology has been afforded by a comparative study of Permo-Carboniferous coals which were typical of Laurasia and Gondwanaland. The two coals chosen were the Top Hard Seam of the Nottinghamshire coalfield and the No.2 or Main Seam of the Witbank coal field of South Africa. The Top Hard Seam is the highest seam in the middle Coal Measures and varies from 4 ½ to 5 ½ ft. in thickness, while the Main Seam of the Witbank occurs in the middle Ecca Stage of the Karroo System. The failure of these bituminous coals to furnish distillates, even in a molecular still, at temperatures below which decomposition starts, indicates that they are either networks of polymeric structures connected by primary bonds, or they are mainly mixtures of large high boiling point molecules. Consequently, several attempts have been made to distil coal in vacuo. Extremely variable results have been obtained and on the few occasions when the individual constituents, vitrain, durain and fusain have been distilled it was found that with the exception of fusain, the composition changed gradually. Since such observations are at variance with the petrographic differences between these macerals they were subjected to pyrolysis in a stream of argon and the volatiles analysed in a vapour-phase chromatogram. At instantaneous temperatures of about 800°C in a pyrolysis unit the volatile content of the coal was swiftly inserted into the stream of argon gas. From experience, using standard compounds of high boiling points, there was remarkably little pyrogenic break-down. From earlier work in vacuum distillation, it was expected that large volumes of methane, CO and water would emerge (HOWARD, 1947). On the contrary, the chromatogram revealed the direct emergence of paraffins, along with saturated and unsaturated hydrocarbons (see Fig. 114). At this stage in the use of this new technique it has not proved possible to identify all these compounds, but with use of known compounds and mixtures thereof, some of them can be established with reasonable certainty. Since the samples of vitrain, durain and fusain were not in excess of 0.3 mg it was evident that the pyrochromatograms indicated the dominant components of each maceral. The surprising result obtained in this way was that the macerals of the Top Hard Seam were remarkably similar to those obtained from the Main Coal of Witbank and were capable of being classified into eight groups (see Fig. 114):

Group 1. Low boiling point paraffins diluted in a methane-ethane eluate, with peak No. 2 indicative of trimethylpentane. Included are the low boiling point cyclopentanes and olefines.

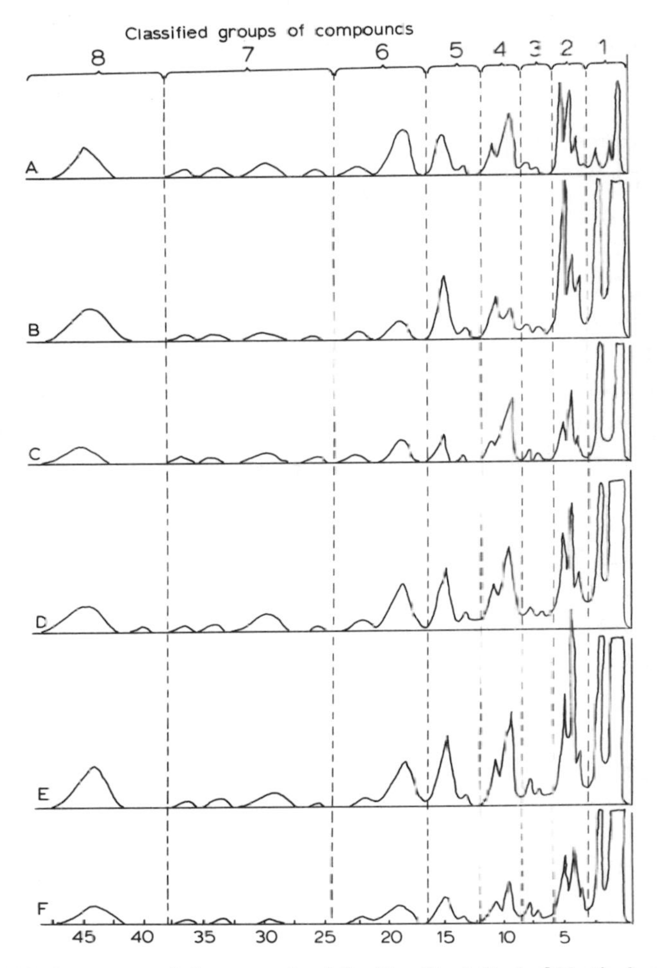

Fig.114. Pyrograms of the macerals of the Top Hard Coal of Nottinghamshire and the Witbank Coal of South Africa. A. Vitrain from the Top Hard Coal. B. Durain from the Top Hard Coal. C. Fusain from the Top Hard Coal. D. Vitrain from the Witbank Coal. E. Durain from the Witbank Coal. F. Fusain from the Witbank Coal.

Group 2. A characteristic group with boiling points in excess of 65°C and including ethylcyclohexane.

Group 3. Two groups which have not yet been resolved with boiling points in excess of 80°C.

Group 4. Principally composed of decane and phenol.

Group 5. Represented by peaks No.10 and 11 which is probably xylene.

Group 6. Represented by peaks No.12 and 13 which have not been resolved.

Group 7. A variable group of compounds with boiling points ranging between 140°C and 200°C and probably consisting of phenols, cresols and fatty acids.

Group 8. This group contains naphthalene and compounds with boiling points in excess of 218°C.

From this it will be seen that it will soon be possible to place this form of pyrochromatography of coal on a quantitative basis and throw more light on the origin and utilisation of coal.

Pyrochromatography is in its infancy but it will certainly penetrate a wide variety of geochemical areas of research in the near future. The wealth of knowledge which now exists in the orthodox use of gas chromatography will unquestionably assist the development of these high temperature techniques. For the direct evaluation of the characteristics of various types of petroleum it offers considerable advantages. From work in progress it seems likely that in the near future it will be possible to distinguish the organic content of source rocks from reservoir rocks in an oilfield by pyrochromatography. In a entirely different field of research it can be applied to the identification of the organic content of sea-water and in an equally spectacular way pyrochromatography is a natural technique for the examination of the organic content of meteorites.

References

CHAPTER 1

ABELSON, P.H., 1957. *Ann. N. Y. Acad. Sci.*, 69: 276.
BARTH, T., 1962. *Norsk Geol. Tidsskr.*, 42 (2): 330.
BIEN, G.S., CONTOIS, D.E. and THOMAS, W.H., 1958. *Geochim. Cosmochim. Acta*, 14: 35.
BIERMANN, W.J. and GESSER, H., 1960. *Anal. Chem.*, 32: 1525.
BRISCOE, H.V.A., MATTHEWS, J.W., HOLT, P.F. and SANDERSON, P.M., 1937. *Trans. Inst. Mining Met.*, 46: 145.
BURRI, C., 1964. *Petrochemical Calculations*. Siran Press, Jerusalem.
CLARKE, F.W., 1924. *U.S., Geol. Surv., Bull.*, 770: 80.
COULTER, E.E., 1956. *Trans. A.S.M.E. (Am. Soc. Mech. Engrs.)*, 78: 869.
CROSS, C.W., IDDINGS, J.P., PIRSSON, L.V. and WASHINGTON, H.A., 1902. *J. Geol.*, 10: 555.
DAVIDSON, J.F., 1960. *Trans. Inst. Chem. Engrs. (London)*, 38: 197.
ELLIOTT, R.B. and COWAN, D.R., 1965. *Norsk Geol. Tidsskr.*, 45: 1.
ELLIOTT, R.B. and MORTON, R.D., 1966. *Norsk Geol. Tidsskr.*, 46: 309.
EVANS, W.D., 1948. *Proc. S. Wales Inst. Engrs.*, 64: 2–110.
EVANS, W.D., 1951. *Colliery Eng.*, 28: 333–334.
EVANS, W.D., 1955. *Trans. Inst. Mining Met.*, 65: 13.
EVANS, W.D., 1962. *Proc. Intern. Conf. Organic Geochem., Milan.*
EVANS, W.D., 1963. *Proc. Intern. Conf. Organic Geochem., Milan,* p.33.
EVANS, W.D., 1964. *Experimental Pedology*. Proc. Easter School Agr. Sci., Univ. Nottingham.
EVANS, W.D. and AMOS, D.H., 1961. *Proc. Geologists' Assoc. (Engl.)*, 72: 445.
FORMAN, S.A., WHITING, F. and CONNELL, R. 1959. *Can. J. Comp. Med. Vet. Sci.*, 23: 157.
FRANKENBERG, T.T., 1959. *Proc. Am. Power Conf., 21 (1959)*, p.169.
FREISER, H., 1959. *Anal. Chem.*, 31: 1440.
GARDENER, L.U., 1938. *Am. Inst. Mining Met. Engrs., Spec. Publ.*, 929: 1.
GARRELS, R.M., 1959. In: *Researches in Geochemistry*. Wiley, New York, N.Y.
HARINGTON, J.S., 1962. *Nature,* 193: 43.
HOLST, S., 1958. *Nord. Med.*, 60: 1169.
JOHANNSEN, A., 1931. *A Descriptive Petrography of the Igneous Rocks*. Univ. Chicago Press, Chicago, Ill., 1.
JORGENSON, E.G., 1953. *Physiol. Plantarum,* 6: 301.
JUVET, R.S. and WACHI, F.M., 1960. *Anal. Chem.*, 32: 290.
KAGI, J., 1961. *Sulzer Tech. Rev. (Switz.),* 1: 29.

KING, E.J., 1945. *Med. Res. Council, Special Rept. Ser.,* 250: 69.
KOSTRIKIN, U.M. and FILIMONOV, A.I. 1955. *Teploenerg.,* 1: 34.
KRAUSKOPF, K.B., 1959. In: *Researches in Geochemistry.* Wiley, New York, N.Y.
KULLERUD, G., 1959. In: *Researches in Geochemistry.* Wiley, New York, N.Y., pp.301–335.
LOUW, H.A. and WEBLEY, D.M., 1959. *J. Appl. Bacteriol.,* 22 (2): 227.
MANDL, I., GRAUER, A. and NEUBERG, C., 1953. *Biochim. Biophys., Acta,* 8: 654.
MANDL, I., GRAUER, A. and NEUBERG, C., 1953. *Biochim. Biophys. Acta,* 10: 540.
MOREY, G.W. and HESSELGESSER, G.M., 1951. *J. Econ. Geol.,* 46: 821.
NASH, V.E. and MARSHALL, C.E., 1956. *Missouri Univ., Agr. Expt. Sta., Res. Bull.,* 614: 36.
NEUBERG, C. and MANDL, I., 1948. *Z. Vitamin, Hormon Fermentforsch.,* 2: 480.
NEWMAN, P.C., 1959. *Flame of Power.* Longmans, Toronto, Ont.
NIGGLI, P., 1920a. *Lehrbuch der Mineralogie.* Berlin, 476 pp.
NIGGLI, P., 1920b. *Systematik der Eruptivgesteine.* C. Bf. Min., p.161.
NIGGLI, P., 1936. *Schweiz. Mineral. Petrog. Mitt.,* 16: 285.
NIGGLI, P., 1938. *Schweiz. Mineral. Petrog. Mitt.,* 18: 610.
NIGGLI, P., 1950. *Intern. Geol. Congr., 18th, Great Britain, 1948, Pt. II., Proc. Sect. A.,* p.101.
OSANN, A., 1899. *Tschermaks Mineral. Petrog. Mitt.,* 19: 100.
POCOCK, F.J. and STEWART, J.F., 1962. *Am. Soc. Mech. Engrs., Paper,* 61: 140.
QUENSEL, P., 1918. *Bull. Geol. Inst. Univ. Uppsala,* 16: 10.
RAKOV, K.A., 1956. *Combustion,* 28 (3): 42.
RITCHIE, A.S., 1964. *Chromatography in Geology.* Elsevier, Amsterdam, 158 pp.
ROSE, P.N., 1965. *Sci. J.,* 1 (4): 59.
SHORT, M.N., 1940. *U.S., Geol. Surv., Bull.,* 914: 311.
SIEVER, R., 1957. *Soc. Econ. Palaeontologists Mineralogists, Spec. Publ.,* 7: 55.
STIEFEL, W., 1961. *Sulzer Tech. Rev. (Switz.),* 3: 21.
STRAUB, F.G., 1946. *Illinois Univ., Eng. Expt. Sta., Bull.,* 364: 54 pp.
STYRIKOVICH, M.A., 1956. *Combustion,* 28 (3): 49.
SWINGLE, K.F., 1953. *Am. J. Vet. Res.,* 14: 493.
TARR, W.A., 1927. *Am. J. Sci.,* 194: 409.
TURNER, R., 1958. *Chemical Engineering Practice.* Butterworths, London, 6: 302 pp.
TWENHOFEL, A., 1950. *Principles of Sedimentation,* 2nd ed. McGraw-Hill, New York, N.Y., 402 pp.
ULMER, R.C., 1960. *Trans. A.S.M.E. (Am. Soc. Mech. Engrs.),* 82: 264.
WEBLEY, D.M., DUFF, R.B. and MITCHELL, W.A., 1960. *Nature,* 188: 766.
WILLIAMS, D. and NAKHLA, F.M., 1950. *Trans. Inst. Mining Met.,* 60 (7): 284.

CHAPTER 2

BYRON, A.S., 1925. *J. Am. Chem. Soc.,* 47: 981.
DICKSON, J.A.D., 1966. *Nature,* 205: 587
DWYER, F.P., 1938. *Australian Chem. Inst., Proc.,* 5: 37.
EVANS, W.D., 1966. *Trans. Inst. Mining Met., Sect.B,* 75: 165.

FEIGL, F., 1946. *Spot Tests*. Elsevier, Amsterdam, 600 pp.
FRITZ, H., 1929. *Z. Anal. Chem.*, 78: 418.
GLAZUNOV, A., 1929. *Chim. Ind. (Paris), Spec. No.*, 425:[illegible]
GLAZUNOV, A., 1938. *Österr. Chemiker - Ztg.*, 217: 41.
GUTZEIT, G., 1942. *Am. Inst. Mining Engrs., Tech. Publ.*, 1457: 87 pp.
GUTZEIT, G., GYSIN, M. and GALOPIN, R., 1933. *Compt. Rend. Soc. Phys. Hist. Nat. Genève*, 50: 192.
JONES, M.P. and FLEMING, M.G., 1965. *Identification of Mineral Grains*. Elsevier, Amsterdam, 102 pp.
KODAMA, K., 1963. *Methods of Quantitative Inorganic Analysis*. Wiley, New York, N.Y., 507 pp.
LINGANE, J.J., 1958. *Electroanalytical Chemistry*. Interscience, New York, N.Y., 350 pp.
SHORT, M.N., 1940. *U.S., Geol. Surv., Bull.*, 914: 305 pp.
WEINIG, A.J. and SCHODER, W.F., 1947. *Technical Methods of Ore Analysis*. Wiley, New York, N.Y., 205 pp.
WELCHER, F.J., 1947. *J. Am. Chem. Soc.*, 69: 682.
WELCHER, F.J., 1948. *Organic Analytical Reagents*. Van Nostrand, New York, N.Y., 3: 152 pp., 4: 161 pp.
WILLIAMS, D. and NAKHLA, F.M., 1950. *Trans. Inst. Mining Met.*, 60: 284
YOE, A. and ROBERT, T.H., 1937. *J. Am. Chem. Soc.*, 59: [illegible]72.

CHAPTER 3

BARREDS, J.M.G. and TAYLOR, J.K., 1947. *Electrochem. Soc. Trans.*, 92: 437.
BETZ, J. and NOLL, C.A., 1950. *J. Am. Water Works Assoc.*, 42: 49.
BRANNOCK, W.W. and BERTHOLD, S.M., 1953. *U.S., Geol. Surv., Bull.*, 992: 1.
GROVES, A.W., 1951. *Silicate Analysis*, 2nd ed. Allen and Unwin, London, 336 pp.
HEGEMANN, F. and ZOELLNER, H., 1952. *Ber. Deut. Keram. Ges.*, 29: 68.
HILLEBRAND, W.F., 1900. *U.S., Geol. Surv., Bull.*, 176: 52 pp.
KITSON, R.E. and MELLON, M.G., 1944. *Ind. Eng. Chem., Anal. Ed.*, 16: 379.
KRALHEIM, A., 1947. *J. Opt. Soc. Am.*, 37: 585.
PARKER, C.A. and GODDARD, A.P., 1950. *Anal. Chim. Acta* 4: 577.
RILEY, J.P., 1958a. *Analyst*, 83: 42.
RILEY, J.P., 1958b. *Anal. Chim. Acta*, 19: 413.
SARVER, L.A., 1927. *J. Am. Chem. Soc.*, 49: 1472.
SCHOLES, P.H. and SMITH, D.V., 1958. *Analyst*, 83: 615.
SHAPIRO, L. and BRANNOCK, W.W., 1952. *U.S., Geol. Surv., Bull.*, 1036 C: 42 pp.
SHAPIRO, L. and BRANNOCK, W.W., 1956. *U.S., Geol. Surv., Bull.*, 1636-C: 56 pp.
SHAPIRO, L. and BRANNOCK, W.W., 1962. *U.S., Geol. Surv., Bull.*, 1144-A: 62 pp.
VOGEL, A.L., 1962. *Quantitative Inorganic Analysis*, 3rd ed. Longmans, London, 1216 pp.
WASHINGTON, H.S., 1932. *The Chemical Analysis of Rocks*. Wiley, New York, N.Y., 506 pp.
YOE, J.H. and ARMSTRONG, A.R., 1947. *Anal. Chem.*, 19: 100.

CHAPTER 4

AGTERDENBOS, J. and TELLINGEN, E.J.V., 1961. *Talanta,* 8: 532.
ALIMARIN, I.P. and YUNG-SCHAING, T., 1959. *Zh. Anal. Khim.,* 14: 5.
ALIMARIN, I.P. and YUNG-SCHAING, T., 1961. *Talanta,* 8: 317.
AYRES, G.H. and YOUNG, F., 1950. *Anal. Chem.,* 22: 1277.
BALLCZO, H. and DOPPLER, G., 1956. *Z. Anal. Chem.,* 152: 321.
BANDISCH, O., 1909. *Chem. Z.,* 33: 1298.
BANKS, C.V. and EDWARDS, R.E., 1955. *Anal. Chem.,* 27: 947.
BARKOVSKII, Y.A. and LOBANOVA, E.F., 1959. *Zh. Anal. Khim.,* 14: 523.
BEAMISH, F.E., 1958. *Talanta,* 1: 3.
BELCHER, R. and WILSON, C.L., 1964. *New Methods in Analytical Chemistry.* Chapman and Hall, London, 304 pp.
BENEDETTI-PICHLER, A., 1932. *J. Ind. Eng. Chem., Anal. Ed.,* 4: 336.
BERKA, A. and ZYKA, J., 1956. *Chem. Listy,* 50: 829.
BERKA, A., DVORAK, V., NEMIC, J. and ZYKA, J., 1960. *Anal. Chim. Acta,* 23: 380.
CARTWRIGHT, P.F.S., 1960. *Analyst,* 85: 216.
CAYLEY, E.R. and KAHLE, G.R., 1959. *Anal. Chem.,* 31: 1880.
CHENG, K.L., 1955. *Anal. Chem.,* 27: 1594.
CHENG, K.L. and WILLIAMS, T.R., 1955. *Chem. Anal.,* 44: 96, 98.
CHRETIEN, A. and LONGI, Y., 1944. *Bull. Soc. Chim. France,* 11: 241, 245.
CUNNINGHAM, T.R. and COTTMAN, R.W., 1924. *J. Ind. Eng. Chem.,* 16: 58.
DAS, J.D. and SHOME, S.C., 1962. *Anal. Chim. Acta,* 27: 58.
DE VOE, J.R. and MEINKE, W.W., 1959. *Anal. Chem.,* 31: 1428.
ERDEY, L. and BUZAS, H., 1954. *Acta Chim. Acad. Sci. Hung.,* 4: 195.
ERDEY, L. and RADY, G., 1958. *Talanta,* 1: 159.
FEIGL, F., 1924. *Mikrochemie,* 2: 187.
FEIGL, F. and NEUBER, F., 1923. *Z. Anal. Chem.,* 62: 373.
FURMAN, N.H., 1942. *Ind. Eng. Chem., Anal. Ed.,* 14: 367.
FURMAN, N.H. and MUSTAFIN, I.S., 1960. *Zh. Anal. Khim.,* 15: 671.
GANTIER, J.A. and PIGNARD, P., 1951. *Mikrochem. Mikrochim. Acta,* 36/37: 793.
HAHN, R.B. and BAGINSKI, E.S., 1956. *Anal. Chim. Acta,* 14: 45.
HARVEY, A.E. and YOE, J.H., 1953. *Anal. Chim. Acta,* 8: 246.
HILLEBRAND, W.F. and LUNDELL, G.E.F., 1929. *Applied Inorganic Analysis.* Wiley, New York, N.Y., 929 pp.
HOLZBECHER, L., 1958. *Chem. Listy,* 52: 430.
ISHIBASHI, M. and HARA, T., 1953. *Japan Analyst,* 2: 300.
JOHNSON, R.A. and KWAN, F.P., 1951. *Anal. Chem.,* 23: 651.
KAIMAL, V.R.M. and SHOME, S.C., 1962. *Anal. Chim. Acta,* 27: 298.
KOLTHOFF, I.M. and SANDELL, E.B., 1929. *Ind. Eng. Chem., Anal. Ed.,* 1: 181.
KOLTHOFF, I.M. and SANDELL, E.B., 1952. *Textbook of Quantitative Inorganic Analysis,* 3rd ed. Macmillan, New York, N.Y., 759 pp.
KOLTHOFF, I.M., SANDELL, E.B. and MOSKOVITZ, B., 1933. *J. Am. Chem. Soc.,* 55: 1454.
KOMAROWSKY, A.S. and POLUEKTOFF, N.S., 1933/34. *Microchemie,* 14: 317.
KONONENKO, L.I., LANER, R.S. and POLUEKTOV, N.S., 1960. *Ref. Zh. Khim., Abstr.,* 47061.

KOPANICA, M. and DOLEZAL, J., 1956. *Chem. Listy*, 50: 1225.
KROUPA, E., 1944. *Mikrochemie*, 32: 245.
LEONARD, M.A. and WEST, T.S., 1960. *J. Chem. Soc.*, 98: 3577.
LUCCHESI, P.J., 1954. *Anal. Chem.*, 26: 521.
MAHR, C. and OHLE, H., 1937. *Z. Anal. Chem.*, 109: 1.
MATSUURA, J., 1953. *J. Chem. Soc. Japan*, 74: 337.
MCKAVENEY, J.P. and FREISER, H., 1958. *Anal. Chem.*, 30: 1955.
MINCZEWSKI, J. and GLABISZ, V., 1960. *Talanta*, 5: 179.
MOORE, C.E. and ROBINSON, T.A., 1960. *Anal. Chim. Acta*, 23: 533.
MOTOGIMA, K., 1956a. *J. Chem. Soc. Japan*, 77: 95, 97, 100.
MOTOGIMA, K., 1956b. *Bull. Chem. Soc. Japan*, 29: 29, 71, 75.
ONISHI, H., 1957. *Bull. Chem. Soc. Japan*, 30: 567.
PILIPENKO, A.T. and OBOLONCHIK, A.T., 1959. *Ref. Zh. Khim.. Abstr.*, 23058.
PLYUSHCHEV, V.I. and KORSHUNOV, B. G., 1955. *Zh. Anal. Khim.*, 10: 119.
POWELL, A.R. and SCHOELLER, W.R., 1925. *Analyst*, 50: 485.
PRIBIL, R. and VESELY, V. 1963. *Talanta*, 10: 383.
REMINGTON, W.J. and MOYER, H.V., 1937.*Dissertation Abstr.* Ohio State Univ. Press, Columbus, Ohio.
RYAN, D.E., 1950. *Anal. Chem.*, 22: 599.
RYAN, D.E., 1960. *Can. J. Chem.*, 38: 2488.
RYAN, D.E. and LUTWICK, G.D., 1953. *Can. J. Chem.*, 31: 9.
SCHOELLER, W.R., 1932. *Analyst*, 57: 750.
SCOTT, W.S., 1944. *Standard Methods of Chemical Analysis*, *1*, 5th ed. Van Nostrand, New York, N.Y., 805 pp.
SEATH, J. and BEAMISH, F.E., 1937. *Ind. Eng. Chem., Anal. Ed.*, 9: 373.
SINHA, S.K. and SHOME, S.C., 1959. *Anal. Chim. Acta*, 21: 45.
SOGANI, N.C. and BHATTACHARYA, S.C., 1956. *Anal. Chem.*, 28: 1616.
TANDON, S.G. and BHATTACHARYA, S.C., 1960. *Anal. Chem.*, 32: 194.
WALTER, J.L. and FREISER, H., 1953. *Anal. Chem.*, 25: 127.
WESTLAND, A.D. and BEAMISH, F.E., 1956. *Mickrochim. Acta*, 41: 1474.
WHITE, J.C. and GOLDBERG, G., 1955. *Anal. Chem.*, 27: 1188.
WILLIS, V.F., 1942. *Analyst*, 67: 219.

CHAPTER 5

AHRENS, L.H., 1950. *Spectrochemical Analysis*. Addison-Wesley, New York, N.Y., 270 pp.
FELDMAN, C., 1949. *Anal. Chem.*, 21: 1041.
GATTERNER, A. and JUNKES, J., 1947. *Specola Vaticaria*, 2nd ed., pp.1–20.
GERLACH, W., 1925. *Z. Anorg. Chem.*, 142: 393.
GOLDSCHMIDT, V.M., 1954. *Geochemistry*. Clarendon, Oxford, 604 pp.
GRABOWSKI, R.J. and UNICE, R.C., 1958. *Anal. Chem.*, 30: 1374.
HARRISON, G.R., 1939. *M.I.T. Wavelength Tables*. Wiley, New York, N.Y., 429 pp.
HARRISON, G.R., LORD, R.C. and LOOFBOUROW, J.R., 1948. *Practical Spectroscopy*. Prentice-Hall, New York, N.Y., 605 pp.

HARVEY, C.E., 1950. *Spectrochemical Procedures*. Appl. Res. Lab. Glendale, Glendale, Calif., 30 pp.
KING, P.B., FERGUSON, H.W., CRAIG, L.C. and RODGERS, J., 1944. *Tenn. Dept. Conserv. Div. Geol., Bull.*, 52: 275.
LEWIS, S.J., 1936. *Spectroscopy in Science and Industry*. Blackie, London, 94 pp.
MANNKOPF, R. and PETERS, C., 1931. *Z. Physik*, 70: 444.
MEGGERS, W.F., 1941. *J. Opt. Soc. Am.*, 31: 39.
MELOCHE, V.W. and SHAPIRO, R., 1958. *Anal. Chem.*, 30: 1374.
MITCHELL, R.L., 1948. *Commonwealth Bur. Soil Sci. (Gt. Brit.), Tech. Comm.*, 44.
MITTLEDORF, A.J. and LANDON, D.O., 1956. *Appl. Spectry.*, 10: 12.
MOSHER, R.E., 1958. *Anal. Chem.*, 30: 1374.
MOSHER, R.E., BIRD, E.J. and BOYLE, A.J., 1951. *Anal. Chem.*, 23: 1514.
NACHTRIEB, N.H., 1950. *Principles and Practice of Spectrochemical Analysis*. McGraw-Hill, New York, N.Y., 324 pp.
NEUMANN, H., 1956. *Norsk Geol. Tidsskr.*, 36: 52.
SCOTT, R.O. and URE, A.M., 1958. *Analyst*, 83: 561.
YOUNG, L.G., 1962. *Analyst*, 87: 6.

CHAPTER 6

BREALEY, L., GARRATT, D.C. and PROCTOR, K.A., 1952. *J. Pharm. Pharmacol.*, 4: 717.
BRYAN, H.A. and DEAN, J.A., 1957. *Anal. Chem.*, 29: 1289.
CHOW, T.J. and THOMPSON, T.G., 1955a. *Anal. Chem.*, 27: 18.
CHOW, T.J. and THOMPSON, T.G., 1955b. *Anal. Chem.*, 27: 910.
DEAN, J.A., 1960. *Flame Photometry*. McGraw-Hill, New York, N.Y., 205 pp.
DEAN, J.A. and CAIN, C., 1957. *Anal. Chem.*, 29: 530.
DEAN, J.A. and LADY, J.H., 1955. *Anal. Chem.*, 27: 1533.
DEAN, J.A. and LADY, J.H., 1956. *Anal. Chem.*, 28: 1887.
DIAMOND, J.J., 1955. *Anal. Chem.*, 27: 913.
EDER, J.M. and VALENTA, E., 1893. *Denkschr. Math. Naturwiss. Kl. Kais. Akad. Wiss. Wien*, 467.
EGGERTON, F.T., WYLD, G. and LYKKEN, L., 1951. *Am. Soc. Testing Mater., Spec. Tech. Bull.*, 116: 52.
ESHELMAN, H.C., DEAN, J.A., MENIS, O. and RAINS, T.C., 1959. *Anal. Chem.*, 31: 183.
GEHRKE, C.W., AFFSPRUNG, H.W. and WOOD, E.L., 1955. *J. Agr. Food Chem.*, 3: 48.
GILBERT, P.T., HAWES, R.C. and BECKMAN, A.O., 1950. *Anal. Chem.*, 22: 772.
GOUY, C.L., 1879. *Ann. Chim. Phys.*, 18: 5.
IKEDA, S., 1956. *Sci. Rept. Res. Inst. Tohoku Univ.*, 8: 134.
KIRCHHOFF, G. and BUNSEN, R., 1860. *Ann. Phys. Chem.*, 110: 161.
LUNDEGARDH, H., 1929. *Die Quantitative Spektralanalyse der Elemente, 1*. Fischer, Jena, 201 pp.

LUNDEGARDH, H., 1934. *Die Quantitative Spektralanalyse der Elemente, 2.* Fischer, Jena, 156 pp.
MENIS, O., RAINS, T.C. and DEAN, J.A., 1959. *Anal. Chem.*, 31: 187.
MITCHELL, R.L. and ROBERTSON, I.M., 1936. *J. Soc. Chem. Engrs.*, 55: 269.
MITSCHERLICH, A., 1862. *Ann. Phys. Chem.*, 116: 499.
PINTA, M., 1951. *J. Rech. Centre Natl., Lab. Bellevue (Paris)*, 4: 210.
PRO, M.J. and MATHERS, A.P., 1954. *J. Assoc. Offic. Agr. Chemists*, 37: 945.
SERVIGNE, M. and DE MONTGAREUIL, P.G., 1954. *Chim. Anal.*, 36: 115.
TALBOT, W.H.F., 1826. *Brewsters J. Sci.*, 5: 77.
VALLEE, B.L. and BARTHOLOMAY, A.F., 1956. *Anal. Chem.*, 28: 1753.
WIRTSCHAFTER, J.D., 1957. *Science*, 125: 603.

CHAPTER 7

ADLER, I. and AXELROD, J.M., 1953. *J. Opt. Soc. Am.*, 43: 769.
ADLER, I. and AXELROD, J.M., 1954. *Anal. Chem.*, 26: 931
ADLER, I. and AXELROD, J.M., 1955. *Spectrochim. Acta*, 7: 91.
ADLER, I. and AXELROD, J.M., 1957a. *Anal. Chem.*, 29: 1280.
ADLER, I. and AXELROD, J.M., 1957b. *Econ. Geol.*, 52: 694.
BIRKS, L.S., 1959. *X-Ray Spectrochemical Analysis*. Interscience, New York, N.Y., 137 pp.
BIRKS, L.S. and BROOKS, E.J., 1950. *Anal. Chem.*, 22: 1017.
BIRKS, L.S. and BROOKS, E.J., 1957. *Denver Res. Conf.*
BIRKS, L.S., BROOKS, E.J., FRIEDMAN, H. and ROE, R.M., 1950. *Anal. Chem.*, 22: 1258.
BRAGG, W.L., 1912. *Nature*, 90: 410.
BRISSEY, R.M., 1952. *Anal. Chem.*, 24: 1034.
CAMPBELL, W.J. and CARL, H.F., 1954. *Anal. Chem.*, 26: 800.
CAMPBELL, W.J. and THATCHER, W.J., 1958. *R.I. Bur. Mines*, 5416: 1.
CARL, H.F. and CAMPBELL, W.J., 1955. *Anal. Chem.*, 27: 1884.
CASTAING, R., 1951. *Application des Sondes Electroniques à une Méthode d'Analyse Ponctuelle Chimique et Cristallographique*. Thesis, Univ. Paris, 140 pp.
CHODOS, A.A. and ENGEL, C.G., 1960. *Advan. X-Ray Anal.*, 4: 401.
CLAISSE, F., 1956. *Can. Dept. Mines Tech. Surv., Rept.*, 327.
COSTER, D. and VON HEVESY, G., 1923. *Nature*, 111: 79. *Chem. News*, 127: 65.
DESPUJOIS, J., 1952. *J. Phys. Radium*, 13: 31.
FRIEDMAN, H. and BIRKS, L.S., 1948. *Rev. Sci. Instr.*, 19: 323.
GULBRANSEN, L.B., 1955. *Anal. Chem.*, 27: 1181.
HEIDEL, A.H. and FASSEL, V.A., 1961. *Anal. Chem.*, 33: 9_3.
KOH, P.K. and CAUGHERTY, B., 1952. *J. Appl. Phys.*, 23: 427.
MORTIMORE, D., ROMANS, P.A. and TEWS, J.L., 1954. *Norelco Reptr.*, 1: 107.
MOSELEY, H.G.J., 1913. *Phil. Mag.*, 26: 1024.
MOSELEY, H.G.J., 1914. *Phil. Mag.*, 27: 703.
PATRICK, R.F., 1952. *J. Am. Ceram. Soc.*, 35: 189.
PFEIFFER, H.G. and ZEMANY, P.D., 1954. *Nature*, 174: 397.
SHALGOSKY, H.I., 1960. *Methods in Geochemistry*. Interscience, New York, N.Y., 240 pp.

STEVENSON, J.S., 1954. *Am. Mineralogist,* 39: 436.
VON HEVESY, G., 1932. *Chemical Analysis by X-Rays and its Applications.* McGraw-Hill, New York, N.Y., 333 pp.
VON LAUE, M., FRIEDRICH, W. and KNIPPING, P., 1913. *Ann. Physik,* 41: 971.
WATLING, J., 1961. *The Determination of Uranium in Low Grade Ores by an X-Ray Spectrographic Technique. Pamphlet AM.81.* H.M. Stationary Office, London, 7 pp.
WEBBER, G.R., 1957. *Trans. Can. Inst. Mining. Met.,* 60: 138.
WILSON, H.M. and WHEELER, G.V., 1958. *Appl. Spectry.,* 11: 128.

CHAPTER 8

ALEXANDER, L. and KLUG, H.P., 1948. *Anal. Chem.,* 20: 886.
BERRY, L.G. and THOMPSON, R.M., 1962. *X-Ray Powder Data for Ore Minerals.* Geol. Soc. Am., New York, N.Y., 281 pp.
BRAGG, W.L., 1912. *Nature,* 90:410.
BRINDLEY, G.W., 1951. *X-Ray Identification and Crystal Structures of Clay Minerals.* Mineral. Soc., London, 345 pp.
KOSSENBERG, M., 1955. *J. Sci. Instr.,*32: 117.
TAYLOR, A., 1951. *J. Sci. Instr.* 28: 200.
WEISSENBERG, K., 1924. *Z. Physik,* 23: 229.
WYCKOFF, H.W.G., 1935. *The Structure of Crystals,* 2nd ed. Chem. Catalog. Co., New York, N.Y., 150 pp.

CHAPTER 9

BOWEN, E.J. and WOKES, F., 1953. *Fluorescence of Solutions.* Longmans, London, 201 pp.
BREALEY, L. and ROSS, R.E., 1957. *Analyst,* 82: 769.
EVANS, W.D., COOPER, B.S., CORBETT, D.W. and GOUGH, K., 1962. *Quart. J. Geol. Soc. London,* 118: 23.
SANDELL, E.B. 1950. *Colorimetric Determination of Traces of Metals.* Interscience, New York, N.Y., 487 pp.
STOKES, G. 1852, *Phil. Trans. Roy. Soc. London,* 143: 463.

CHAPTER 10

AMBROSE, D. and AMBROSE, B.A., 1961. *Gas Chromatography.* Newnes, London, 176 pp.
BREALEY, L., ELVIDGE, D.A. and PROCTOR, K.A., 1959. *Analyst,* 84: 221.
BRITISH DRUG HOUSES, 1958. *Spot Tests.* British Drug Houses, Poole, 84 pp.
CONDON, R.D., 1959. *Anal. Chem.,* 31: 1717.

DESTY, D.H., GOLDUP, A. and SWANTON, B.H.F., 1959. *J. Inst. Petrol.*, 45: 624.
ELLIOTT, R.B. and EVANS, W.D., 1963. *Nature*, 199: 686.
EVANS, W.D., 1963. *Proc. Geol. Am.* 1608: 119.
EVANS, W.D., 1964. *Experimental Pedology.*
EVANS, W.D., COOPER, B.S. and GUNN, R.K., 1963. *Proc. Intern. Conf. Organic Geochem., Milan*, p. 417.
FEIGL, F., 1954. *Spot Tests*. Elsevier, Amsterdan, 600 pp.
FORD, T.D., 1955. *Proc. Yorkshire Geol. Soc.*, 30: 35.
GOLAY, M.J.E., 1958. *Gas Chromatography*. Butterworth, London, 406 pp.
GUNN, R.K. and COOPER, B.S., 1963. *Proc. Intern. Conf., Org. Geochem., Milan*, p. 147.
HARINGTON, J.S., 1962. *Nature*, 193: 43.
HARRISON, G.F., 1956. *Vapour Phase Chromatography*. Butterworths, London, 206 pp.
HARTLEY, J. and PRETORIUS, V., 1958. *Nature*, 181: 177.
HOPKINS and WILLIAMS Ltd., 1964. *Organic Reagents for Metals*. Hopkins and Williams, Essex, 1: 199 pp., 2: 275 pp.
HOWARD, H.C., 1947. *Chemistry of Coal Utilisation*. Wiley, New York, N.Y., 1: 506 pp.
JAMES, A.T. and MARTIN, A.J.P., 1952. *Biochem. J.*, 50: 679.
JAMES, A.T., MARTIN, A.J.P. and SMITH, G.H., 1952. *Biochem. J.*, 52: 238.
JONES, C.E.R. and MOYLES, A.F., 1961. *Nature*,191: 663.
JOST, W., 1960. *Diffusion*. Academic Press, New York, N.Y., 282 pp.
KEULEMANS, A.I.M., 1959. *Gas Chromatography*, 2nd ed. Reinhold, New York, N.Y., 305 pp.
KIRCHNER, J.G., MILLER, J.M. and KELLER, G.J., 1951. *Anal. Chem.*, 23: 420.
KUHN, R., WINTERSTEIN, A. and LEDERER, E., 1931. *Hopper-Seyler's Physiol. Chem.*, 197: 141.
LEDERER, E. and LEDERER, M., 1957. *Chromatography*. Elsevier, Amsterdam, 712 pp.
LOVELOCK, J.E., 1958. *J. Chromatog.*, 1: 35.
MARTIN, A.J.P. and JAMES, A.T., 1956. *Biochem. J.*, 63: 138.
MARTIN, A.J.P. and SYNGE, R.L.M., 1941. *Biochem. J.*, 35: 1358.
MCWILLIAM, I.G. and DEWAR, R.A., 1958a. *Nature*, 182: 1664.
MCWILLIAM, I.G. and DEWAR, R.A., 1958b. *Gas Chromatography*. Butterworths, London, 210 pp.
MOTT, R.A., 1942. *Fuel*, 21: 129.
PARRISS, W.H. and HOLLAND, P.D., 1960. *Brit. Plastics*, M 9228: 10463.
PLUMSTEAD, E.P., 1957. *Coal in Southern Africa*. Witwatersrand Univ. Press, Johannesburg, 24 pp.
PURNELL, H., 1962. *Gas Chromatography*. Wiley, New York, N.Y., 305 pp.
RICE, S.A. and BRYCE, W.A., 1957. *Can. J. Chem.*, 35: 1293.
RITCHIE, A.S., 1964. *Chromatography in Geology*. Elsevier, Amsterdam, 185 pp.
ROUSSEAU, P.E., 1961. The conversion of South African low-grade coal to oil and chemicals. *Commonwealth Mining Met. Congr., 7th, Congress Paper*, 15 pp.
SCOTT, R.P.W., 1955. *Nature*, 176: 793.
SHAKESPEAR, G.A., 1916. *Brit. Pat.*, 124: 453.

STAHL, E., SCHRÖTER, G., KRAFT, G. and RENZ, R. 1956. *Pharmazie*, 11: 633.
STOPES, M.C., 1919. *Proc. Roy. Soc. London*, 90 B: 470.
TENNEY, H.M. and HARRIS, R.J., 1957. *Anal. Chem.*, 29: 317.
TIMMS, D.G., KONRATH, H.J. and CHIRNSIDE, R.C., 1958. *Analyst*, 83: 603.
TSWETT, M., 1906. *Ber. Deut. Botan. Ges.*, 24: 316.
TUEY, G.A.P., 1960. *Materials for Gas Chromatography*. May and Baker, *198 pp.*
WIRTH, M.M., 1957. In: *Vapour Phase Chromatography*. Butterworths, London, p.203.
WITHAM, B.T., 1957. In: *Vapour Phase Chromatography*. Butterworths, London, p.198.

Index